Vol. 29. **The Analytic[illegible]** [illegible] *in three parts).* By J. H. Karchmer

Vol. 30. **Ultra[illegible]**

Vol. 31. **Phot[illegible]**

Vol. 32. **Determ[illegible] of [illegible]** [illegible]Frederick T. Weiss

Vol. 33. **Masking and Demasking of Chemical Reactions.** By D. D. Perrin

Vol. 34. **Neutron Activation Analysis.** By D. De Soete, R. Gijbels, and J. Hoste

Vol. 35. **Laser Raman Spectroscopy.** By Marvin C. Tobin

Vol. 36. **Emission Spectrochemical Analysis.** By Morris Slavin

Vol. 37. **Analytical Chemistry of Phosphorus Compounds.** Edited by M. Halmann

Vol. 38. **Luminescence Spectrometry in Analytical Chemistry.** By J. D. Winefordner, S. G. Schulman, and T. C. O'Haver

Vol. 39. **Activation Analysis with Neutron Generators.** By Sam S. Nargolwalla and Edwin P. Przybylowicz

Vol. 40. **Determination of Gaseous Elements in Metals.** Edited by Lynn L. Lewis, Laben M. Melnick, and Ben D. Holt

Vol. 41. **Analysis of Silicones.** Edited by A. Lee Smith

Vol. 42. **Foundations of Ultracentrifugal Analysis.** By H. Fujita

Vol. 43. **Chemical Infrared Fourier Transform Spectroscopy.** By Peter R. Griffiths

Vol. 44. **Microscale Manipulations in Chemistry.** By T. S. Ma and V. Horak

Vol. 45. **Thermometric Titrations.** By J. Barthel

Vol. 46. **Trace Analysis: Spectroscopic Methods for Elements.** Edited by J. D. Winefordner

Vol. 47. **Contamination Control in Trace Element Analysis.** By Morris Zief and James W. Mitchell

Vol. 48. **Analytical Applications of NMR.** By D. E. Leyden and R. H. Cox

Vol. 49. **Measurement of Dissolved Oxygen.** By Michael L. Hitchman

Vol. 50. **Analytical Laser Spectroscopy.** Edited by Nicolo Omenetto

Vol. 51. **Trace Element Analysis of Geological Materials.** By Roger D. Reeves and Robert R. Brooks

Vol. 52. **Chemical Analysis by Microwave Rotational Spectroscopy.** By Ravi Varma and Lawrence W. Hrubesh

Vol. 53. **Information Theory As Applied to Chemical Analysis.** By Karel Eckschlager and Vladimir Stepánek

Vol. 54. **Applied Infrared Spectroscopy: Fundamentals, Techniques, and Analytical Problem-solving.** By A. Lee Smith

Vol. 55. **Archaeological Chemistry.** By Zvi Goffer

Vol. 56. **Immobilized Enzymes in Analytical and Clinical Chemistry.** By P. W. Carr and L. D. Bowers

Vol. 57. **Photoacoustics and Photoacoustic Spectroscopy.** By Allan Rosencwaig

Vol. 58. **Analysis of Pesticide Residues.** Edited by H. Anson Moye

Vol. 59. **Affinity Chromatography.** By William H. Scouten

Vol. 60. **Quality Control in Analytical Chemistry.** By G. Kateman and F. W. Pijpers

Vol. 61. **Direct Characterization of Fineparticles.** By Brian H. Kaye

Vol. 62. **Flow Injection Analysis.** By J. Ruzicka and E. H. Hansen

Spot Test Analysis

CHEMICAL ANALYSIS

A SERIES OF MONOGRAPHS ON
ANALYTICAL CHEMISTRY AND ITS APPLICATIONS

VOLUME 75

A WILEY-INTERSCIENCE PUBLICATION

JOHN WILEY & SONS

New York / Chichester / Brisbane / Toronto / Singapore

Spot Test Analysis

CLINICAL, ENVIRONMENTAL, FORENSIC, AND GEOCHEMICAL APPLICATIONS

ERVIN JUNGREIS

The Hebrew University of Jerusalem
Jerusalem, Israel

A WILEY-INTERSCIENCE PUBLICATION

JOHN WILEY & SONS

New York / Chichester / Brisbane / Toronto / Singapore

Library of Congress Cataloging in Publication Data:

Jungreis, Ervin, 1926–
Spot test analysis.

(Chemical analysis, 0069-2883; v. 75)
"A Wiley-Interscience publication."
Includes index.
1. Spot tests (Chemistry) I. Title. II. Series.

QD98.S6J86 1984 543'.08134 84-15176
ISBN 0-471-86524-9

Printed in the United States of America

10 9 8 7 6 5 4 3

To Nancy

PREFACE

In the past several decades chemical analysis has undergone a formidable process of sophistication. Advanced instrumental tools have improved the generation of the signal given by the analyte and transduced it to a more conveniently measured one, and with the tremendous developments in the field of electronics, enormous amplification of the resulting signal has become feasible.

Parallel with this trend toward instrumental sophistication, a trend toward simplification was seen in some marginal selected areas in the form of simple, rapid, and inexpensive spot and screening tests. A screening test is defined as a simple method that provides a sufficient answer to the analytical question with a minimum expenditure of time and money. Rapid methods in various areas of application register gross deviation from a norm, and although positive results generally require confirmation, the negative findings are conclusive. It would be wasteful to devote resources to attaining precision or accuracy several magnitudes beyond that necessary in the clinical urine analysis, for example, when pathological changes in the composition of urine are evidenced by a simple spot test.

It must be emphasized that the applicability of spot test methodology is limited even in the areas discussed in this book, and that for the exact determination of most chemical substances, complex analytical procedures are unavoidable. Although the use of screening tests is a marginal one; the margin, however, is quite significant.

This book deals with the contemporary uses of spot and screening tests in clinical, forensic, geochemical, soil, and water testing and in occupational safety protection. It is intended to be a laboratory reference book providing detailed procedural information.

I am grateful to the following individuals and institutions for their help in supplying the variety of technical literature used in the preparation of this book: Mrs. Nechama Weissman, Chemistry Institute, The Hebrew University, Jerusalem; Dr. I. Brenner, Israel Geological Survey, Jerusalem; Dr. C. O. Rupe, Ames Company, Miles Laboratories, Inc., Elkhart, Indiana; Inspector Y. Leisst, Israeli Police Headquarters; Dr. V. C. O. Schüler, Anglo American Research Laboratories, Crown Mines, South Africa; Mrs. Agnes Mühlrad, Hebrew University Medical Library Service, Jerusalem; Hach Chemical Co., Loveland, Colorado; Macherey-Nagel, Düren, F. R. G.; Mr. I. Günther, Drägerwerk AG, Lübeck, F. R. G.; Hoechst AG, Frankfurt am Main, F. R. G.; Anatole G. Sipin Co., Inc., New York; Boehringer-Mannheim GmbH, Mannheim, F. R. G.; Ecologic Instrument Corp., New York; Taylor Chemicals, Inc., Baltimore, Maryland; E. Merck Co., Darmstadt, F. R. G.; Eastman-Kodak Co., Rochester, New York; M. S. A. International Co., Pittsburgh, Pennsylvania; Hellige, Inc., Garden City, New York; and B. D. H. Chemicals Ltd., London. I gratefully acknowledge the kindness of the Executive Committee of the Association of Official Analytical Chemists (AOAC) in permitting me to make free mention of the relevant AOAC methods.

Jerusalem, Israel
November 1984

Ervin Jungreis

CONTENTS

Spot Test Analysis

CHAPTER 1
INTRODUCTION

In the last three decades the classical spot reactions have been successfully applied in large-scale analytical problems in clinical analysis, in control tests of air quality, in food and water analysis, in geochemical prospecting, in crime laboratories, in soil testing, and in juridical chemical studies.

Commercial companies in both the United States and Europe are selling vast quantities of compact spot test systems for the qualitative analysis of urine and water, in which the goal is the rapid establishment of the presence or absence of certain materials. The extreme simplicity of these tests, their time- and money-saving nature, and their ultimate reliability make these tests very useful, and the analytical problems solved using simple reagent strips and preprepared reagent pillows or tablets constitute a significant part of the total spectrum of analytical problems.

Aquatic ecology tests are based on spot test analysis. Water quality, whether the criteria are based on human consumption, irrigation, or feed lot or farmland run off, can be examined easily and reliably and soil characteristics can be established by unsophisticated spot tests.

Around the middle of the last century single-chemical tests were already being performed on filter paper matrices. Runge was one of the important pioneers in those days in the use of absorbent papers as the seat of the chemical reaction (1). In a sense, he can be regarded as the creator of paper chromatography and spot test analysis. For testing for free chlorine, he impregnated paper with starch and potassium iodide, as described in Volume II of his "Farbenchemie." When the paper was spotted with bleaching solution a blue fleck appeared because of the liberation of iodine and its reaction with starch. This was probably the earliest scientifically checked authentic spot reaction with impregnated paper; even more amazing, this reagent paper has been sold commercially ever since. This test antedated Schiff's test for uric acid in urine on silver carbonate by 25 years.

The modern development of spot test methods of analysis occupied Fritz Feigl for half a century. Even before their systematic development began, spot tests were used in isolated organic and inorganic applications.

According to the writings of Pliny the Elder, early Romans detected iron in vinegar by means of a reagent paper prepared by soaking papyrus in an extract obtained from gall nuts.

Modern instrumental analytical chemistry, the application of microprocessors, and the rapid advances in spectral, chromatographic, nuclear, mass spectroscopic, and electroanalytic research changed analytical chemistry spectacularly, dramatically reducing the complexity of the wet analytical procedures. Nearly any physical property characteristic of a particular element or compound was made the basis of an analytical method. Together with this rapid sophistication took place another development in certain distinct areas of application, in exactly the opposite direction. This was the search for simple, compact, and inexpensive analytical devices for semiquantitative evaluation of certain elements or compounds where such an approximate evaluation has diagnostic value in at least the first stage of the examination. A new look was taken at the old spot tests; sometimes they were refined, and sometimes new ones were elaborated. In many of the rapid analytical procedures the human eye was used as a detector to measure color changes; others employed simple reflectometric instruments. Major manufacturers of chemical reagents concentrated their efforts mainly on three applicative areas, namely, clinical analysis, water quality control tests, and air quality control tests, because of the vast number of screening tests needed in these areas.

Urine analysis test strips for such preliminary screening are offered by several reagent firms. The several simultaneous spot reactions carried out by just dipping the strip into the urine provide reliable evidence of pathological changes in the composition of the urine. This testing for complete chemical urinalysis in both hospital and private practice is the first step toward establishing a diagnosis that can then be corroborated by clinical examination and quantitative laboratory tests.

An important development is the application of spot test procedures for specific, visual determination of important components such as glucose and urea in whole blood. A semipermeable membrane on the impregnated reagent area serves as a barrier, and quantitation is established by giving the soluble component a predetermined time period to diffuse to the chromogenic reagent zone. The simplicity of this procedure allows the rapid, semiquantitative bedside measurement of blood urea nitrogen and blood glucose. Because no deproteinization or exact volume measurement of the blood is necessary, self-monitoring by diabetics or patients having an elevated or depressed renal threshold is feasible. Such "instant" blood sugar and blood urea nitrogen measurements can be carried out by simple visual comparison with a color chart or, for more exact measurement, by a portable, battery-operated reflectance meter that measures the light reflected from the reagent strip and converts this to a reading.

The very same principle of rapid separation of low- and high-molecular-weight components in blood through semipermeable membranes in a spot reaction combined with automated colorimetric and potentiometric assays is used by the Eastman-Kodak Company in a very significant system that considerably simplifies and speeds up the routine analysis of substances in human body fluids. Thin-film coating technology, which is basic to modern color photography, was elaborated to create a system consisting of a spreading layer, a semipermeable layer, and a layer containing the pigment precursors. The potential uses of this thin-film, multilayer approach to rapid clinical colorimetric assays seem extensive. A variety of operations can be performed on serum components as they diffuse through the multilayer structure. These layers may contain enzymes, buffers, ion-exchange materials, semipermeable membranes, and mordant and masking layers (to separate dyed layers optically).

Besides using exact analytical methods, geochemical prospecting uses spot colorimetry, qualitative tests for pathfinders, and simple field methods for rough differentiation of ore bodies.

The testing of tap water, lake and stream limnology studies, and the determination of seasonal changes in water supplies are also feasible with simple tests. Semiquantitative measurement is achieved either by visual comparison with color discs or with battery-operated portable photometers. The cooling water used in industry and the quality of water used in agriculture, of boiler feed water, and of swimming pool water can all be monitored with spot-test-based chemical reactions.

In police investigations, detection of traces of metallic lead in certain portions of substrates indicates firearm discharge. Blood, sperm, explosive, and drug tests are a very important part of crime detection.

Air quality standards are often monitored by direct reading of colorimetric indicators. Reagent papers can be used to detect gases present in the atmosphere. A classical example is the Gutzeit method, in which arsine turns a mercuric-bromide-impregnated paper strip black. Semiquantitative screening is achieved by using a simple sampling device that passes a measured volume of air through a preprepared detector tube. The first indicating tubes containing solid chemicals for quick, direct reading were detection tubes for carbon monoxide, hydrogen sulfide, and benzene. A sample of the workroom atmosphere being tested is drawn through a glass tube containing the reagent system. The appearance of a color change indicates the presence of the particular contaminant, and comparison of the hue produced against a set of standard colors, or measurement of the length of a stain, indicates the approximate level of the airborne contaminant.

In the last decade interest in methods for checking air quality in working areas has grown tremendously. Several commercial companies (Bacharach Industrial Instrument Co.; Davis Engineering Equipment Co. Inc.; Union Industrial Equipment Corp.; Mine Safety Appliances Co.; Acme Protection Equipment Corp.; Drägerwerk AG) have made impressive progress in constructing indicating tubes for rapid, convenient, and inexpensive analysis of toxic gases in air. These tests can be performed easily by semiskilled operators.

REFERENCE

1. B. Anft, *J. Chem. Educ.* **32,** 566 (1953).

CHAPTER 2
TECHNIQUES

The simplest methods of spot test analysis involve the mere mixing of a drop of the unknown substance and a drop of the reagent solution. The reaction medium is either a porous or a nonporous supporting surface such as filter paper, glass, or porcelain. Use of a porous medium (e.g., filter paper, asbestos, or gelatin) impregnated with water-soluble or water-insoluble reagents raises the sensitivity of the test. Applied spot test analyses in clinical chemistry and in compact water analysis systems use strips of absorbent cellulose, one end of which is impregnated with the reagent system. The impregnated reagent area is sometimes covered with a semipermeable membrane to prevent staining by a colored matrix (e.g., red blood cells).

Another type of spot test employs one of the reactants in the solid form. In many clinical and technical appications a stable reagent tablet is simply dropped into the test solution and the extent of precipitation or color formation established.

The equipment required for spot test analysis is made of glass, porcelain, plastic, and metal. A classical spot test laboratory setup consists of assorted sizes of beakers, volumetric flasks, Erlenmeyer flasks, suction flasks, round-bottom flasks, distillation heads, Conway cells, crystallizing dishes, evaporating dishes, filter sticks, separating funnels, extraction pipets, fritted glass crucibles, graduated cylinders, pipets, burets, weighing bottles, storage bottles, vials, test tubes, centrifuge tubes, microscope slides, cover glasses, and spot plates.

White and black porcelain crucibles, dishes, and spot plates should be on hand. Platinum equipment, including dishes, foils, crucibles, and boats, is standard for many spot test operations. Crucibles made of nickel and beakers of stainless steel are also valuable. The use of aluminum pans and dishes is very practical, since, owing to their low cost, they are discardable. Tweezers and spatulas made of nickel-plated steel are essential too. A good assortment of microburners and blast lamps should be on hand. Other essential metalware includes sand baths, water baths, tripods, ring stands, clamps, rings, and buret holders.

The following devices and equipment are essential in the classical spot test laboratory: hand balance, microanalytical balance, pH meter, ultraviolet lamp, infrared lamp, centrifuge, hot closet, drying apparatus, ignition furnace, hot plates, water baths, microdistillation equipment, cooling block, spectrophotometer, stereomicroscope, chemical microscope, and melting point apparatus.

The simplest spot test analysis procedure is the mixing of a drop of the test solution with a drop of the reagent. This extremely simple procedure can be carried out in various ways. It is generally not essential to control the drop size, but in some special cases equal drop sizes should be applied to enable semiquantitative estimation. For the accurate measurement of drop size a Gilmont micropipet–buret can be used (1). This consists of a glass container and a delivery tip and can be made to deliver from 0.001 ml to 1 ml of solution. The liquid volume is controlled with the aid of an adjustable plunger made of synthetic rubber, which is located in a Teflon guide. When measuring a volume of liquid, a micrometer screw is turned to move the rubber piston. The pipet is mounted on a stand, and the delivered volume is read directly from a microgauge. Use of this pipet is important in the development of new spot tests, when precise measurement of the test solution volume is needed to establish the limit of dilution of the test.

Filter paper impregnated with a stable reagent is the most common configuration of spot test applied in clinical screening and water quality control tests. Many commercial companies dealing with qualitative and semiquantitative control tests have developed strip tests in which a square of absorbent cellulose that has been impregnated with reagents, buffers, sequestering agents, and other substances and affixed to a plastic strip serves as the reaction zone. The test is performed by immersing the test zone of the strip in the sample solution and comparing the resulting tint with a color scale.

Stable reagents must be available for preimpregnation of spot test papers. To prevent "bleeding" of the reagent from the site of the reaction, minimally water-soluble agents are used advantageously. Most reagents applied in "dip and read" sticks are organic reagents, which are only slightly soluble in water, and the impregnation is carried out in an organic solvent solution.

Test papers impregnated with substances that are only barely soluble in water react with an acceptable velocity, even though insoluble compounds applied in *macro*analysis would react extremely slowly. The high dispersion of the extremely fine solid particles in the capillaries of the filter paper leads to a very high surface area, so the chemical reaction proceeds quickly. The reaction product does not diffuse, and thus is more localized. This feature improves the sensitivity of the reaction greatly.

Insolubility of the reagent in water contributes to its stability. Spot test paper impregnated with alkali sulfide for the detection of heavy metals, for example, has a very limited stability, due to oxidation to sulfate in air. Moreover, when such test paper is immersed in the aqueous test solution, the extremely soluble alkali sulfide is leached. In contrast, when the test paper is impregnated with scarcely soluble sulfides, such as zinc sulfide, cadmium sulfide, antimony sulfide, and so forth, relatively stable papers result. The concentration of the active sulfide ion depends on the solubility product of the particular compound; this fact enables selective precipitation of metal ions. The detection of silver, copper, or mercury is possible, for example, with antimony sulfide paper in the presence of lead, cadmium, tin, iron, nickel, cobalt, and zinc. The detection of copper ions with the soluble reagent sodium diethyldithiocarbamate reaches a limit of identification at 0.2 μg, whereas the sensitivity of detection of the same ion increases 100-fold when filter paper impregnated with insoluble zinc diethyldithiocarbamate is used as the reagent.

Since the majority of spot test reactions lead to the formation of distinctive colors, simple color comparing or estimating devices belong in the applicative spot test laboratory. The simplest of these are calibrated color charts, which are usually printed on the test kit container. Another handy device for rapid screening of color intensity is a system of colored cubes molded of acrylic plastic and matched to laboratory standards. The sample and the reagent are mixed in a cavity next to the five-step color scale, and after visual matching the concentration is read directly (Hach Chemical Co., Loveland, Colorado; Visocholor®, Macherey-Nagel Co., Düren, F. R. G.). A series of nonfading glass color standards mounted on a circular disc may be used in a comparator box (Hellige Inc., Garden City, New York), or continuous color discs may be applied for visual color comparison (Hach). Use of the latter eliminates the necessity of interpolating between individual color standards.

For titrimetric tests, so-called drop count measurements are carried out using a dropper bottle. The use of digital titrators (Hach) fits the requirements of a rapid analytical procedure in the field while still providing a good relative accuracy (±1%). The titrant is packaged in disposable polypropylene plastic containers with Teflon-covered neoprene seals and resealable vinyl closures to cover the cartridge tips. These cartridges contain sufficient solution for approximately 50–100 average titrations.

The range of instruments used in applicative spot test analysis today encompasses simple portable instruments for the measurement of the reflected light of the reacted reagent area of a test strip (Ames™ Reflective Meter), portable turbidimeters, battery operated colorimeters, conductivity meters, nephelometers, dissolved oxygen meters, low-cost field gas chromatographs, and pH meters.

For the semiquantitative testing of airborne contaminants in a workroom atmosphere the detector tube technique has gained ground in the applicative spot test laboratory (Drägerwerk AG, Lübeck, F. R. G.; MSA International, Pittsburgh, Pennsylvania; Anatole S. Sipin Co. Inc., New York; and others). The air to be tested is drawn through the tube with a hand pump, squeeze pump, or continuously operated sampler pump. In the presence of the contaminant to be measured the contents of the tube change color. The length or the intensity of the color change indicates the concentration of the gas being measured.

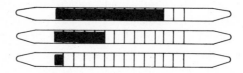

Sampling is one of the most important operations in the test procedure. The sample naturally must be representative; it is totally useless to waste one's efforts on other than representative samples. For spot test operations applied in geochemical screening, soil testing, or food chemistry, the first task should be the detection of inhomogeneities in the sample. Sometimes, mainly in geochemical prospecting or in forensic types of investigations, selected parts of the sample have to be examined. In this case a wide-angle, low-power microscope is generally used to sort out the desired particles. Fine glass fibers moistened with glycerol can be useful in this sorting procedure.

Prior to testing, the representative solid sample is dissolved in deionized water, acid, or alkali, or disintegrated by fusing it with solid sodium carbonate, sodium peroxide, or sodium pyrosulfate. For many silicate-bearing minerals, fusing of the sample with hydrofluoric, hydrochloric, or sulfuric acid is necessary to effect complete dissolution. Extraction with organic solvents is a useful means of separating certain ingredients. In any of the above treatments, the reaction rate is function of the maximum reactive surface of the solid. Freshly precipitated water-insoluble compounds that have been gently dried have a high surface area, but most natural or technical products should be pulverized carefully before testing.

The pulverizing of hard samples is generally carried out in agate mortars. Exceptionally hard specimens should first be crushed in a steel diamond mortar and then pulverized in an agate micromortar. The crushed and pulverized samples should be sieved through very fine silk, which may be prepared by stretching the silk over a beaker. The pulverized sample is screened through the cloth, and the retained coarser particles are ground again until the total sample passes through the silk barrier.

To remove excess solvent or to evaporate the test solution completely to dryness, a simple water bath, air bath, or sand bath heated with an infrared lamp, hot plate, or gas burner may be applied. A simple air bath can easily be built in the spot test laboratory from a nickel crucible. Lateral slits in an asbestos triangle hold the microcrucible, in which the test solution to be concentrated is placed. Platinum spoons and silica casseroles also serve as useful tools for evaporation of small volumes of samples. Platinum spoons are used also for the fusing of insoluble materials, and the melts may be subjected to oxidizing or reducing flames as desired.

Barnes-type reagent bottles are important tools in the spot test laboratory. These bottles are stoppered with a dropper pipet, and the rubber cap controls the delivery of the drop. The dropper pipet is held at right angles to the horizontal receiving surface 1 or 2 cm above the recipient area. Another way to deliver drops of reagent solutions is with all-glass dropping bottles equipped with turn caps. When the drop size cannot be controlled by any of the above methods, solutions may be transferred with single glass rods. The volume transferred is a function of the diameter of the rod. For transfer of very small liquid volumes, loops made of fine platinum wire are very useful. The size of the loop determines the size of the drop, and such loops can be calibrated with an acceptable accuracy.

In the simple techniques of spot test analysis, separation processes are avoided as much as possible. Interfering substances are eliminated by sequestering procedures using masking reactions. These reactions produce stable soluble complexes or pseudosalts. Such masking agents are often part of a compact reagent tablet system or of the reactive zone of a "dip and read" reagent strip. Use of masking procedures yields the tremendous advantage that separation of phases is not required.

One of the advantages of classical spot test analysis is the extremely simple and elegant way that precipitation and filtration are accomplished on the surface of filter paper. The insoluble reaction product is retained in the capillaries of the paper after a certain particle size is reached, while the excess reagent solution, together with any nonreacting cosolutes, passes through by capillary diffusion. This microprecipitation and microfiltration makes possible a very simple technique for washing the product: One simply places drops of water on the center of the spot or hangs the spotted paper in an appropriate wash liquid.

The technique of separation of small water-soluble molecules from high-molecular-weight proteins by means of a semipermeable layer is utilized in several screening tests for glucose and urea in blood (Haemoglucotest®—Boehringer-Mannheim. Dextrostix®—Ames Co. Azostix®—Ames Co. Reflotest-Urea—Boehringer-Mannheim).

A special branch of spot test analysis was developed for the nondestructive screening of metals and alloys. Sampling of metals and alloys is generally carried out by drilling holes through a representative ingot at selected points; all the material from the holes is then collected and mixed, and a sample of suitable size is used for analysis.

This technique cannot be applied in the analytical screening of finished tools or parts of instruments, where sampling would destroy the further practical use of the sample or greatly decrease its commercial value. In these cases the sample should be taken in the nondestructive way. In this method, a few drops of the solvent (generally a suitable acid) are placed on the surface of the sample article for such a duration that the amount of dissolved material lost will not damage the tool's further practical use. However, the sensitivities of the applied tests must be enormously high. A long series of such highly sensitive spot tests with identification limits of 10^{-7}–10^{-8} g/ml using a drop size of 0.002–0.005 ml were worked out and are now applied for both qualitative and semiquantitative purposes. The tests are generally carried out on filter papers or reagent strips, using color charts, Lovibond comparators, or continuous color discs for screening.

A nondestructive metal test generally starts with the mechanical grinding, chemical cleaning (e.g., with nitric acid), deionizing with distilled water, and drying of the surface. Before the solvent is placed on the metal, a circular paraffin barrier is prepared by dropping hot melted paraffin on the surface and cutting a 0.5- to 1.0-cm-diameter depression into the paraffin. Alternatively, a circle may be drawn on the metal with a petroleum-jelly-wetted drinking glass; the acid is then dropped in the middle of the circle.

In such nondestructive screening tests of metals and alloys, 1 : 1 hydrochloric acid, 1 : 1 nitric acid, or a mixture of the two is used. Hydrochloric acid dissolves cast iron, carbon steels with low chromium contents, alloyed steels, and various aluminum alloys, but does not dissolve the copper-containing alloys such as brass, bronze, and gun metals. Nitric acid, on the other hand, attacks readily the copper, iron, and steel alloys, but does not dissolve the aluminum alloys. Steels with high chromium contents, carbon steels, cast iron, and tin–lead alloys are easily dissolved in hydrochloric acid–nitric acid mixtures.

The nondestructive screening of metals and alloys starts with the placing of 1 or 2 drops of solvent on the barriered surface of the object. The course of the reaction, as judged from the rate of gas bubble formation, can be followed visually with a magnifying glass. The end of the reaction is signaled by the stoppage of microbubble formation. The solution formed on the alloy surface can be used directly for testing, but it is advisable to include a further simple step in the procedure. This step is necessary because at the end of the dissolution reaction the insoluble particles of the alloy [carbides, certain oxides, and hydroxides (e.g., metastannic acid)] may appear as fine precipitates in the solution. The whole solution containing the insoluble particles is sucked from the surface of the alloys into a micro test tube by means of a capillary tube, and the dissolution is completed by heating the tube and adding more acid.

A simple spot test (2) may provide a qualitative, but fairly accurate, indication of the carbon content of a sintered hard metal. The test also serves for rapid discrimination between hard metals with cobalt or nickel interfaces. It is based on the rapid dissolution of the binder phase by a concentrated mixture of acids and the formation of either green hydrated nickel salts or pinkish hydrated cobalt salts. The carbon content is indicated by the richness of the final color. As the carbon content rises, the color brightens, because the binder phase contains less tungsten, leading to a better acid solubility.

PROCEDURE. Place a drop of a freshly prepared mixture of conc. nitric acid and hydrochloric acid on the surface of the specimen with the end of a narrow (2 mm diameter) glass rod. Photograph the progress of the reaction at 1.5, 3, and 9 min. The clarity of the green or pink colors may be calibrated using hard metals with known carbon contents.

Extraction methods of separation are often integral parts of spot test analyses. The unique value of solvent extraction as a means of isolation and concentration has to be emphasized. Trace amounts of materials can be selectively extracted and estimated: Auric chloride, for example, can be isolated and concentrated by extraction into *n*-butyl alcohol and the metal then detected or determined with any selective gold test.

A simple procedure for separation from a single drop of sample is made possible by the Weisz ring oven (3). Soluble salts or salts extracted through the action of acids, bases, or complexing agents diffuse through paper to an outer ring, where the solvent is evaporated by a heated ring. The diffused materials are thus concentrated in a thin ring and an efficient, rapid separation from the main residue is obtained. The main advantage of the method is that the solvent is quickly and easily evaporated to leave the extract in a concentrated form immediately available for direct spot test analysis.

Concentration of trace amounts of materials by precipitation often makes use of collecting agents. Numerous collectors are described in the literature and may successfully be applied to spot test analysis. For example, silver oxide is used as a collector for metals forming hydrated oxides, whereas copper sulfides "gather" metals precipitable as sulfides. Copper sulfide collectors can concentrate 0.02 μg mercury from an enormous dilution of 1 : 2,500,000 (4).

Trace amounts of heavy metals can be enriched very simply and effectively on chemically modified filter papers (5). Sample solutions containing part per billion concentrations of these metals are passed through a modified paper. Ten milliliters of the sample solution is adjusted to pH 5.3 with acetate buffer and then sucked 10 times at a 10-ml/min velocity through filter paper containing silica gel (Whatman SC_T 81) to which dithiocarbamate groups have been bonded.

Comparison of the speed of decolorization on filter paper may serve for quick quantitation (6). A spot of the sample solution and also spots of the third and sixth standard of a series of six standards having concentrations in the ratio 1 : 2 : 4 : 6 : 8 : 10 are placed on filter paper and the color is developed with a suitable reagent. The color is then bleached with a second reagent, and the order in which the spots are decolorized is noted. The procedure is repeated once or twice with two more of the standards and then with the actual sample diluted. The concentration is finally calculated by a weighted average method.

A simple pretreatment using selective volatilization is occasionally used to effect easy separation. As little as 0.008 μg osmium can be detected selectively by exploiting the volatility of perosmic acid. Germanium can also be simply separated, by distillation of germanium chloride, and the volatilization of arsenic hydride makes the arsenic test highly selective.

Stable reagent tablets containing the reagent material, buffering agents, and inert, powdered, water-soluble or insoluble carriers constitute the reagent system in many spot reactions, taking the place of reagent solutions. The carriers are impregnated with the reagent system and then dried. The tablets are often superior to reagent solutions in stability and economy, and eliminate the tedium of reagent preparation. Premeasured reagent powder pillows (Hach) have better solubility features than do tablets.

REFERENCES

1. R. Gilmont, *Anal. Chem.* **20,** 1109 (1948).
2. B. Roebuck and A. T. May, *Z. Prakt. Metall.* **18,** 31 (1981).
3. H. Weisz, *Mikrochem. Acta,* 140 (1954).
4. A. H. Stock, F. Lux Cucuel, and H. Kohle, *Z. Angew. Chem.* **46,** 62 (1933)
5. G. Gendre, W. Haerdi, H. R. Linder, B. Schreiber, and R. W. Frei, *Int. J. Environ. Anal. Chem.* **5,** 63 (1977).
6. H. Weisz, S. Pantel, and R. Giesin, *Anal. Chem. Acta* **101**(1), 187 (1978).

CHAPTER

3

THE APPLICATION OF SPOT TEST IN CLINICAL ANALYSIS

3.1. GENERAL

The urine continuously formed in and excreted from the human body provides important information—at least in the screening stage—with regard to many health anomalies. The general profile of health indicated by the analysis of urine is used in the preliminary examination of individuals in hospitals and in assessing new applicants for insurance policies and inductees into the army.

The history of urine study reaches back to the Babylonians and Sumerians (1), who tried to relate the physical appearance of the urine to various sicknesses. Tasting of urine was used in primitive medicine to detect diabetes mellitus. One method practiced by early Hindu physicians was ingenious. These doctors observed that urine containing high levels of glucose attracts ants and they reached the conclusion that such urine was excreted by individuals suffering from carbuncles.

In the Middle Ages simple sensory and visual tests on urine were carried out, with observations being made on color, consistency, transparency, quantity, sediments, odor, and froth. It is highly interesting that in those days, the appearance of a peculiar odor in children's urine indicated mental retardation. Was the peculiar odor due to phenylpyruvic acid, and was the disease detected phenylketonuria?

Exact scientific methods for the examination of urine were not practiced until the 18th century, when the black urine of alkaptonuria was described, protein was detected in urine by deproteinization by means of boiling the sample with acids, the correlation of sweet-tasting urine and diabetes was discovered, and the specific gravity of urine was measured. The first two spot tests employed in urinalysis were the measurement of the pH of urine by litmus paper and the detection of uric acid in urine by means of silver-carbonate-impregnated filter paper strip, both by William Prout in 1859. Trommer, in 1841, detected glucose in urine by the reduction of glucose by cupric ions in hot alkaline solution. The modification of this test devised by Fehling a decade ago is still in use in some clinical laboratories.

Although literally thousands of compounds are present in urine in very small quantities, the presence or absence of most of these substances is not known to be associated with disease states. However, substantial changes in the concentrations of glucose, protein, ketone bodies, bilirubin, urobilinogen, nitrites, leukocytes, erythrocytes and in the pH of the urine may have significant diagnostic value.

3.2. SCREENING FOR GLUCOSE IN URINE

Glycosuria, the presence of glucose in the urine, is the oldest known symptom of diabetes mellitus, and is still frequently the first to be discovered. This is why a simple, rapid test for urinary glucose is the most important method of screening for unidentified diabetics. Early diagnosis of diabetes is of decisive importance, for early treatment can prevent or at least delay the development of sequelae. Normal urine contains a minute quantity of glucose, insufficient to give a positive reaction with most of the tests.

The chemical basis of practically all of the commonly used tests for recognizing glucose in the urine is either the specific enzymatic catalysis of glucose oxidation by oxygen in air (2), or the reduction of copper by the sugar in a hot alkaline solution (Benedict's reaction) (3).

The compact spot test configuration of the Benedict's reaction is the Clinitest® reagent tablets (Ames Company, Elkhart, Indiana), which serve for the quick, simple, semiquantitative estimation of sugar in urine. They are an ingenious adaptation of the alkaline copper reduction test in self-heating tablet form. Each tablet contains copper sulfate, sodium hydroxide, sodium carbonate, and citric acid. Sodium carbonate and citric acid form an effervescent couple, which facilitates the rapid solution of the tablet and generates a little heat. Much more heat is liberated by the solution of sodium hydroxide and its partial neutralization by citric acid. In the alkaline medium sugar reduces the blue copper sulfate solution to reddish, insoluble cuprous oxide. The carbon dioxide displaces the air from above the reaction and prevents reoxidation of the cuprous oxide during the test. The color of the mixture indicates the proportion of sugar in the urine.

The Clinitest tablets are not specific for glucose. They react with the reducing sugars, such as glucose, lactose, galactose, and fructose; with nonsugar reducing substances such as ascorbic acid; and with reducing drugs such as salicylates.

According to a clinical study (4), the tablet test carried out in the usual recommended way (i.e., with 5 drops of urine) is accurate in the 0–2%

sugar concentration range. Above 4% there is a reversal of color that can be mistaken as indicating a 0.75–1% concentration and which is referred to as "pass through." When 2 drops of urine are used instead of 5, the accuracy range of the Clinitest is extended to 0–5% and the "pass through" delayed until a concentration over 10% sugar is reached. In this case, the color changes are as follows:

0%	Trace Amount	0.5%	1%	2%	3%	5%
Blue	Blue-green	Light green	Olive green	Dark brown	Light brown	Orange

Because the Clinitest, based on the alkaline reduction of copper, is not specific for glucose, meliturias involving sugars other than glucose affect the outcome of the reaction. These include pentosuria, fructosuria, galactosuria, lactosuria, sucrosuria, and manniheptulosuria, and are relatively rare. In a series of 352 urine samples from healthy subjects only one was found to give a positive reducing test with Clinitest, but in this case a negative result given by a glucose-specific enzymatic test (2) suggested by the presence of nonglucose reducing substances. This is a rate of occurrence of approximately 0.3%. In 1423 urine samples from hospital patients, the number of positive reducing tests resulting from the presence of nonglucose substances increased to about 1%. Jenson, in discussing the meliturias, has indicated that even though diabetes mellitus is the melituria of major importance, other meliturias are significant and not always benign (5). It is important to differentiate the other sugars from glucose to avoid unnecessary treatment for diabetes (6).

Although the Clinitest tablets gave a false positive reaction with saturated solutions of methyldopa, the urine of hypertensive patients receiving methyldopa has shown no false reactions.

In 1957 a specific enzymatic glucose test was published (2) that bypassed the classical principle of alkaline reduction of copper ions. The test was based on the activity of the enzyme glucose oxidase, which was identified by Muller (7). It was reported (8) that although the enzyme had a high specificity for glucose, it would nevertheless react slowly with mannose, xylose, maltose, and galactose. It turned out that the concentrations of these other sugars encountered in the urine are nonreactive with the reagent system.

The reagent system for the enzymatic test is composed of a buffered mixture of glucose oxidase, peroxidase, and *o*-tolidine. Glucose oxidase catalyzes the oxidation of β-D-glucose, but has no effect on the oxidation of α-D-glucose. A freshly prepared solution of crystalline glucose contains only the α-form and is essentially nonreactive with the glucose oxidase test. However, in solution an equilibrium mixture between the α- and β-forms is readily established. In blood and urine containing glucose, equilibrium mixtures of α- and β-glucose are present, so β-D-glucose is always available for immediate reaction with the enzyme.

The reaction consists of two consecutive steps. First, glucose is oxidized by atomospheric oxygen in the presence of glucose oxidase to gluconic acid and hydrogen peroxide:

$$\text{Glucose} + O_2 \xrightarrow[\text{oxidase}]{\text{glucose}} \text{gluconic acid} + \text{hydrogen peroxide}$$

Second, hydrogen peroxide in the presence of the enzyme peroxidase oxidizes *o*-tolidine to a blue quinoidal oxidation product:

$$\text{Hydrogen peroxide} + \underset{\text{(white)}}{o\text{-tolidine}} \xrightarrow{\text{peroxidase}} \underset{\text{(blue)}}{\text{oxidized } o\text{-tolidine}} + H_2O$$

In a newer test configuration, Diastix® (Ames Co., Elkhart, Indiana), the *o*-tolidine in the second step of the reaction is replaced by potassium iodide, which is oxidized by hydrogen peroxide to form free iodine. This oxidation is also catalyzed by the peroxidase.

In spite of the fact that it is not possible to quantitate the amount of glucose in urine by the enzymatic test because of the different pH values, temperatures, and concentrations of ascorbic acid in urine, the test has become very useful in diabetes-detection screening programs, in routine urinalysis in hospitals and physicians' offices, in the differential diagnosis of diabetic coma from insulin shock or other coma, in regular urine testing on known diabetics by both physicians and patients, and in differentiation of glucosuria from other meliturias.

A self-containing spot test configuration of the enzymatic glucose test employing dry-reagent chemical technology appeared in 1956—Clinistix® (Ames Co., Elkhart, Indiana). A firm plastic strip (about 8 cm long) to which a stiff absorbent cellulose area was affixed was impregnated with a buffered mixture of glucose oxidase, peroxidase, and *o*-tolidine. These reagent strips were especially convenient for use in testing urine because they were stable and disposable. Their existence not only stimulated the incorporation of known and new spot tests for urine into very simple and comfortable "dip and read" stick configurations, but also generated rapid developments in the same direction for other diagnostic tests. Nowadays, several companies are offering plastic strips with eight or nine separate reagent areas affixed that fulfil in a single minute the function of a routine urinalysis laboratory (N-Multistix, Ames Co., Elkhart, Indiana; Combur 9 Test, Boehringer-Mannheim; Rapignost, Hoechst AG).

Many clinical studies have confirmed the "dipstick tests" as simple and reliable procedures. Comparative studies between N-Multistix (Ames Co.) and Glucotest (Boehringer-Mannheim) (9, 10) indicated a marginally higher sensitivity for the latter. False negative results were obtained only in the presence of levodopa, ascorbic acid, glutathione and dipyrone (11).

To eliminate interpersonnel variation in interpretation of the dipstick colors a reflectance scanning instrument (Clini-Tek Analyzer, Ames Co.) may be used (12). It irradiates the dipstick with polychromatic light and measures the reflected radiation with use of an appropriate filter. In tests of this system, reference urine samples containing glucose, protein, ketones, occult blood, nitrite, bilirubin, and urobilinogen (but not including drug metabolites or pathological substances) were measured at different concentrations of each substance. Good reproducibility was achieved for glucose, ketones, urobilinogen, and nitrite; values for occult blood and bilirubin were the least reproducible.

3.3. SCREENING FOR PROTEIN IN URINE

Protein tests are, next to glucose tests, the most frequently performed tests in routine urinalysis. In 1827, Richard Bright of Guy's Hospital in London made the first known clinical correlation of urinary excretion of albumin with kidney disease. His coagulation of urine protein with heat has since been modified to include the addition of acetic acid to give the common laboratory procedure known as the heat and acetic acid test. The coagulation of protein with various types of acids has led to a series of turbidimetric tests (13–16). These tests use sulfosalicylic acid for deproteinization and turbidimetric measurement for estimation of the amount of coagulated protein particles.

The first spot test configuration of the sulfosalicylic acid test was introduced in 1950 as the Bumintest® tablet (Ames Co.), which contained a premeasured amount of sulfosalicylic acid. The tablets were dissolved in water and the resulting solution was mixed with urine to precipitate protein; the amount of protein was then visually estimated by the degree of turbidity of the solution.

This turbidimetric method of estimating urine protein is not specific, nor is it designed to quantitate the amount of protein present. X-ray contrast media and metabolites of tolbutamide may give false positive results, and highly buffered alkaline urines may give false negative results. The interpretation of results is also difficult when the urine is initially turbid.

The main advantage of the sulfosalicyclic method is that it reacts with all the proteins tested, including γ-globulin and Bence-Jones protein (17).

The most important spot test for native albumin, widely applied today by manufacturers of diagnostic products, is based on the "protein error" discovered by Sørenson (18) in 1909. Sørenson was the first to show that proteins and their decomposition products often interfere with colorimetric pH determinations. In the presence of these colloidal materials, the colorimetric values are higher than those obtained by potentiometric methods. This influence of proteins is due to their ability to form soluble compounds, and the extent of the error depends on the indicator used.

The analytical use of anomalies in color reactions carried out in the presence of protective colors was demonstrated by Feigl and Anger (19) in their spot test for native albumin in urine, but was only utilized by a commercial company 20 years later in a system applying dry-reagent chemical technology (Albustix®, Ames Co.). This test exploits the fact that the blue aqueous solution of the potassium salt of tetrabromophenolphthalein ethyl ester becomes yellow on addition of dilute acetic acid. The blue can be restored by adding alkali. Consequently, such solutions behave like dyestuff indicators. If, however, a solution or suspension of native albumin is added to a dilute blue solution, a great deal of acetic acid can be introduced without causing the blue to yellow change. Only concentrated acetic acid or mineral acids are capable of breaking the bond between the protein and the ester to cause the color change. This test is sensitive and selective for native albumin, since its degradation products do not bring about colloidal masking.

PROCEDURE. Mix a drop of the test solution on a spot plate with a drop of the blue reagent solution and then acidify with a drop of 0.2 *N* acetic acid. A blank turns yellow, but the blue or greenish color persists if the sample contains protein. The reagent is a 0.1% solution of the potassium salt of tetrabromophenolphthalein ethyl ester in alcohol.

The test revealed 0.5 μg albumin, 0.5 μg casein, 0.5 μg hemoglobin, 5 μg edestin, 0.35 μg serum albumin, 0.5 μg clupeine, 0.5 μg salmine, and 1 μg gliadin.

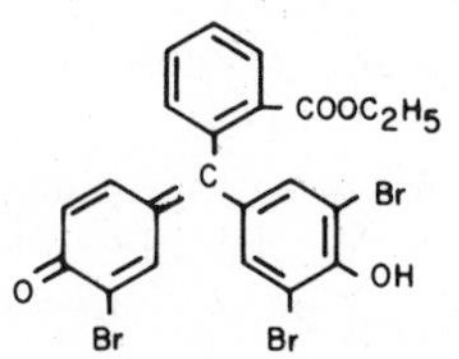

$COOC_2H_5$ C Br OH O Br Br

Tetrabromophenolphthalein ethyl ester

The same protein error principle is used in the solid-state "dip and read" reagent system Albustix® (Ames Co.), using 3′,3″,5′,5″-tetrabromophenolsulfophthalein, and a similar configuration called Albym-Test® (Boehringer-Mannheim) uses 3′,3″,5′,5″-tetrachloro-3,4,5,6-tetrabromosulfophthalein. These standardized color tests for the elimination of protein in urine are composed of a strip of absorbent cellulose one end of which is impregnated with the indicator and buffered at pH 3.

These dipsticks have been used for many years in routine urinalysis and their convenience and clinical usefulness are well established. There are, however, some shortcomings. The test is most sensitive to albumins, and appreciably less sensitive to γ-globulins, Bence-Jones proteins, and muco-proteins. These same proteins do not interfere with the reaction of the dipsticks with albumin (17).

Addition of detergent to urine samples influences the results of the dipsticks. However, the concentrations of detergents that produce these effects are relatively high, so the detergent effect should not ordinarily be observed in the common laboratory setting. However, contamination of the screening by quaternary ammonium compounds used to clean the container and not thoroughly rinsed out may cause false positive results.

Although the reagent strips are buffered for a wide range of urinary pH values, specimens that are exceptionally alkaline may give positive results in the absence of protein, due to the buffer of the strip being overcome. Such urines are obtained from patients receiving alkaline medication. Stale or fermented urines may also be extremely alkaline.

On the other hand, the readings from the sticks are unaffected by urine turbidity, x-ray contrast media, tolbutamide, *p*-amino salicylic acid, sulfafurazole, and penicillin, which cause positive interference with the Bumintest tablet. The paper strip test for urine protein does give false positive reactions in the presence of chlordiazepoxide, quinidine, and chlorpromazine, whereas the presence of benzyl-pennicilin, and iodipamide resulted in false negative responses (11). The broader specificity of the sulfosalicylic acid method makes it an important additional method for routine use. Mutual confirmation of readings can add certainty to reporting and eliminate diagnoses of Bence-Jones proteinuria and globulinuria.

3.4. SCREENING FOR KETONE BODIES IN URINE

The detection of ketone bodies in urine is especially important for the timely recognition of metabolic decompensation in diabetics. "Ketone bodies" is a term applied to three compounds that appear in blood and urine as a result of excessive lipid metabolism or minimal carbohydrate metabolism: acetone, acetoacetic acid, and β-hydroxybutyric acid.

Coma and precomatic states are almost always accompanied by ketoacidosis and ketonuria. Ketonuria is also found in starvation and in the course of weight-reducing diets in which carbohydrate intake is reduced and replaced by protein-rich food.

In diabetic ketoacidosis, insulin deficiency shifts the balance between lipolysis and lipogenesis in favor of the former. As a result, fatty acid catabolism is accelerated, liberating larger amounts of acetoacetic acid, part of which is then converted to β-hydroxybutyric acid and acetone. Diabetics whose disorder has never been diagnosed can become ketoacidotic suddenly. Testing for ketonuria is very important in the monitoring of the management of both juvenile and maturity-onset diabetics, and should be carried out quite often.

The two types of reactions most commonly employed for ketone detection are the nitroprusside tests (20), which react with acetone and acetoacetic acid, and the ferric chloride tests, which react with acetoacetic acid but not with acetone.

The basis of the nitroprusside color reaction is that the NO group of the nitroprusside reacts with the ketone to give isonitrosoketone, which remains in the complex anion. At the same time, the iron (III) is reduced to iron (II).

$$[Fe(CN)_5NO]^{2-} + CH_3COCH_3 + 2OH^- \rightarrow [Fe(CN)_5ON{=}CHCOCH_3]^{4-} + 2H_2O$$

Other methyl ketones that contain an enolizable CO group give analogous color reactions, whereas ketones that lack methyl or methylene groups bound to CO groups are not active in this respect.

Ferric chloride tests are very insensitive, give false positive reactions with such common drugs as salicylate, and develop, in the presence of certain biological compounds, other colors that mask the true acetoacetic acid reaction.

Free and Free (21) have found that in the nitroprusside reaction the whole response is given by acetoacetic acid, acetone making no significant contribution to such ketone body reactions. This also applies to the ketone body analysis of serum. In both serum and urine, acetoacetic acid is present in ketoacidosis in much greater quantities than acetone, and it is about 10 times as reactive with nitroprusside.

The first tablet-form spot test configuration of the nitroprusside reaction was the Acetest® (Ames Co.), in which the reagents are compounded into a single tablet containing sodium nitroprusside, disodium phosphate, and aminoacetic acid. The test procedure is to place a tablet on a clean white surface, such as typing paper, then place a drop of urine on the tablet, and 30 sec later compare the color of the top of the tablet with the color chart provided, which has three color blocks with various shades of purple representing trace, moderate, and strongly positive reactions. The tablet remains white or turns cream colored if the urine is negative for ketone bodies. The sensitivity of the test is 5 mg acetoacetate/100 ml, but the color blocks are calibrated for 10–20 mg/100 ml, designated as the trace amount.

Color Block	Acetoacetic Acid (mg/100 ml)	Acetone (mg/100 ml)
Trace (+)	10–20	25–50
Moderate (++)	25–40	200–250
Strong (+++)	>50	400–1000

Reagent strips for testing of ketone bodies with similar compositions to that of Acetest® are in commercial distribution in single or multiple test systems (N-Multistix; Ames Co.; Combur 9 Test, Boehringer-Mannheim; Rapignost®, Hoechst AG). The spot test for ketone bodies in the strip form consists of stiff absorbent cellulose impregnated with a buffered mixture of sodium nitroprusside and glycine. The procedure for use is to dip the coated end of the strip in urine, remove it 1 min later, and compare the color of the dipped portion with a color chart. Three color blocks from pink to purple represent small, moderate, or large amounts of ketone bodies. The test is practically nonreactive to acetone, since concentrations of acetone of 1–2% in urine give no color after 1 min.

The active ketone in this reaction—acetoacetic acid—is prone to decomposition, and acetone is volatile. Both processes are speeded up by heat. The presence of yeast or bacteria can also cause the rapid removal of acetoacetic acid from urine.

There are some substances besides the ketones in the body fluids that give color reactions with the reagent system. The first group of such compounds is the phenylketones, which give colors ranging from brick red to greenish yellow unlike the shades on the chart. These occur only rarely, for example, in the urine of untreated phenyketonuria. The other potentially disturbing compounds in this test are three indicators: bromsulfalein, phenolsulfonphthalein and phenolphthalein (11). At the alkaline pH prevailing on the strip, these indicators turn shades of red or purple, which, although not matching the colors on the chart, could cause confusion. Bromsulfalein is administered as a test of liver function, phenolsulfonphthalein is occasionally injected as a test of renal function, and phenolphthalein is an ingredient of some laxative preparations; thus only phenolphthalein might be self-administered. A suspected positive result from one of these three can be confirmed by adding alkali to the specimen, when the color-change will appear.

No drug or drug metabolite in urine is known to inhibit the strip test for ketone bodies.

3.5. SCREENING FOR BILIRUBIN IN URINE

Bilirubin, a product of hemoglobin catabolism, is formed in the reticuloendothelial system, the spleen, and Kupffer cells in the liver. It is present in the circulating blood at all times, and is one of the important components of bile. Bilirubin is not detectable in the urine of healthy persons. The bilirubin excreted in the bile is conjugated in the liver as the mono- or diglucuronide, and this conjugated bilirubin is the precursor for urobilinogen, which is formed in the intestine. The conjugated bilirubin is lipid soluble and water insoluble. Conjugation with glucuronic acid converts the water-insoluble bilirubin to water-soluble compounds that can be excreted via the urine.

In the blood, bilirubin is normally in the free form and is bound to the serum proteins. If this normal pathway of bilirubin metabolism is interfered with in any way due to an intra- or extrahepatic obstruction, excess quantities of the glucuronide may appear in the blood. If there is inflammation of the liver that prevents the normal excretion of conjugated bilirubin into the bile, the conjugated bilirubin backs up into the bloodstream and appears in the urine, since the glucuronide is soluble and can be excreted by the kidney. This occurs in liver damage caused by infectious agents or hepatotoxic agents, such as carbon tetrachloride and other organic solvents, and in biliary obstructions.

Bilirubinuria is encountered in intra- and extrahepatic obstructive jaundice, hepatocellular jaundice, acute and chronic hepatitis, and hepatocirrhosis.

There are a large number of tests for bilirubin in urine. These are mainly based on the following procedures:

1. dye dilution procedures involving the blending of the yellow bilirubin color with a dye such as methylene blue or methyl violet;
2. the oxidation of bilirubin to give colored derivatives, most often green biliverdin; and

3. diazotization procedures in which bilirubin is coupled to a reagent to produce a colored compound.

The need for a quick and a reliable spot test for urinary bilirubin received added emphasis from the increased incidence of liver disease after World War II. There was an urgency for a simple, sensitive, and reliable bilirubin test that could be used routinely by the physician for the detection of bile in the urine and employed in the army and in industry for mass screening examinations to detect latent liver disease.

Harrison (22) employed barium chloride in his test for bilirubin, which has come to be widely known as the Harrison test. In this procedure urine and barium chloride solution are mixed, after which the voluminous precipitate of insoluble barium salts is filtered off onto filter paper. Fouchet's reagent (10% $FeCl_3$, 25% trichloroacetic acid, in water) is then dropped on this precipitate, and if bilirubin was present in the initial urine sample, the characteristic green color of biliverdin is noted.

This relatively simple procedure was further simplified to a point where it was much more applicable to mass usage, as in the armed service or industry. In this simplified bilirubin spot test (23) barium-chloride-impregnated dry paper strips are dipped into the urine samples for 30–120 sec and then spotted with two drops of Fouchet's reagent. A positive test is denoted by the appearance of a green color, which varies in intensity with the amount of bilirubin present. It is claimed that barium acts not only as an efficient adsorbent, but also may serve a catalytic function in this reaction. Calcium ions appear to adsorb bilirubin from the urine just as efficiently as does barium, but the sensitivity of this reaction to the calcium-adsorbed bilirubin is considerably lowered.

This barium strip modification of the Harrison test has best met the requirements of sensitivity, simplicity, and applicability to mass and serial use. With this test, higher concentrations of bilirubin give a definite green color. However, at low concentrations of bilirubin (<0.1 mg/100 ml) the color becomes blurred and indistinct.

A reagent tablet that can be prepared simply and inexpensively was described by Franklin (24). This kit makes use of the same principles of concentration of biliary pigment and subsequent oxidation of the bile pigment by Fouchet's reagent as in the Harrison spot test. The tablets used in this test concentrate bilirubin by the processes of filtration, adsorption by gypsum particles, and barium chloride adsorption.

PREPARATION OF TABLETS. Pour a mix of commercial plaster of Paris (350 g) and saturated barium chloride solution (250 ml) into a rubber mold with appropriate indentations. After they become firm, dry the tablets in an oven.

PROCEDURE. Drop 10 drops of urine slowly on the surface of the tablet so as not to cause any overflow. If bilirubin is present in the urine, a fine layer of yellow bile pigment will be adsorbed on the surface of the tablet. The remainder of the urine will filter through. Then drop one or two drops of Fouchet's reagent directly on the yellow area. A positive test is denoted by a blue-green color, which varies in intensity with the amount of bilirubin present. Normal urinary pigments give neither the yellowish discoloration on the tablet surface nor a positive test with the use of Fouchet's reagent.

In another tablet-configurated reagent system Free and Free (25) described a diazotization procedure for identifying bilirubin in the urine under acidic conditions using a diazonium salt of the formula

$$\rho - AC_6H_4N : N^+ - 0_2SB$$

where A is a -COOH, sulfo, or nitro group and B is a phenyl, ρ-tolyl, or 5-nitro-o-tolyl group. The diazonium salt couples with the bilirubin to give a characteristic purple color.

The reaction may be carried out on a fiber test mat using a tablet containing 0.6 parts *p.* nitrobenzene diazonium *p.* toluenesulfonate, 10 parts $NaHCO_3$, 100 parts sulfosalycilic acid and 20 parts diluent such as boric acid. The sensitivity of the reaction is 0.05–0.1 mg bilirubin/100 ml urine, which coincides with the lower limit of accepted pathological significance.

Since 1970 dip-and-read tests for bilirubin have been available from the major companies dealing with diagnostics. These tests are also based on the coupling of bilirubin with a stable diazonium compound. To carry out the test, the strip is merely dipped into the urine and then compared with the color chart after 20 sec. Actually, this strip test for bilirubin is a part of various multiple dip-and-read systems (N-Multistix, Ames Co.; Rapignost, Hoechst AG; Combur 9 Test, Boehringer-Mannheim). The bilirubin strip test exhibits abnormal colors in the presence of methyldopa and adrenochrome (11).

3.6. SCREENING FOR UROBILINOGEN IN URINE

Urobilinogen and stercobilinogen are formed from bilirubin to a small extent in the efferent biliary ducts, but are formed predominantly in the intestines by bacterial reduction. As analytical differentiation between urobilinogen and stercobilinogen is very difficult and has no diagnostic significance, the two are called collectively "urobilinogen." It has been found (26) that antibiotics may interfere with the bacterial reduction of bilirubin to urobilinogen. The amount of urobilinogen formed is a function of the bilirubin excreted via the bile into the intestine.

Urinary excretion of urobilinogen is increased when the capacity of the liver for enterohepatic circulation of bile pigments is restricted or overburdened, or when the liver is bypassed. If excessive bilirubin concentrates in the bile, then more urobilinogen will be formed. If bilirubin excretion by the liver is decreased, then less urobilinogen than normal will be formed. The major portion of the urobilinogen formed in the intestine is excreted in the stool. Some of the color of the stool is due to urobilin, and most of the darkening of fecal specimens in air is due to urobilin formation from the urobilinogen present.

A small proportion of the urobilinogen is excreted by the kidney into the urine. Small amounts of urobilinogen are normally present in urine. If the liver's excreting capacity for urobilinogen is impaired, then a proportionally greater amount of urobilinogen will be excreted in the urine. If some blockage of bilirubin metabolism occurs, so that less bilirubin enters the intestine, there will be a corresponding decrease in urobilinogen formation, reabsorption, and excretion by the kidney. If there is excessive bilirubin production, there will be increased excretion of bilirubin into the gut and increased urobilinogen production, reabsorption, and urinary excretion.

Urobilinogen was first estimated by a spot test leading to fluorescence (27). Urobilinogen was converted to urobilin, which fluoresces markedly in a slightly alkaline alcoholic solution of zinc acetate. In working with pure solutions of urobilin, the sensitivity reached was 0.002% (1 : 50,000 dilution limit).

The classical reagent for urobilinogen in urine, the Ehrlich reagent, was introduced by Otto Neubauer in 1903, and is still part of some of the most modern dip-and-read sticks. The Ehrlich reagent is a *p*-dimethylaminobenzaldehyde in hydrochloric acid solution that forms a red Schiff base when reacted with urobilinogen. The reaction is extremely sensitive, having a lower limit of 1.5 ppm urobilinogen. A rough but clinically very useful index of the amount of elevated urine urobilinogen was derived by Wallace and Diamond (28), who, using the Ehrlich reagent, carried out serial dilutions of the urine and noted the actual dilution that still gave a detectable pink color.

In 1969 the Ames Company introduced the first dip-and-read solid state reagent for urobilinogen. It gave the added advantage that the test could be done quickly on spot specimens and thus urobilinogen was less apt to deteriorate. The test consists of 2.9% (w/w) *p*-dimethylaminobenzaldehyde and 97.1% (w/w) acid buffer. The stick is dipped in the urine specimen and compared with a color chart 60 sec later. The color blocks are yellow for normal values (0.1 and 1 Ehrlich unit/100 ml) and darker orange to brown-red for values of 4, 8, and 12 Ehrlich units/100 ml.

Another strip test [Ugen Test®, Boehringer-Mannheim (29, 30)] is based not on principles of condensation of urobilinogen with *p*-aminobenzaldehyde, but rather on the reaction of the urobilinogen with polysaturated benzenediazonium salts to produce red to blue condensation products in the presence of acids. A typical impregnating solution for test papers contains 0.3 g 4-methoxy benzenediazonium fluoroborate, 10 g metaphosphoric acid, 0.4 g sodium dodecyl benzenesulfonate, and 5 ml methanol diluted to 100 ml with water.

This test is so sensitive that it reveals even physiologic levels of urobilinogen in urine by a weak pink coloration. The practical sensitivity limit is about 7 μmol/liter (0.4 mg/100 ml), whereas the upper limit of physiologic urobilinogenuria is 17 μmol/liter (1 mg/100 ml).

This test is not affected by the presence of porphobilinogen, indican, sulfonamides, or sulforylurea, which occasionally occur in urine. 4-Aminosalicyclic acid, adrenochrome, chlorprozamine, isoniazid, and methyldopa, on the other hand, interfere with the urobilinogen strip color reaction (11).

3.7. SCREENING FOR NITRITE IN URINE

Testing of urine for nitrite is important, because it can be used to detect urinary tract infection. Pathogenic bacteria in urine, such as *Escherichia coli,* reduce the nitrate present to nitrite, and thus its presence indicates bacterial infection.

Urinary tract infection is generally a disease of women. Bacteriuria can be determined in about 1–4.5% of girls between 5 and 16 years of age (31). Continual screening for urinary tract infection and infectious nephropathia, especially among women and persons falling into the high-risk category, makes possible the early discovery and treatment of incipient pathological processes.

The nitrite test is so easy to carry out and evaluate that follow-up tests to detect relapses can be performed by the patient himself.

The basis of the spot test used in the Nitur-Test® (Boehringer-Mannheim) and in the Rapignost® strip system (Hoechst AG) is the classical Griess test for nitrite. The Griess test (32) uses naphthylamine for diazotization with nitrous acid to form diazonium cations:

$$[Ar{-}NH_3]^+ + HNO_2 \rightarrow [Ar{-}N{\equiv}N]^+ + 2H_2O$$

These cations may condense with primary aromatic amines such as sulfanilic acid to form the red-colored *p*-benzenesulfonic acid-azo-α-naphthyl-amine. The diazonium cation formed between the nitrite ion and sulfanilic acid may be coupled with a hydroxypyridine (33). The azo dye produced serves as a basis of a highly sensitive spot test for nitrite.

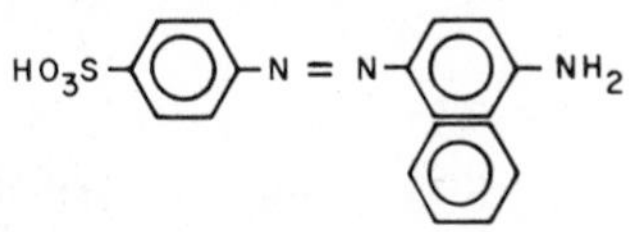

HO_3S—C₆H₄—N = N—C₁₀H₆—NH_2

PROCEDURE. Add to one drop of urine two to three drops of aqueous ammonia (1 : 1, v/v) and one drop each of 1% sulfanilic acid solution dissolved in 30% aqueous acetic acid and 0.5% alcoholic solution of *N*-amino-3-hydroxypyridine. The appearance of an intensely colored red azo dye indicates the presence of nitrite in the urine sample.

In the strip formulation of the Nitur-Test®, the aromatic amine sulfanilamide reacts with nitrite in acid buffer medium to form a diazonium salt, which then couples with 3-hydroxy-1,2,3,4-tetrahydrobenzo-quinoline to give a red azo dye. The intensity of the red coloration is a measure of the nitrite concentration in the urine.

The nitrite test of the Merck Co. (34) detects as little as 0.01 mg nitrite/liter. It uses filter paper impregnated with 100 ml of an aqueous solution containing methanol, 1–2 g of a diazotable amine (e.g., the sulfate of 4[2-amino-(2-hydroxyethylsulfonyl)-phenyl]benzoic acid), 0.1–0.5 g of a coupling component (e.g., the dichloride of *N*-1-naphthyl ethylenediamine), and 2–8 g of a solid organic acid. (e.g., tartaric acid). The paper is dried and cut into strips. This test is specific for nitrite and independent of urinary pH. A reddish coloration of the test paper after brief immersion in the specimen indicates that nitrite is present. The test reveals significant bacteriuria when the bacterial count reaches 1×10^7/ml urine.

3.8. SCREENING FOR OCCULT BLOOD IN URINE

Occult blood can occur in urine in nearly every type of renal disease. Pollak and Mendoza (35) have indicated that hematuria is one of the four critical findings in patients with rapidly progressing glomerulonephritis.

An abundance of data indicates that hematuria is one of the most common symptoms of hemorrhagic complications associated with anticoagulant therapy. It is one of the earliest signs of inadequate control and can be an early warning of impending serious hemorrhage. Routine testing of urine for occult blood has been urged for patients receiving anticoagulants (36). James (37) has reported that positive screening tests for hematuria may reveal cancer of the prostate or bladder in patients without symptoms. Urine testing is a simple procedure for preliminary screening, because hematuria must always be regarded as indicative of a tumor until the possibility can be excluded.

Spot tests for occult blood in urine are based on the peroxidase-like activity of hemoglobin, myoglobin, and some of their degradation products. The reaction makes use of the catalytic properties of these compounds in the oxidation by peroxide of various reagents such as benzidine or *o*-tolidine. The pseudoperoxidase activity of hemoglobin transfers an oxygen atom from the peroxide to the appropriate chromogen.

The adaptation of this spot test to the tablet configuration is the Occultest® (Ames Co.). The tablet contains *o*-tolidine, strontium peroxide, tartaric acid, calcium acetate, and sodium bicarbonate. The tartaric acid, calcium acetate, and strontium peroxide react in solution to produce hydrogen peroxide, which in the presence of hemoglobin or certain of its iron-containing degradation products oxidizes *o*-tolidine to a blue quinoidal oxidation product. The sodium bicarbonate forms an effervescent couple with a little of the tartaric acid and assists the solution of the reagents.

PROCEDURE. Place one drop of well-mixed, uncentrifuged urine in the middle of a filter paper square; then place the reagent tablet in the middle of the moist area. Pour two drops of water over the tablet and observe the color of the filter paper around the tablet exactly 2 min later. The presence of occult blood is indicated by the appearance of a blue color around the tablet.

These tablets will detect 1 part of blood in 100,000 parts of urine, equivalent to approximately 50 red blood cells/mm^3. Unlike microscopic examinations, this spot test will detect the presence of blood whether or not the cells are hemolyzed.

A more convenient development was the dip-and-read spot test for occult blood (Hemastix®, Ames Co.; Sangur-Test®, Boehringer Mannheim) (38). Both configurations use absorbent cellulose, one end of which is impregnated with the reagent system. This consists of an organic peroxide [cumene hydroperoxide and 2,5-dimethylhexane-2,5-di(hydroperoxide), respectively], *o*-tolidine, and a buffer system.

Both strips react with specimens containing intact red cells or free hemoglobin, but they are more sensitive to hemolyzed than unhemolyzed blood. When the urine contains a small amount of blood almost entirely in the form of red cells, the blue color on the strip may be rather mottled. The use of the strip is indicated in the case of hemolyzed blood as after hemolyzis the microscopic detection fails.

The practical sensitivity limit of this test for intact erythrocytes is approximately 5 erythrocytes/μl.

False positive reactions are found only with urine samples collected in containers contaminated by strong oxidizing substances such as hypochlorites. The test is not affected by other cellular constituents of urine such as epithelial cells, leukocytes, and spermatozoa. Excessively low or even false negative results are obtained in the presence of large amounts of ascorbic acid.

The presence of occult blood in urine may also be tested for with the leucomalachite green reagent, into which isoquinoline has been incorporated to increase sensitivity (39). The reagent is prepared by mixing 1% ethanolic leucomalachite green, 10% isoquinoline solution in 30% acetic acid, and 3.5% aqueous hydrogen peroxide solution in the ratio 3 : 1 : 1. Filter paper strips are then impregnated with the reagent. The detection limit of the test is 0.5 mg hemoglobin/100 ml urine. Clinical results agree well with those obtained by use of the *o*-tolidine test.

The classical benzidine test for occult blood is no longer in use because of the carcinogenic nature of this compound. 3,3′,5,5′-Tetramethylbenzidine, however, has turned out to be a safe and sensitive reagent (40) in an acetic acid–hydrogen peroxide system.

3.9. URINARY pH SCREENING

The pH of urine is dependent not only on nutritional habits but also on individual metabolic balance, and its measurement is part of most routine urinalysis. The pH of fresh urine from healthy individuals can vary considerably (pH 4.5–8), but it is generally found to lie between 5 and 6. If fresh urine is persistently alkaline (pH 7–8) throughout the day, the possibility of urinary tract infection must be considered. The pH can also contribute information in the monitoring of drug therapy, in the prophylaxis of renal calculi, and in the diagnosis and monitoring of renal tubular alkalosis and acidosis. Persistently alkaline urine may be of importance in identifying some respiratory disorders (41), and persistently acidic urine may be of importance in recognizing respiratory situations in which there is carbon dioxide retention.

The first spot test applied for urinary pH indication used litmus paper, which identified only whether the urine was acidic or alkaline. Recently several major manufacturers in the clinical diagnostics industry incorporated pH measuring strips into their multiple dip-and-read systems. The pH strips contain a combination of methyl red and bromothymol blue that produces distinct color changes from orange to blue over the pH range 5–9. They give rather consistent results that correlate well with pH meter readings. There is no possibility, however, of reading to a greater accuracy than plus or minus half a unit of pH, because slight variations in shade at any one pH may be caused by the pigment of the urine.

3.10. SCREENING FOR PHENYLKETONES IN URINE

In 1934 Fölling discovered the metabolic disease phenolketonuria (PKU) (42). He found that two mentally retarded children in the same family excreted urine containing a substance that gave a green color with ferric chloride. The substance was later identified as phenylpyruvic acid.

It was found that the chemical defect of a PKU-affected person is reflected in an inability to oxidize adequate amounts of phenylalanine to tyrosine and therefore in the production of increased amounts of other metabolic end products, for example, phenylpyruvic acid. Wallace, Moldave, and Meister (43) showed that the deficiency may be due to the absence of at least one component of the enzyme system that effects the conversion of phenylalanine to tyrosine. This enzyme is phenylalanine hydroxylase. With the increase of the concentration of serum phenylalanine, irreversible injury occurs to the brain. In the absence of phenylalanine hydroxylase, a portion of the phenylalanine is deaminated and oxidized to phenylpyruvic acid, which is excreted into the urine.

$$C_6H_5{-}CH_2{-}CH(NH_2){-}COOH \longrightarrow C_6H_5{-}CH_2{-}C({=}O){-}COOH$$

Phenylalanine → Phenylpyruvic acid

Because effective dietary management started early in life can prevent severe mental retardation, the early detection of this abnormality is crucial.

The first spot test for phenylpyruvic acid in urine was suggested by Baird (44), who impregnated paper sticks with ferric chloride and glacial acetic acid. The test was used and improved by the Ames Company in their strip test called Phenistix® (45). This consists of a stiff strip of cellulose impregnated with ferric and magnesium ions; the desired acidity (pH 2.3) is provided by the presence of cyclohexylsulfuric acid. The ferric ion forms the colored chelates with phenylpyruvic acid, whereas the magnesium ions are actually masking agents to minimize interference in the color development by urinary phosphate. The strip is dipped in the urine sample to be tested, and the reaction on the strip is checked 30 sec later against a color chart on the bottle label. If the reaction is negative, the strip remains yellow. If the reaction is positive, a gray to blue color develops, the intensity of which is proportional to the amount of phenylpyruvic acid present.

The strips are able to detect phenylpyruvic acid in urine in amounts greater than 8–10 mg/100 ml urine. Phenylketonuric children commonly excrete about 100 mg/100 ml.

In addition to phenylpyruvic acid, the reagent strips will give a green-colored product with the following compounds:

1. *p-Hydroxyphenylpyruvic acid.* The color due to this compound fades in 2–3 sec, so it does not interfere if the test reading is done after 30 sec. The compound appears mainly in the urine of premature babies.
2. *o-Hydroxyphenylpyruvic acid.* This substance is liable to occur with phenylpyruvic acid in the urine of phenylketonuric patients, and there is no necessity to distinguish between them.
3. *Imidazolepyruvic acid.* This occurs in the urine of patients suffering from histidinemia, a very rare metabolic disease.
4. *Phenothiazine tranquilizers or their metabolites.*
5. *Tetracycline.*

The urinary metabolites of aspirin form brownish-red ferric salicylates with the reagent strip, and this may mask the positive result of PKU. Because it is very unlikely that young babies would be receiving aspirin, tetracycline, or phenothiazines, this potential interference does not bear practical significance.

3.11. SCREEENING FOR CHLORIDE IN URINE

The chloride content of urine is in most instances closely related to the salt content, and thus to the sodium content. Ordinarily, the amount of chloride per volume of urine is measured, but in some instances the amount of chloride excreted over a period of time may be the preferred measurement. Determination of chloride in urine can provide important information about salt metabolism.

The two classical determinations of chloride with wet methods are the direct Mohr titration of the sample with a silver salt and the indirect procedure of Volhard, in which an excess of silver nitrate is added to the urine and the unprecipitated silver is titrated with thiocyanate solution.

Free and Cook (46) described a tablet configuration for urinary chloride testing utilizing the Mohr reaction. This semiquantitative spot test eliminates the use of liquid reagents: All of the compounds needed for the determination are combined into a single tablet. By varying the number of tablets and the amount of urine, it is possible to determine urinary chloride over a wide range of concentrations. The tablet contains silver nitrate, potassium chromate, sodium nitrate, citric acid, and sodium bicarbonate; thus, when the tablet is added to a solution, it rapidly dissolves with effervescence. An excess of silver ions after the interaction of silver and chloride leads to the precipitation of brick-red silver chromate. The tablet is 130 mg in weight, and reacts with 0.15 mg sodium chloride or 0.09 mg of chloride.

The tablet is standardized during manufacture and remains stable when kept in a closed bottle. The simplicity of the technique and of the equipment needed—a test tube and a dropper—makes it possible for an untrained person to carry out the test anywhere, enabling the physician to put the test in the hands of patients on low-salt diets for daily checks of their urine.

The other solid reagent spot test for chloride is the Quantab® Chloride Titrator (Ames Co.) described by Tompkins, Draft, and Zollinger (47), which is a compact and convenient semiquantitative measuring device for chloride in aqueous solution or dilute aqueous extractions of solids in the 0.005–12% sodium chloride range. This chloride titrator is a chemically inert plastic strip, laminated within which is a capillary column impregnated with silver dichromate. Fluid rises in the column by capillary action and continues to progress as long as chloride solutions enter the column. The reaction of silver dichromate with chloride produces a white color in the capillary column:

$$\underset{\text{(brown)}}{Ag_2Cr_2O_7} + 2NaCl \rightarrow Na_2Cr_2O_7 + \underset{\text{(white)}}{2AgCl}$$

When the capillary column is completely saturated a moisture-sensitive zone at the top of the column turns dark blue. The chloride concentration has then been measured, because the length of the white color change is proportional to the chloride concentration.

By measuring urinary chloride, a convenient self-monitoring procedure for low-salt diets can be established. Because the majority of diets nowadays contain processed food, a convenient monitoring test for salt intake for use by patients on low-salt diets is quite useful.

3.12. SCREENING FOR GLUCOSE IN BLOOD

Blood glucose tests are required in a variety of emergency ward, outpatient, and general practice situations. The number of blood glucose tests is larger than that for most biochemical tests, and on occasion the urgency with which results are needed exceeds the speed at which even the on-call service operates. There is an increasing demand for quick and inexpensive screening tests for mass surveys as well as for use in the general practitioner's surgery. These spot tests may also prove useful in small hospitals without modern laboratory facilities.

Abnormalities of glucose metabolism are among the most common encountered in clinical practice, and measurement of blood glucose is one of the most common laboratory investigations. Accurate measurement of blood glucose requires both specialized apparatus and technical skill, neither of which is immediately available to the physician called urgently to see a patient in coma or to the general practitioner caring for an ambulatory patient with brittle diabetes.

The major companies dealing with medical diagnostics (Ames Co.; Boehringer-Mannheim; Biodynamic, Inc., Indianapolis, Indiana) have devised rapid spot tests based on enzymatic oxidation of β-D-glucose (Dextrostix®, Ames Co.; Haemo-Glutotest®, Boehringer-Mannheim; B-G Chemstrips, Biodynamic). The testing apparatus consists of a firm plastic strip to which an impregnated reagent area is affixed. A semipermeable membrane serves as a barrier that retains the high-molecular-weight and cellular components on its surface, and can be wiped clean after the specified reaction time.

The determination is based on the specific glucose oxidase–peroxidase reaction. In a reaction catalyzed by glucose oxidase, D-glucose is oxidized by atmospheric oxygen in the presence of water to β-D-gluconolactone:

$$\text{D-glucose} + O_2 \xrightarrow[\text{oxidase}]{\text{glucose}} \beta\text{-D-gluconolactone} + H_2O_2$$

The hydrogen peroxide obtained in this reaction oxidizes the chromogenic system in the presence of peroxidase.

The strips provide a rapid and convenient method for obtaining approximate blood glucose values. The procedure uses only one drop of whole blood, which eliminates the time loss, venipuncture apparatus, and subsequent serum or plasma preparations and volume measurements inherent in other analytical procedures.

PROCEDURE. Quickly spread a large drop of capillary blood on the active end of the strip. After exactly 60 sec wash the drop off with a jet of cold water; then immediately compare the color of the active end with the color chart. Although significantly hypoglycemic results can be rapidly recognized with the visual test, small deviations from normal values should be confirmed by reflectance measurements.

Clinical evaluations of this test showed (48, 49) the accurate timing of the 60 sec in the procedure to be of crucial importance. The thickness of the blood film is of less importance, provided it is even.

To avoid observer error in judging color charts, a reflectance meter capable of quantitating the degree of the color change can be introduced. It enables the efficient and convenient performance of simple blood glucose determinations in the physician's office, at the patient's bedside, in the hospital outpatient department, or in a mobile screening operation.

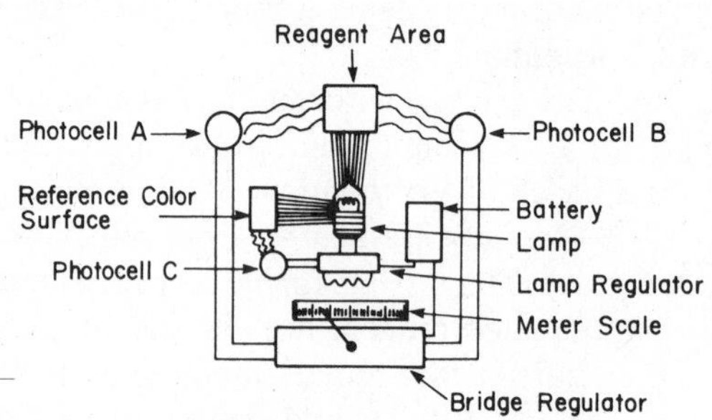

Operational Block Diagram

Reproduced with permission. © 1970 Ames Division, Miles Laboratories, Inc., Elkhart, Indiana

The reflectance meter measures the reflected light from the surface of the reacted reagent area of the strip by means of electronic circuitry and shows the reading on a calibrated meter. The lamp provides a predetermined level of illumination through a fixed aperture onto the reagent area of the strip. The light reflected from this area is a function of the color density and is transmitted to photocells A and B. The higher the blood glucose level, the darker the strip, and the less light is reflected. The light source is stabilized to maintain a constant light. Calibration is checked against standard color chips.

A clinical study (50) showed that the reflectance measurement adds significant quantitative precision to the enzyme strip technique for glucose determination. According to the evaluation of Vickers (51), however, the strip–reflectance meter combination somewhat underread the results as compared with those obtained by means of the Auto Analyzer:* A blood glucose sample measured as 20 mg/100 ml in the Auto Analyzer will read as 18 mg/100 ml; 100 mg/100 ml, as 93 mg/100 ml; 250 mg/100 ml, as 235 mg/100 ml; and 600 mg/100 ml, as 510 mg/100 ml. Thus, it is suggested that laboratories using this simplified technique should apply an appropriate correction formula.

Diabetic patients may apply for home use another simple test for glucose, which uses dried blood spots on filter paper (52).

PROCEDURE. Punch a 6.5-mm-diameter disc from the center of the dried blood spot. Elute the disc with 0.33 *M* perchloric acid solution; then determine the glucose in the eluate with any test strip based on the glucose oxidase principle.

3.13. SCREENING FOR UREA IN BLOOD

Urea is quantitatively the major end product of protein metabolism. It is formed in the liver in the process known as the Krebs cycle and is excreted predominantly by the kidneys by means of glomerular filtration. Depending on the volume of urine excreted, 30–60% of the urea is reabsorbed by the blood via the tubules.

The clinical significance of the determination of blood urea owes to the fact that permanently elevated blood urea indicates the presence of renal insufficiency. Organic and functional nephropathies may impair renal function. Some extrarenal factors may also affect the urea concentration. A high-protein diet or increased protein catabolism due to pyrexia, infectious diseases, surgery, injuries, starvation, corticosteroid therapy, burns, wasting disease, collagenoses, or hemorrhage causes the blood urea to rise. A low-protein diet causes the blood urea concentration to fall.

Rapid bedside methods for measuring blood urea include the Azostix® (Ames Co.), Reflotest®-Urea (Boehringer-Mannheim), and Merckognost® (E. Merck Co., Darmstadt F.R.G.) reagent strips. These are based on the

* Manufactured by Technicon Co. Tarrytown, N.Y.

same chemical principle: Urease catalyzes the hydrolysis of blood urea to carbon dioxide and ammonium hydroxide. The ammonium hydroxide concentration is measured by the color change of the indicator.

The Azostix strip consists of a reagent zone impregnated with urease and the pH indicator bromothymol blue. It is coated with a membrane that is freely permeable to urea but prevents staining of the paper by blood pigments. Estimations are performed by leaving a drop of whole blood on the reagent zone for 60 sec, after which it is quickly washed away with cold water and the color of reagent zone compared with color chart blocks. The color change of the indicator from yellow to green varies with the blood urea concentration, since the increase in pH is due to ammonia liberated by the action of urease on urea that has diffused through the membrane. The color chart blocks were originally claimed to represent blood urea nitrogen concentrations of 10, 20, 30, and 50 mg/100 ml, but the results of clinical evaluations showed that with this scale the strips consistently underestimated blood urea concentration (53). The manufacturer subsequently assigned new values to the blocks and the scale now reads 20, 45, 85, and 130 mg/100 ml.

Visual color matching limits the performance of this spot test. Although Azostix strips cannot replace laboratory methods, if the proper technique is observed, they should provide a screening test for azotemia when laboratory facilities are not readily available. In general and domestic practice they provide a convenient screening test for azotemia, provided that readings greater than 20 mg/100 ml are taken as indications for further investigation.

Reflotest-Urea (Boehringer-Mannheim) is a very similar test strip that uses a reflectance photometer for the quantitative measurement of the reaction product. The strip consists of two layers, one a urease paper and one an indicator paper, separated by a hydrophobic mesh laminate layer. The sample is applied to the test patch and permeates the urease paper. The gaseous ammonia liberated diffuses through the hydrophobic mesh laminate into the indicator paper, whereas the liquid components are retained. The ammonia then reacts with the indicator to produce a bluish-violet dye.

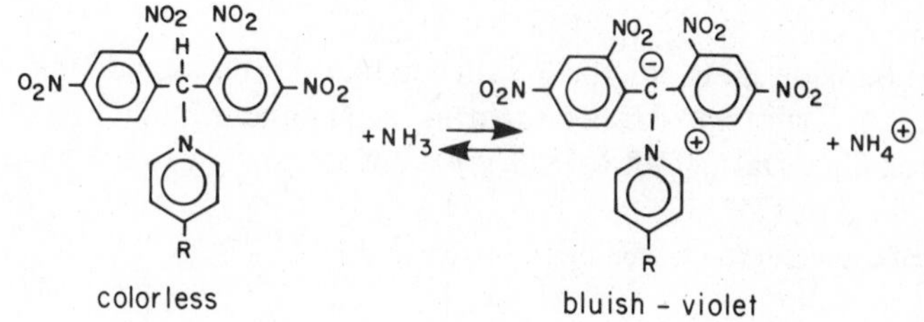

$+ NH_3 \rightleftharpoons$ $+ NH_4^{\oplus}$

colorless

bluish - violet

Although physiological and pathological concentrations of ammonia in the blood are also detected, these are in fact so small that they do not affect the urea concentration measurement. The test is not affected by hemolysis.

Clinical studies (54–56) comparing the performance of the urea test strips against alternative spectrophotometric methods showed a high correlation coefficient between the results (0.985). It was concluded that the strips are useful for routine and emergency diagnosis.

Miscellaneous Screening Tests of Clinical Importance

3.14. OCCULT BLOOD DETECTION IN FECES

Testing for occult blood in feces may lead to early detection of carcinoma, and the passage of more than 3 ml of blood in a 24-hour period is taken as an important sign of gastrointestinal disease. Loss of larger quantities of blood (50–75 ml) during a day generally imparts a dark red to black appearance to the stool, but smaller increases in blood content may not alter its appearance. Such stools are said to contain "occult blood." It is in the detection of occult blood that examination of feces can be most useful in uncovering or localizing disease.

The applied spot tests for fecal occult blood depend on the determination of peroxidase activity as an indication of hemoglobin content. Reagents used include guaiac, benzidine, *o*-tolidine, *o*-dianisidine, and 2,6-dichlorophenolindo-phenol (57). Peroxidases (including hemoglobin, which can act as either a catalase or a peroxidase) catalyze the oxidation of the test substances by peroxidase, leading to the development of various shades and intensities of blue. The chromogenic agents differ very much in sensitivity: *o*-Tolidine is about 10 times more sensitive than benzidine, and benzidine is several hundred times more sensitive than guaiac.

The classical test with benzidine is rarely performed today because of the proven carcinogenic nature of this reagent. Before this test was carried out, the patient was kept on a meat-free and green-vegetable-free diet for at least 3 days.

PROCEDURE. Mix a small fragment of the stool with ~5 ml water in a test tube and heat it in a boiling water bath for ~5 min. After cooling, add the solution to ~5 ml of a saturated solution of benzidine in glacial acetic acid to which 10 drops of 10% hydrogen peroxide solution have been added.

The presence of occult blood is indicated by the appearance of a blue color. The presence of ascorbic acid interferes with the test.

The guaiac test does not require a meat-free diet preceding the test. It detects 0.5–1 mg hemoglobin/ml aqueous solution. The serious shortcoming of this screening test is the short world supply of powdered gum guaiac.

PROCEDURE. Place ~0.5 g of feces in a test tube and add about 2 ml tap water. After mixing, add 0.5 ml glacial acetic acid and ~2 ml of a 1 : 60 (w/v) solution of gum guaiac in 95% ethyl alcohol. Add ~2 ml 3% (v/v) hydrogen peroxide, mix, and start the timer. Observe for 2 min and record the maximal development of color. In the presence of occult blood a blue color appears.

Some commercial reagent systems such as Fecatest (Finnpipette, Ky, Helsinki, Finland) (58) and Hemoccult (Smith Kline Diagnostics, Sunnivale, California) are based on the guaiacum resin, whereas other self-contained systems contain *o*-tolidine (Hemastix®, Hematest®, and Occultest®, Ames Co.), which is oxidized by the peroxidase-like reaction.

The Hematest reagent tablets (Ames Co., Stoke Pages, Bucks, England) contain *o*-tolidine, strontium peroxide, tartaric acid, and calcium acetate. In contact with water, tartaric acid, calcium acetate, and strontium peroxide produce instantly hydrogen peroxide, which in the presence of hemoglobin and certain of its iron-containing degradation products oxidizes *o*-tolidine to a blue compound.

PROCEDURE. Make a thin fecal smear on a filter paper square and place the reagent tablet across the edge of the smear. Flow one drop of water on top of tablet, wait 5–10 sec and flow a second drop on the tablet so that it runs down the sides on to the filter paper. Observe the color of the filter paper around the tablet exactly 2 min later. The presence of occult blood in the stool is indicated by the appearance of a blue color on the paper, the intensity of which is proportional to the blood concentration.

Because this test is relatively insensitive it can be applied without dietary preparation. A similar tablet configuration (Occultest, Ames Co.) has a much higher sensitivity, but cannot be used without dietary preparation of the patient.

Fecal occult blood can be screened using 2,2′-azinodi-(3-ethyl benzothiazoline-6-sulfonic acid) (ABTS) as a reagent (59).

PROCEDURE. Place a pea-size suspension of feces in 2 ml water in a boiling water bath for 2 min; then cool. Mix two drops of fecal suspension, 10 drops of reagent solution, one drop of 0.2% ethylenediaminetetraacetate (EDTA) solution, and one drop 3% aqueous hydrogen peroxide solution. In the presence of occult blood a green color is produced.

To prepare the reagent solution, dissolve 50 mg ABTS in 50 ml water and 50 ml anhydrous acetic acid and discharge the slight green color with a just sufficient amount of ascorbic acid solution.

Various phenothiazines have recently been applied for the detection of fecal occult blood (60).

PROCEDURE. Suspend 2 g feces in 5 ml water and heat for 2–3 min in a boiling water bath. Mix on a spot plate two drops of the reagent solution (0.2% EDTA, 1% prochlorperazine maleate or chlorpromazine hydrochloride) with two drops of anhydrous acetic acid, two drops of fecal suspension, and one drop of 10% aqueous hydrogen peroxide solution. The presence of occult blood is indicated by the appearance of a pink or red color at room temperature 5 min after the mixing.

3.15. SPOT TESTS FOR URINARY CALCULI

Analysis of the constituents of urinary calculi is the first step in the management of patients with calculous disease. The chemical composition of the calculi depends on the character of the urine in which they form. Since the urine may frequently change its character during the slow growth of the stone, the end result is usually a laminated calculus with layers of different color, consistency and chemical composition. The most common varieties of urinary calculi are the calcium oxalate, uric acid, ammonium urate, and phosphatic calculi.

PRELIMINARY EXAMINATION. Heat a small portion of the powdered calculus on platinum foil. This will show whether the bulk of the calculus is chiefly organic or if it contains a high proportion of ammonium salt. The presence of cystine is shown by a blue flame and a pronounced mercaptanlike odor.

PLATINUM WIRE TEST. Heat a small portion of powdered calculus in a platinum wire loop in a burner flame. After moistening it with a drop of conc. hydrochloric acid, heat it again in the flame. Calcium that was present in the original powder as oxalate or carbonate will yield a brick-red flame.

To decide whether a calculus consists chiefly of uric acid, of phosphate, or of oxalate, the followinig test is useful.

PROCEDURE. Boil a small amount of the sample with 50% hydrochloric acid. This will dissolve everything but uric acid. This insoluble uric acid is readily soluble in a solution of lithium carbonate. The presence of uric acid must be confirmed by the murexide test (below). Flocculent organic debris should not be mistaken for the insoluble uric acid. Add conc. ammonia to the filtrate. Any precipitate will indicate the presence of either phosphate or oxalate. Addition of a slight excess of glacial acetic acid will dissolve phosphate but not calcium oxalate. Confirm the presence of calcium oxalate by the flame test and of phosphate by the molybdate reaction (below). Cystine is precipitated from the ammonia solution by acetic acid. Confirm its presence by the lead test (below).

Procedure for the Murexide Test for the Presence of Uric Acid. Heat a pinch of the sample in an evaporating dish with three or four drops of conc. nitric acid. The presence of uric acid is indicated by the appearance of a red color.

Procedure for the Molybdate Test for Phosphates. Warm a pinch of the sample material with a few drops of conc. nitric acid in a test tube, add 3–4 ml 10% ammonium molybdate solution, and warm. The presence of phosphate will be indicated by the appearance of a canary-yellow precipitate.

Procedure for the Lead Test for Cystine. Boil a pinch of the powdered sample with three to four drops of 1% lead acetate solution and five to six drops of 20% sodium hydroxide. In the presence of cystine, a black precipitate of lead sulfide will form.

Rare stones may contain sulfonamide derivatives. These may be detected by the diazotization of a free amino group. The excess nitrite is destroyed by ammonium sulfamate, and a purplish-red azo dye is formed by the coupling of the diazotized sulfanilamide with *N*-(1-naphthyl)-ethylenediamine dihydrochloride.

Procedure. Add to several centigrams of the pulverized stone in a test tube two drops of 10% hydrochloric acid and wait 30 sec. Then add two drops of 0.5% ammonium sulfamate and two to three drops of 0.1% *N*-(1-naphthyl)-ethylenediamine dihydrochloride solution. In the presence of sulfonamide a brownish-pink to magenta color develops.

3.16. SCREENING FOR URINARY LACTIC ACID

Various serious disorders such as shock, hepatic insufficiency, and biguanide intoxication lead to hyperlactatemia and lactic acidosis. This diagnosis may indicate a dangerous situation with a high mortality (61). Early detection of lactic acidosis is possible by means of the tedious and time-consuming blood lactate determination. The known correlation between the blood lactate concentration and urinary lactate excretion enabled the elaboration of a simple test for urinary lactic acid, which is indicative of the blood lactate content.

One simple test for the detection of increased urinary lactic acid (62) transforms the lactic acid to acetaldehyde, which reacts with sodium nitroprusside to produce a blue color. The test easily detects urinary lactic acid concentrations of greater than 2 m*M*. Clinically relevant hyperlactatemia is detected easily and with a high degree of probability.

PROCEDURE. Add 200 μl urine to a test tube together with 30 mg calcium chloride (pressed into a tablet for convenience) and ~30 mg cerium (IV) sulfate. Cover the test tube with a piece of dried filter paper that has been impregnated with a 2.5% aqueous sodium nitroprusside solution and wetted with a drop of 10% aqueous morpholine solution. The presence of lactic acid is indicated by the appearance of a blue color on the indicator paper within 5–10 min.

3.17. SPOT TESTS FOR DETECTION OF CYSTIC FIBROSIS

Cystic fibrosis (CF) is one of the most common of the genetic diseases that lead to early death. The classical clinical indications for the diagnosis of CF are chronic pulmonary disease, pancreatic deficiency and an abnormally high concentration of sweat electrolytes.

Several screening tests are used for early detection of CF. The "palm print" test (63) estimates the concentration of chloride ions in the sweat from a palm imprint on an agar plate. The test medium is composed of silver nitrate and potassium chromate suspended in agar. The agar plates are bulky and must be refrigerated during storage.

One rapid diagnostic method for detecting CF has a strip test configuration (64). The test strip shows a characteristic color reaction at chloride concentrations larger than 60 mEq/liter when laid on the clean skin of a patient.

PROCEDURE. Dissolve 50 mg agar in 100 ml boiling water and add 0.85 g silver nitrate and 0.425 g potassium chromate in hot water. Pour 200-μl aliquot of the warm solution onto a strip of filter paper and dry. Attach the strip to a small plate covered with polymer-coated paper with a hole to allow access to the test paper. Application of the strip to the clean skin of person suffering from CF leads to the appearance of a gray or silver-gray color.

An indicator system for CF consists of three filter paper strips impregnated with 0.1 *M* potassium bichromate. These are soaked in a 0.05 *N*, a 0.066 *N*, and a 0.1 *N* silver nitrate solution, respectively (65). The strips are covered with waxed paper for protection.

PROCEDURE. Attach the indicator system to a suitable part of the body of the child for 15 min. Remove the indicator strips and read them. A positive result is signaled by the decoloration of the originally red-brown papers. Zero or one indicator showing a positive reaction means that the sweat chloride concentration is less than or equal to 35 mEq/liter, and the child is considered normal. Children showing chloride concentrations of 35–55 mEq/liter (two of three papers reacting positively) need to be checked more thoroughly, and concentrations of more than 55 mEq chloride/liter (all three indicators reacting positively) indicate a high probability of CF.

Another spot test for CF in newborn babies is based on increased albumin content in the meconium (66, 67). The test strip is a piece of filter paper soaked in a citrate buffer solution of a "protein error" indicator (preferably tetrabromophenolphthalein ethyl ester), dried, and sealed onto a resin strip for support.

PROCEDURE. Cover the strip with meconium and then place it in a small vessel containing water. The presence of albumin (and hence, a positive diagnosis for CF) is indicated by the appearance of a blue color on the test paper.

REFERENCES

1. L. Gershenfeld, *Urine and Urinalysis,* 3rd ed., W. B. Saunders, Philadelphia (1948).
2. A. H. Free, C. A. Adams, M. L. Kercher, H. M. Free, and M. H. Cook, *Clin. Chem* **3,** 163 (1957).
3. S. Benedict, *JAMA* **57,** 1193 (1911).
4. M.M. Belmonte, E. Sarkozy, and E. R. Harpus, *Diabetes* **16,** 557 (1967).
5. W. K. Jenson, "Melituria," in *Diseases of Metabolism,* 5th ed., G. Duncan, editor, W. B. Saunders, Philadelphia (1964), p. 912.
6. G. R. Constam, "Diagnosis: Meliturias other than Diabetes Mellitus," in *Diabetes Mellitus: Diagnosis and Treatment,* T. S. Danowsky, editor, American Diabetes Association, New York (1964), p. 59.
7. D. Muller, *Biochem. Z.* **199,** 136 (1928).
8. D. Keilin and E. T. Hartree, *Biochem. J.* **42,** 221 (1948).
9. B. C. Smith, M.J. Peake, and C. G. Frazer, *Clin. Chem.* **23** (12), 2337 (1977).
10. J. Dyerberg, L. Pedersen, and O. Aagard, *Clin. Chem.* **22** (2), 205 (1976).

11. T. Aoki and I. Shinonosane, *Eisei Kensa* **26** (12), 1117 (1977).
12. J. P. Peele, Jr., R.H. Gadsen, and R. Crews, *Clini. Chem.* **23** (12), 2238 (1977).
13. G. Esbach, *C. R. Soc. Biol., (Paris)* **89,** 417 (187)
14. W. G. Exton, JAMA **80,** 529, (1923).
15. W. G. Exton, *J. Lab. Clin. Med.* **10,** 722 (1925)
16. F. B. Kingsbury, C. P. Clark, G. Williams, and A. L. Post, *J. Lab. Clin. Med.* **11,** 981 (1926)
17. W. L. Gyure, *Clin. Chem.* **23** (5), 876 (1977).
18. S. P. L. Sørensen, *Biochem. Z.* **21,** 131 (1909).
19. F. Feigl and V. Anger, *Mikrochim. Acta* **2,** 107 (1937).
20. A. C. H. Rothera, *J. Physiol.* (*Lond.*) **37,** 491 (1908).
21. A. H. Free and H. M. Free, *Am. J. Clin. Pathol.* **30,** 7 (1958).
22. G. A. Harrison, *Chemical Methods in Clinical Medicine,* J. and A. Churchill, London (1937).
23. V. Hawkinson, C. J. Watson, and R. H. Turner, JAMA **129,** 514 (1945).
24. M. Franklin, *J. Lab. Clin. Med.* **34,** 1145 (1949).
25. A. H. Free and M. H. Free, 2.854.317 (1958) and German Patent 1,011.775 (1957).
26. G. W. Osbaldiston, *Clinical Biochemistry of Domestic Animals,* Academic Press, New York (1971), p. 55.
27. N. Schlesinger, *Dtsch. Med. Wochenschr.* **32,** 561 (1903).
28. G. B. Wallace and J. G. Diamond, *Arch. Intern. Med.* **35,** 698 (1925).
29. D. Kutter et al., *Dtsch. Med. Wochenschr.* **98,** 112 (1973).
30. Boehringer-Mannheim GmbH, British Patent 1,343.247, applied for June 14, 1972.
31. H. Zunkley, *Med. Welt,* **25,** 1683 (1974).
32. P. Griess, "Berichte der deutschen chemischen Gesellschaff," *Ber.* **12,** 427 (1879).
33. B. S. Garg, Y. L. Mehta, and M. Katyal, *Talanta,* **23** (1), 71 (1976).
34. Merck Patent Gesellschaft, British Patent 1,361.523, applied for July 21, 1972.
35. V. E. Pollak and N. Mendoza, *Med. Clin. North Am.* **55,** 1417 (1971).
36. M. A. Peyman, Lancet 2, 486 (1956).
37. R. M. James, *Lancet* **1,** 1164 (1969).
38. D. Kutter et al., *Dtsch. Med. Wochenschr.* **99,** 2332 (1974).
39. K. Sumi, Y. Shinomiya, Sh. Shimizu, Sh. Kagawa, K. Mimura, T. Tozawa, and A. Matsuoka, *Rinsho Byori* **25** (6), 518 (1977).
40. V. R. Holland, B. C. Saunders, F. L. Rose, and A. L. Walpole, *Tetrahedron* **30** (18), 3299 (1974).
41. R. M. Kark, J. R. Lawrence, V. E. Pollak, C. L. Pirani, R. C. Muehrcke, and H. Silva, *A Primer of Urinalysis,* 2nd ed., Harper and Row, New York (1963), p. 13.
42. A. Fölling, *Z. Physiol. Chem.* **227,** 169 (1934).
43. H. W. Wallace, K. Moldave, and A. Meister, *Proc. Soc. Exp. Biol. Med.* **94,** 632 (1957)

44. H. Baird, *J. Pediatr.* **52,** 715 (1958).
45. C. O. Rupe and A. H. Free, *Clin. Chem.* **5,** 405 (1959).
46. A. H. Free and M. H. Cook, *Am. J. Clin. Pathol.* **22,** 588 (1952).
47. R. K. Tompkins, A. R. Draft, and R. M. Zollinger, *Surg. Gynecol. Obstet.* **131,** 500 (1970)
48. B. Scherster, *Acta Med. Scand.* **178,** 583 (1965).
49. A. E. Davis, *Am. J. Med. Technol.* **42** (1), 18 (1976).
50. E. L. Mazzaferri, T. G. Skillman, R. R. Lanesc, and M. P. Keller, *Lancet* **1,** 331 (1970).
51. M. F. Vickers, *Med. Lab. Technol.* **30,** 235 (1973).
52. A. S. Abiholm, *Scand. J. Clin. Lab. Invest.* **41** (3), 269 (1981).
53. A. M. Bold, I. S. Menzies, and G. Walker, *J. Clin. Pathol.* **23,** 85 (1970).
54. J. D. Kruse-Jarres, F. Dunsbach, O. Grossler, F. Kaltwasser, W. Minder, V. Sasse, W. Bablock, P. Koller, and W. A. Poppe, *Dtsch. Med. Wochenschr.* **105** (21), 756 (1980).
55. A. Scholer, A. Pianezzi, D. Vonderschmidt, and W. Goetz, *J. Clin. Biochem.* **15** (10), 583 (1977).
56. F. Dunsbach and V. Seifert, *Mad. Lab.* (*Stuttg.*) **27** (11), 263 (1974).
57. N. J. Dent, *Lab. Pract.* **22** (11), 674 (1973).
58. H. Adlerkreutz, K. Liewendahl, and P. Virkola, *Clin. Chem.* **24** (5), 756 (1978).
59. N. M. Deadman, *Clin. Chim. Acta* **48** (4), 433 (1973).
60. S. A. Ahmed and H. S. Gowda, *Clin. Chim. Acta* **111** (2–3), 275 (1981).
61. K. G. Alberti and M. Nathrass, *Lancet* **2,** 25 (1977).
62. H. Bartels and W. Berger, *Intensive Care Med.* **6,** 235 (1980).
63. H. Schwachman and N. Gahm, *N. Engl. J. Med.* **225,** 999 (1956).
64. H. M. Emrich, German Patent 2,631.864, January 19, 1978.
65. E. Jungreis, Israeli Patent Application, March 26, 1981.
66. Boehringer-Mannheim GmbH, British Patent 1,377.463, March 27, 1973.
67. M. Hellsing and H. Kolberg, *Scand. J. Clin. Lab. Invest.* **33,** 333 (1974)

CHAPTER

4

FORENSIC APPLICATION OF SPOT TEST ANALYSIS

4.1. GENERAL

A fairly new member of the law enforcement team is the forensic scientist. Part of his job is to help the police find links between the crime and the person who committed it. The routine work of a modern police laboratory involves, among other disciplines, a great deal of analytical chemistry, and quick, inexpensive screening by spot test analysis still has considerable importance.

Of the objective physical evidence that a forensic scientist collects, a considerable amount is based on simple spot tests. In homicides committed with a gun it is important to find out the shooting range of the firearm and other important facts. A great deal of corroborative evidence can be collected by detecting red blood cells and spermatozoa at the scene of the crime. The detection of strong acids and caustic material by means of indicator papers is another preliminary examination done by the forensic scientist. Differentiation between organic and inorganic substances with the preliminary charring test has, at times, diagnostic value.

The search for illicit drugs is greatly facilitated by selective spot tests. More than 6000 organic compounds are considered to drugs. It is unrealistic to expect the field investigator to differentiate among this overwhelming number of drugs; however, the number actually encountered is much smaller, and spot tests are used extensively for the preliminary identification of drug material.

4.2. SPOT TESTS FOR DETECTING THE RESIDUES OF FIREARM DISCHARGE

Reliable and sensitive methods for demonstrating the presence of residues of firearm discharge on the clothing of a victim or on the skin of a suspect have been sought for many years by forensic investigators. The detection of nitrates and nitrites in such residues can be achieved by spot tests.

Spot tests for the screening of suspects must have the following characteristics: (1) They must be simple enough to be performed by police officers lacking formal training in chemistry; and (2) test outcomes must be simple and decisive, with distinct, relatively stable colors formed.

A very important aspect of these tests is to establish whether a death is a suicide or a homicide. The detection of traces of antimony and barium, which are present in many cartridge priming charges, on the web of the thumb and the back of the hand can supply data relevant to the question of whether a head wound resulting from being shot with a revolver was self-inflicted (1).

The relatively recent discharge of a firearm is indicated by the qualitative detection of the most frequently encountered metallic components of firearm discharge residues, such as lead. Other components of firearm discharge residue are derived from gunpowder, primer, and lubricants.

Primers contain sensitizers, in most cases lead stiphnate or lead trinitroresorcinate, and in older formulations, lead acids or mercury fulminate. They are charged with oxidizing agents such as barium nitrate, barium peroxide, lead nitrate, lead peroxide, and potassium chlorate. Antimony sulfide, calcium silicide, lead thiocyanate, zirconium, or aluminum powder serve as fuel.

In the early days, gunpowder was composed of 75% potassium nitrate, 15% sulfur, and 10% carbon. This formulation was replaced about 40 years ago by smokeless powder, composed of nitrocellulose or a mixture of nitrocellulose and nitroglycerin.

Bullets are composed mostly of lead containing a small percentage of antimony to increase hardness. Lead bullets are covered with a layer of nickel or copper. Spot tests for lead and copper are used in police investigations to detect substrates that have been penetrated or pierced by an uncoated or a copper-coated bullet.

4.2.1. Spot Tests for Lead in Firearm Discharge Residue

The presence of antimony, barium, and lead can be used to characterize the primer and leakage residues of firearm discharge. Lead can be detected by a selective and sensitive spot test. (2). Glossy photographic paper from which the silver salts have been removed and which has been impregnated with an attacking agent is pressed onto areas of the hand in which the weapon would have been held when it was discharged. It is then developed with the sodium rhodizonate reagent. This method is not completely satisfactory for the removal of leakage residues from hands. Another simple method to remove discharge residues from hands uses cotton cloth wetted with 1 *N* HCl; a buffered sodium rhodizonate solution is used to detect any lead (3).

Leisst (4) found, by x-ray diffraction that lead in the discharge residue is actually in metallic form. This previously unreported fact has implications for the effective dissolution and detection of lead-containing residue. Because these tiny (0.1–100 microns), spherical lead particles are easily soluble in sodium acetate solution, removing the residue with sodium-acetate-impregnated Benchkote paper makes possible a remarkable sensitivity in lead detection.

The yellow aqueous solution of sodium rhodizonate produces colored precipitates of basic lead rhodizonates from neutral or slightly acidic lead solutions. Violet $Pb(C_6H_6) \cdot Pb(OH)_2 \cdot H_2O$ is precipitated in neutral solution, and scarlet-red $2Pb(C_6O_6) \cdot Pb(OH)_2 \cdot H_2O$ is formed from weakly acidic solutions.

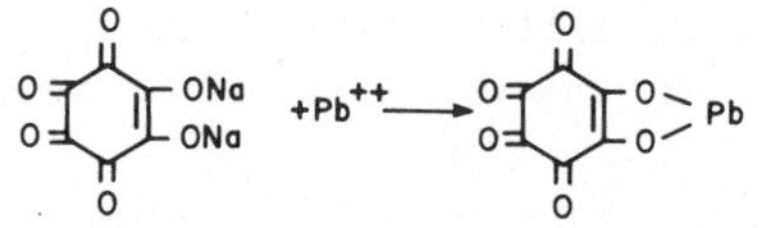

PROCEDURE. Place a drop of the test solution on a piece of filter paper. After the liquid has been soaked up, add a drop of freshly prepared 0.2% sodium rhodizonate. A blue spot or ring is formed if lead is present. When an intense reaction occurs, the blue spot can be transformed to scarlet by spotting with a drop of buffer solution (pH 2.8) containing 1.9 g sodium bitartrate and 1.5 g tartaric acid in 100 ml water.

Limit of identification	0.1 μg lead
Limit of dilution	1:500,000

When a lead bullet has been fired from a revolver, the chances of detecting lead by this method are very high (4). In 100 test firings, 90 definite positives, 8 negatives, and 2 doubtful positives were recorded. When jacketed lead bullets are fired, even though the amounts of lead deposited appear to be rather less, the detection rate is still on the same order. The quantity of detectable lead after discharge varies with the make and the condition of the revolver, but little variation has been noted with different makes of ammunition.

The amount of lead discharged increases when the weapon is fired from a close distance, indicating possible "blow back" of traces from the target. A gradual build-up of lead deposits on the hand occurs during repeated shooting.

Normal hand washing appears to remove traces detectable by the following method. However, this test is considered to be a useful screening test.

PROCEDURE (5). Twist a pea-sized piece of absorbent cotton wool into the center of a swab of cotton cloth to form a pad. Moisten the pad with 1% hydrochloric acid and use to swab residue from the web and back of the thumb and forefinger. Remove the cotton wool and dry the cloth with warm air. Treat the center of the cloth with a drop of freshly prepared sodium rhodizonate solution. A red color that turns blue on the addition of 5% hydrochloric acid indicates the presence of lead.

4.2.2. Detection of Barium in Firearm Discharge Residue

The sodium rhodizonate test for barium meets the requirements of specificity, sensitivity, and formation of a colored reaction product for a forensic screening test.

The yellow aqueous solution of sodium rhodizonate produces red-brown precipitates from neutral solutions of barium salts.

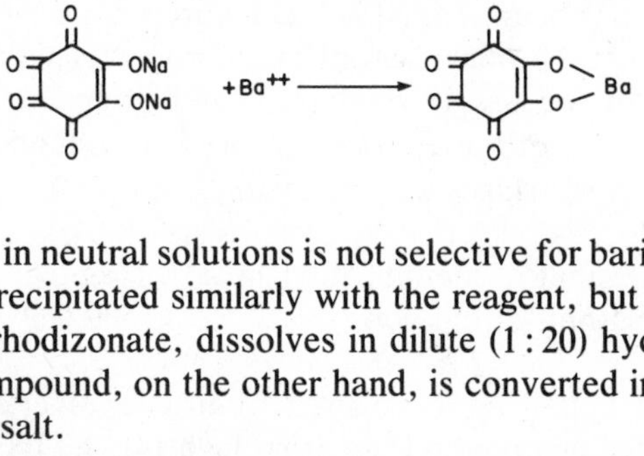

O, O, O, O, ONa, ONa $+ Ba^{++} \longrightarrow$ O, O, O, O, O, O, Ba

The reaction in neutral solutions is not selective for barium ions. Strontium ions are precipitated similarly with the reagent, but the precipitate, unlike barium rhodizonate, dissolves in dilute (1 : 20) hydrochloric acid. The barium compound, on the other hand, is converted into an insoluble bright red acid salt.

PROCEDURE (6). Place a drop of the neutral or slightly acidic test solution on filter paper and then add a drop of an aqueous 0.2% solution of sodium rhodizonate. According to the amount of barium present, a more or less intense red-brown stain is formed.

Limit of identification	0.25 μg
Limit of dilution	1 : 200,000

A relatively stable rhodizonate reagent paper has been developed for barium and lead screening (7). The filter paper is impregnated with a 2% ethylenediamine hydrochloride solution, sprayed with a saturated aqueous solution (~0.2%) of sodium rhodizonate, and dried at a temperature less than 80°C. The resulting test paper is stable for at least 6 months when kept in a dark container.

4.2.3. A Spot Test for Antimony in Firearm Discharge Residue

A specific and sensitive test for antimony (8) involving its precipitation with triphenylmethylarsonium iodide has been developed, and fits the forensic requirements. The huge arsonium cation precipitates the $[SbI_4]'$ anion formed through its weighting effect.

PROCEDURE. Mix a drop of the test solution with a 10% alcoholic solution of trimethylarsonium iodide. In the presence of trivalent antimony, an orange precipitate appears in about 30 sec.

PREPARATION OF THE REAGENT. Triphenylmethylarsonium iodide is not yet available commercially. However, it can be prepared by dissolving 10 g commercially available triphenylarsine and 10 g methyl iodide in ethanol and refluxing for 3 hours. Dry ether is then added to precipitate the salt triphenylmethylarsonium iodide. Purification is achieved by redissolving the salt in alcohol and precipitating with ether.

4.2.4. A Combined Test for Antimony, Barium, and Lead in Firearm Discharge Residue

The three decisive elements can be identified by direct testing of the cloth used to remove the residues. This procedure has the advantage of simplicity, does not require elaborate laboratory equipment, and minimizes the risk of losing the test material.

PROCEDURE. Dry one drop of the test solution containing antimony, barium, and lead on a white cotton cloth. Add a drop of a 10% alcoholic solution of triphenylmethylarsonium iodide. The presence of antimony is indicated by the appearance of an orange ring in approximately 30 sec. Complete color development of the ring is achieved in 2 min. Dry the cloth with the orange ring on a test plate blasted with warm air, under an infrared lamp, or under a hair-drier. Add two drops of a freshly prepared saturated solution of sodium rhodizonate to the center of the ring. A red color within the orange ring indicates the presence of barium, lead, or a combination of both. Dry the cloth again and add one to two drops of 1:20 hydrochloric acid to the red area of the ring. The appearance of a blue spot indicates the presence of lead, whereas if the reddish-brown spot in the center of the ring turns vivid red, the presence of barium is confirmed. These tests should not be done under strong light, since it hastens the decomposition of sodium rhodizonate.

Trace metals in gunshot residue can be estimated by the ring oven technique as well (9). The hydrochloric acid extract is placed in the center of a piece of filter paper. Hydrogen sulfide, bromine water, dilute ammonia, and ammonium sulfide solution are the agents that separate barium, copper, antimony, and lead into different ring zones. This method can also be applied to hand swabs.

4.2.5. Spot Tests for Nitrates and Nitrites in Firearm Discharge Residue

The detection of nitrate and nitrite on the hand is a good indication of the recent use of a firearm. In the so-called dermo-nitrates test (10), melted paraffin is used to gather the nitrate particles from the surface and intradermal layers of the skin. The heat of the melted paraffin dilates the pores of the skin for collection of the trapped nitrate particles, which are then identified with the diphenylamine reagent. Establishing the spatial distribution of the powder residue is an important part of the investigation. Those areas of the hand in which the weapon was held when it was discharged are most likely to be contaminated with leakage residues. The leakage residue contamination of the hand forms a general pattern, with some variation depending on the size and type of the revolver.

There are three types of powder used in the ammunition of small firearms: smokeless, black, and semismokeless. Today, the most common is the smokeless powder. Black gunpowder has been almost completely replaced by the smokeless type, and a mixture of black and smokeless powder is hardly ever used.

Black gunpowder is a mixture of potassium nitrate, charcoal, and elemental sulfur in an approximate ratio of 75 : 15 : 10. The following chemical reaction occurs when the powder is fired in open space:

$$2KNO_3 + 3C + S \rightarrow 3CO_2 + K_2S + N_2$$

This reaction scheme, however, represents only the reaction that occurs in open space. When the explosion occurs in a handgun, a much more complicated reaction takes place, and substances such as potassium carbonate, hydrogen sulfide, potassium thiocyanate, and potassium nitrite are found in the residue.

Smokeless powder is composed of cellulose nitrate or of glycerol nitrate combined with cellulose nitrate. The firing of these compounds in the open leads to the following reactions:

$$C_{12}H_{14}O_4(NO_3)_6 \rightarrow 9CO + 3N_2 + 7H_2O + 3CO_2$$

(Cellulose nitrate)

$$4C_3H_5(NO_3)_3 \rightarrow 12CO_2 + 10H_2O + 6N_2 + O_2$$

(Glycerol nitrate)

The decomposition pattern differs greatly when the explosion occurs in a closed space. Certain organic and inorganic compounds, among them nitrates, are added as stabilizers to regulate the firing rate of the powder and to minimize corrosion of the metal of the gun. Nitrates are present in smokeless powder as well as in black powder; thus, nitrates appear in the firearm discharge residues of both.

The Paraffin–Diphenylamine Test for Nitrates

PROCEDURE. Heat clear, filtered paraffin slowly until just melted. Spray or pour this liquid over the fingers, hand, and wrist of the individual who is to be examined until a coating of paraffin has formed. When the paraffin has cooled and solidified, peel the cast gently from the hand. Slowly add the diphenylamine reagent with a pipet. A positive reaction is indicated by the appearance of dark blue particles on the inner surface of the paraffin cast. The reaction is not at all specific for nitrate, since all oxidizing agents oxidize diphenylamine in a two-stage reaction from diphenylamine through diphenylbenzidine to diphenylbenzidine violet (11).

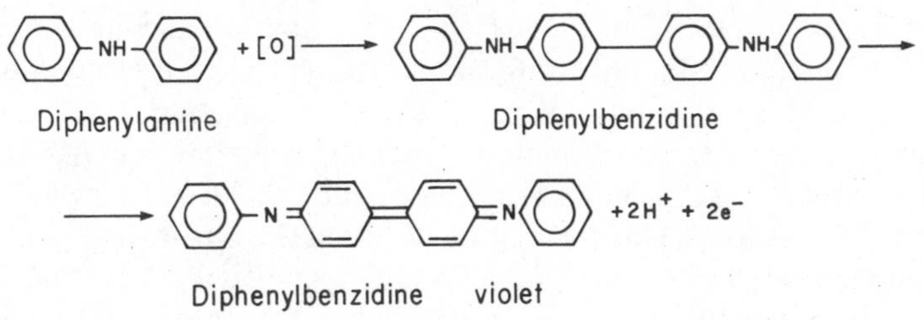

NH + [O] → NH NH →

Diphenylamine Diphenylbenzidine

→ N = = N $+ 2H^+ + 2e^-$

Diphenylbenzidine violet

PREPARATION OF THE REAGENT. Add 10 ml conc. sulfuric acid to 2 ml distilled water under constant stirring. Add to this solution 0.05 g diphenylamine and stir until the solid is completely dissolved.

The Pennsylvania Supreme Court (12) admitted the findings of a positive reaction with the diphenylamine reagent as corroborative evidence that a defendant had recently fired a gun.

Spot Test for Nitrites in Powder Residue (Walker's Test) (13)

The presence of nitrites on cloth or other objects in the line of fire may be detected by this test. The nitrite particles form a distinct pattern of distribution from which it may be possible to estimate the distance between the object and the muzzle of the gun at the time it was discharged.

PREPARATION OF THE TEST PAPER. Remove all the silver salts from unexposed glossy photographic paper by immersing it in sodium thiosulfate solution for 15 min. Wash the paper in running water for 1 hour; then impregnate it with a 10% solution of C acid (2-naphthylamine-4,8-disulfonic acid) and dry. An alternative reagent is the H acid (1-amino-naphthol-3,6-disulfonic acid).

PROCEDURE. Lay a clean bath towel on a worktable; place a piece of the prepared photographic paper face up on it. Then lay the fabric to be examined face down on the photographic paper. Over this arrange, in order, a thin dry towel, a towel moistened with 20–25% acetic acid, and another dry towel. Press with a warm electric iron for 5–10 min. After removal of the photographic paper red-orange spots will appear that correspond exactly with the positions of the nitrite particles in the residue around the bullet hole in the cloth. The paper should be photographed immediately, since the dye may gradually fade.

The basis of the reaction is the classical Griess reaction (Section 3.7.), which consists of consecutive diazotization and coupling. The C or H acid (or any other aromatic amine) reacts in an acidic medium with nitrate to form a diazonium salt. The diazonium salt in turn reacts with some of the remaining unchanged amine to form an azo dye. The Griess reaction is a specific nitrite test; no false positive tests are registered.

This test provides an estimate of the approximate range at which the gun discharged and thus helps to prove whether the deceased was a victim of suicide or murder. The correlation between the nitrite spot pattern and the shooting distance should be calibrated using similar ammunition and the same gun against the same type of target. A common obstacle faced by criminologists is the presence of blood, which can easily mask the specks of nitrates on clothing at the entry hole of a bullet. It can also result in confusion because the color of blood and that of the red azo dye produced by the Griess reaction are similar.

Detecting nitrites by fluorescence under ultraviolet light solves this problem (14). Nitrites react with *o*-phenylenediamine in an acidic medium to form 1,2,3-benzotriazole, which is amphoteric in nature. This reacts readily with alkalis or ammonium hydroxide to produce salts that fluoresce under ultraviolet light. The detection of trace amounts of nitrites in firearm discharge residue is based on this phenomenon.

o-Phenylenediamine → 1,2,3-Benzotriazole

PROCEDURE. Soak Whatman no. 1 filter paper in a freshly prepared 5% solution of *o*-phenylenediamine dihydrochloride (made in 50% HCl) and partially dry. Place the filter paper over the suspected area of the gunshot on the clothing, exerting slight pressure for better contact with all parts of the clothing. Then remove the paper and place it in a Petri dish containing a 25% ammonium hydroxide solution. Illuminate the paper with a 254-nm-wavelength ultraviolet lamp. The appearance of intense yellow spots indicates the presence of nitrite particles. The fluorescence pattern may be photographed as permanent evidence.

The main advantages of the fluorescence test for nitrites are its specificity and the fact that the presence of blood at and around the gunshot hole does not influence the outcome of the reaction. The fluorescent spot fades at temperatures above 40°C. The filter paper with the fluorescent spot on it should not be kept under the ultraviolet light for a long time.

As with the photographic paper method, the distribution of spots in this method suggests the range of firing from a particular firearm.

4.2.6. Spot Tests for Detection of Lead and Copper at the Site of Bullet Penetration

Uncoated lead bullets and copper-coated bullets discharged from firearms and penetrating wood, fabric, paper, and so forth leave 0.1 to 100-micron spherical particles of metallic lead and/or copper. These particles are identifiable for use as evidence in court.

Traces of these metals can be detected by scanning electron microscope, atomic absorption spectroscopy, x-ray fluorescence, and neutron activation analysis. However, unsophisticated spot tests have special importance, since these simple tests can easily be done in the field.

To prove that a hole in a given substrate was caused by a gunshot, forensic laboratories use several instruments to record the lead or copper concentration in the vicinity of the hole and compare it with the background concentration. Spot test kits have not been used systematically by law-enforcement agencies, although the feasibility of a spot test for lead for this purpose was shown (15).

A new field-test kit (4) has been developed that can be used effectively by police officers, who must undergo only brief training to master it. The kit enables the detection of firearm use within 3 min. The reaction picture exactly follows the shape of the hole, and because only a few types of bullets contain copper, this simple test can determine the kind of ammunition used. These tests done in the field can furnish preliminary data, saving time for the investigating team.

Test Kit for the Detection of Lead and Copper in a Gunshot Hole

This kit consists of two polyethylene bottles of different colors. Bottle 1 contains 10 mg solid sodium rhodizonate and a glass ampoule containing 5 ml distilled water; bottle 2 contains 50 mg rubeanic acid and a glass ampoule containing 5 ml ethanol. Neither reagent is very stable in solution, which is why the solutions are prepared just before the tests are carried out. Breaking the ampoules produces a 0.2% sodium rhodizonate solution and a 1% rubeanic acid solution. Two small dropping bottles contain the solvents for the metals: 5% acetic acid for the dissolution of metallic lead particles, and 5 ml 12% ammonium hydroxide solution to dissolve the copper. The tests are carried out on 5×7 cm Benchkote paper sheets.

PROCEDURE FOR THE DETECTION OF LEAD. Moisten the absorptive side of the Benchkote paper with a few drops of acetic acid. Press the wet paper against the bullet hole for 60 sec without moving the paper; then remove the paper. Break the neck of the water ampoule in the sodium rhodizonate bottle and shake the bottle for a few seconds. Pour the reagent solution over the bullet hole area of the paper. The appearance of a red-violet stain indicates the presence of lead.

Traces of copper are detected with the ethanol solution of rubeanic acid, which forms an olive-green precipitate of copper rubeanate from ammonical or weakly acidic solutions of copper salts. The rubeanic acid reacts in its tautomeric diimido form:

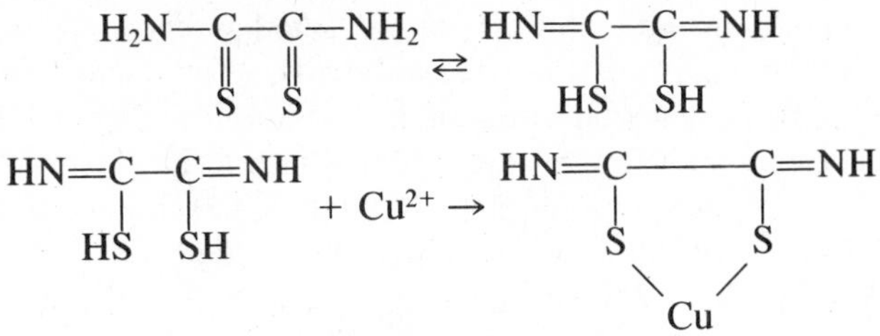

$$
\begin{array}{c}
H_2N{-}\underset{\underset{S}{\|}}{C}{-}\underset{\underset{S}{\|}}{C}{-}NH_2
\end{array}
\rightleftarrows
\begin{array}{c}
HN{=}\underset{\underset{HS}{|}}{C}{-}\underset{\underset{SH}{|}}{C}{=}NH
\end{array}
$$

$$
HN{=}\underset{\underset{HS}{|}}{C}{-}\underset{\underset{SH}{|}}{C}{=}NH + Cu^{2+} \rightarrow
\begin{array}{ccc}
HN{=}C & \text{———} & C{=}NH \\
| & & | \\
S & & S \\
 & \diagdown \quad \diagup & \\
 & Cu &
\end{array}
$$

Procedure for the Detection of Copper. Moisten the Benchkote paper with a few drops of ammonia. Press the wet paper to the bullet hole for 60 sec without moving the paper; then remove the paper. Break the neck of the ethanol ampoule in the rubeanic acid bottle and shake the bottle vigorously for a few seconds. Pour a few drops of the reagent solution over the hole area of the paper. The appearance of an olive-green stain encircling the hole indicates the presence of copper.

It must be noted here that the test for lead must be carried out before the copper test, due to the interfering color produced by the reaction of ammonia and sodium rhodizonate.

An additional advantage of the test is that preliminary conclusions can be drawn about the shooting angle if the colored ring produced in the test does not follow exactly the shape of the hole.

Lead is always detected in the gunshot hole, and copper is detected when copper-coated bullets were used. Since the use of copper coating on bullets is limited to military ammunition, the discovery of copper traces indicates conclusively the source of the bullets.

4.3. SPOT TESTS FOR EXPLOSIVES AND EXPLOSIVE RESIDUES

The task of identifying explosives and explosive residues with certainty is often of outstanding importance for legal or criminalogical purposes. The court may request definite statements regarding the composition of very minute remainders of gunpowder or explosive material. Quick and simple microchemical methods are therefore indispensable. The presence or absence of certain elements or groups as determined by spot test analysis often allows decisive conclusions and judgments to be made.

Procedures for obtaining productive evidence through qualitative testing of explosives appear already in the early work of Feigl (16), and various publications and technical papers from law-enforcement agencies indicate that spot tests are still used extensively for reliable screening.

Black gunpowder, which contains large amounts of nitrates, may be tested for by checking for the presence of nitrate in a drop of aqueous extract. The previously mentioned (Section 4.2.5) color reaction using diphenylbenzidine or diphenylamine may be applied.

According to one relatively recent publication (17), the initial extraction of the explosive and its residue should be done with acetone, which extracts both inorganic ions and explosive organic compounds to a sufficient extent to enable the successful performance of the appropriate spot tests. Nitroglycerin, ethyleneglycol dinitrate, cyclomethylenetrinitramine (RDX), pentaerythrityl tetranitrate (PETN), tetryl, and nitrotoluene are all sufficiently soluble in acetone to pass the sensitivity barrier of their corresponding spot tests. The least soluble of these is RDX, 1 g of which dissolves in approximately 25 ml acetone. Although nitrocellulose and nitrostarch are not truly soluble in acetone, they are completely dispersed in colloidal form in it. After extraction, the extract is allowed to evaporate to dryness, and spot tests are performed on the solid scraped from the evaporating dish. Water extraction is used to improve the performance of the tests of inorganic ions.

The sections immediately below discuss some ions and compounds that may be found in explosive residues and whose presence can be selectively indicated by spot test analysis.

4.3.1. Aluminum

Aluminum is present in powdered elemental form as a constituent of flash powder and in small amounts in a few dynamites. Aluminum is present also in many homemade explosives.

Morin (3,5,7,2′,4′-pentahydroxy flavanol) reacts with aluminum salts in neutral or acetic acid solutions to give an intense green fluorescence in daylight or ultraviolet light. The fluorescence is due to the formation of a colloidally dispersed inner complex aluminum salt of morin (18).

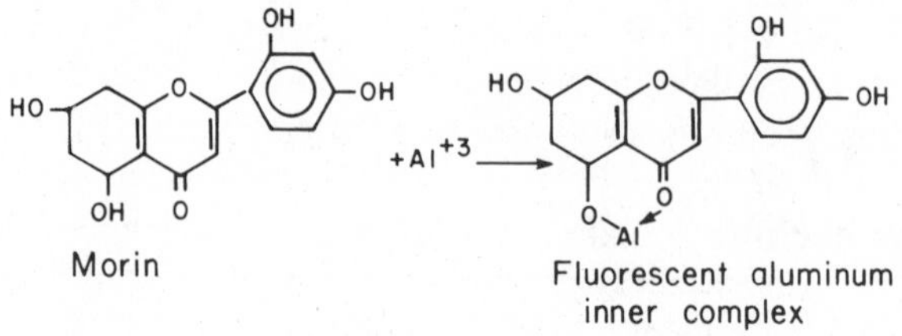

Morin $+ Al^{+3} \longrightarrow$ Fluorescent aluminum inner complex

POSTEXPLOSION TEST PROCEDURE (19). Add a small amount of 1 *N* acetic acid to the water test solution to make it slightly acidic. Place three drops of the acidified solution in a black spot plate and add two drops of the morin reagent. Observe the spot under an ultraviolet light. The presence of aluminum is indicated by a green fluorescence. Always run a known material side by side for comparison.

PREEXPLOSION TEST PROCEDURE. Dissolve the unknown metallic powder in 2 *N* NaOH. Aluminum will dissolve easily; most other metals will not. Place three drops of the solution in a black spot plate. Add two drops of morin solution, and then two or three drops of 2 N HCl. A fluorescent green color under ultraviolet light indicates the presence of aluminum.

REAGENT SOLUTION. Make a saturated solution of morin in methyl alcohol by placing 5 ml alcohol in a small bottle and adding morin until no additional morin will dissolve.

The morin test is not specific for aluminum; beryllium, indium, gallium, thorium, and scandium also form the fluorescent product. However, these ions are not commonly encountered and are not used in explosives.

4.3.2. Ammonium Ion

The ammonium ion in most dynamites appears in the form of ammonium nitrate. It may also be present as a contaminant arising from ammonia-based fertilizers. Ammonium nitrate can also function as an oxidizer for flash-powder-type explosives, although it is inferior to other materials for that purpose. Ammonium perchlorate is sometimes used as an oxidizer for explosives.

A Test for Ammonium Ion with the Nessler Reagent (20)

PROCEDURE. Place a drop of the neutral or weakly acidic test solution in a spot plate and add a drop of the reagent. The formation of an orange-brown precipitate ($HgI_2 \cdot HgNH_2I$) is a positive test result.

PREPARATION OF THE NESSLER REAGENT. Dissolve 10 g mercuric iodide and 5 g potassium iodide in 50 ml water. Add a solution of 20 g potassium hydroxide in 50 ml water. The reagent may be stored for a year.

The presence of ammonium nitrate should be interpreted cautiously in the forensic laboratory. As a primary constituent of fertilizers, it is often found in soil samples. Unless it is accompanied by other materials normally present in explosives, the detection of ammonium nitrate is considered inconclusive evidence for the presence of explosives.

4.3.3. Antimony

Antimony sulfide, a water-insoluble gray powder, is used in some commercial and military pyrotechnic mixtures. After an explosion, the antimony is present in both trivalent and pentavalent oxidation states.

Spot Test with Rhodamine B (21)

Aqueous solutions of xantho dyestuffs, such as Rhodamine B, form violet or blue precipitates. with pentavalent antimony salts in strong hydrochloric acid solutions. In concentrated hydrochloric acid solution an anion of $[SbCl_6]^-$ forms. The large size of both cation and anion precipitates the salt.

$$\left[\text{Rhodamine B cation } (\overset{\oplus}{N}(C_2H_5)_2,\ HOOC)\right] Cl^- \xrightarrow{+[SbCl_6]^-} \left[\text{Rhodamine B cation}\right][SbCl_6]^-$$

red solution   violet precipitate

Gold (III) and thallium (III) form $[AuCl_4]'$ and $[TeCl_4]'$, respectively, in concentrated hydrochloric acid, but since these ions are absent in explosive materials, the test with Rhodamine B is quite selective for antimony. Trivalent antimony is oxidized by nitrous acid to pentavalent antimony salts.

POSTEXPLOSION TEST PROCEDURE. Place several milligrams of the sample in the depression of a spot plate. Add one drop of concentrated hydrochloric acid and a few milligrams of sodium nitrite. Add ten drops of the reagent solution and mix. A change of color from bright red to violet indicates the presence of antimony.

PREEXPLOSION TEST PROCEDURE. Place a few drops of the test solution in a spot plate and add a few milligrams of sodium nitrite. Follow the preceding procedure.

PREPARATION OF REAGENT SOLUTION. Dissolve 0.01 g Rhodamine B in 100 ml distilled water.

4.3.4. Bromide Ion

Bromide ion is not usually present in explosives, but is included here as a possible contaminant of explosive residue. Free bromine converts the yellow dye fluorescein into red tetrabromofluorescein (eosin).

The liberation of bromine from bromides and the formation of eosin occur during prolonged heating with an acetic acid–hydrogen peroxide mixture on filter paper. Any iodide present will be oxidized to iodate under these conditions, not interfere with the test.

PROCEDURE (22). Place one drop of the test solution on a piece of filter paper, let dry, and spot with an acetic acid–hydrogen peroxide solution (2 parts 6% hydrogen peroxide + 1 part glacial acetic acid). Treat the spot with a drop of a 1% alcoholic fluorescein solution. A red stain develops on warming if bromide ion is present.

4.3.5. Chlorates

Chlorates are used in matches and in many explosives, primarily in home-made devices. Unconsumed chlorates may be found after an explosion, although most chlorate will have been converted to chloride.

Test with Manganous Sulfate and Phosphoric Acid (23)

In a strong phosphoric acid solution, chlorates warmed with manganese sulfate form complex, violet-colored manganese (III) phosphate ions:

$$ClO_3^- + 6Mn^{2+} + 12PO_4^{3-} + 6H^+ = 6[Mn(PO_4)_2]^{3-} + Cl^- + 3H_2O$$

This reaction is absolutely unequivocal, even in the presence of large quantities of nitrate; no influence on the testing of explosives need be feared from periodates and persulfates reacting in the same way.

In residues from explosives in which a sugar/chlorate mixture has been used, charring often occurs that masks the color reaction for chlorates. A rapid and simple test (24) avoids this complication.

PROCEDURE. Spot an aqueous solution of the chlorate/sugar mixture onto a small portion of a Polygram Alox N chromatography plate (Camlab Ltd., Milton Road, Cambridge, U.K.) and let dry. Spot a freshly prepared 1 : 1 (v/v) mixture of a saturated solution of manganese sulfate and syrupy phosphoric acid onto the dried area and heat the plate in an oven at 105°C until a violet color develops, indicating the presence of chlorates. Color development usually occurs in 1–3 min, and prolonged heating results in charring.

4.3.6. Cyclotrimethylenetrinitramine

Cyclotrimethylenetrinitramine (also known as Cyclonite, Hexogen, RDX, and T4) is a constituent of the military explosive C-4.

Test for RDX Based on Pyrolytic Oxidation

The colorless compound is not soluble in water or benzene, but does dissolve in acetone. When alone, it is not affected by heating to 180°C, in contrast to Tetryl. However, if heated in the presence of manganese dioxide, which acts as an oxygen donor, RDX yields formaldehyde and nitrous acid (25).

$$(CH_2N\text{-}NO_2)_3 + 6[O] \longrightarrow 3CH_2O + 3N_2O_3$$

PROCEDURE. Place a tiny portion of the solid test material or one drop of its solution in a test tube. Add a few grains of manganese dioxide, and dry if necessary. Cover the tube mouth with filter paper moistened with Nessler reagent or Griess Reagent. Immerse the test tube in a glycerol bath preheated to 180°C. A brown or black stain with the Nessler reagent indicates the evolution of formaldehyde; a red stain with the Griess reagent indicates the presence of nitrous acid.

Spot Test for RDX with J-Acid (26)

RDX is decomposed by concentrated sulfuric acid. One of the products of decomposition is formaldehyde, the production of which serves as the basis for a sensitive spot test for RDX. Formaldehyde reacts with J-acid dissolved in concentrated sulfuric acid, forming a yellow cationic xanthylium dyestuff that fluoresces yellow under ultraviolet light.

J-Acid

xanthylium cation

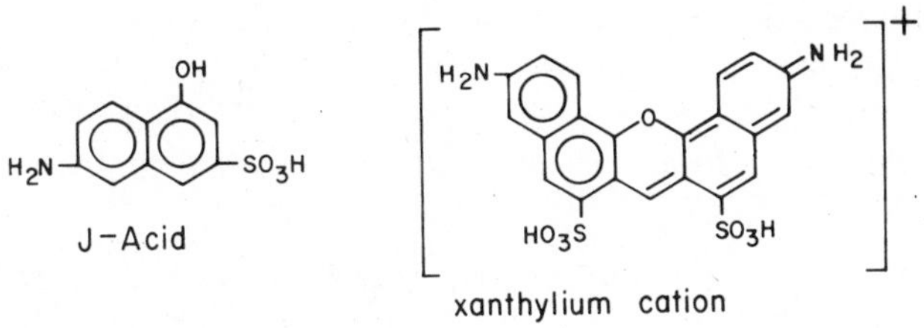

PROCEDURE. Place a drop of a 0.1% solution of J-acid in conc. sulfuric acid on a glass filter paper followed by a drop of the test solution in acetone. The appearance of a yellow fluorescence under ultraviolet light indicates the presence of RDX.

4.3.7. Differentiation of RDX (Hexahydro-1,3,5-trinitro-s-triazine) and HMX (octahydro-1,3,5,7-tetranitro-s-tetrazine)

The dye quinalizarine is destroyed by a nitric acid split-off. Various nitrate and nitramine explosives liberate nitric acid in concentrated sulfuric acid solution, and change the violet color of quinalizarine (1,2,5,8-tetrahydroxyanthraquinone) to yellow after standing for up to an hour. Because the times required for this color change are different for equal weights of HMX and RDX at room temperature, a spot test is possible for the differentiation of these two explosives (27).

PROCEDURE. Place two 50-mg samples of RDX and HMX on two different depressions of a white spot plate. Add enough quinalizarine reagent (3 mg quinalizarine in 40 ml conc. H_2SO_4) to fill the depressions. Agitate the samples for 20 sec, and then allow to stand with only occasional agitation. Observe the color after exactly 20 min, when RDX yields a light yellow color and HMX, a blue color. Observe the color again after another 25–30 min, when HMX yields a yellow color as well.

This test gives clear-cut results when HMX or RDX is present alone. Admixtures of these two explosives make the test results ambiguous.

Spot Test for RDX and HDX with Thymol and Sulfuric Acid

RDX can be identified by the formation of a red color in the presence of thymol and nitrogen-free sulfuric acid (28). The fault of this test is that sugars and aldehydes interfere with it by also producing a red color.

A modification of this test is to heat the reaction partners moderately; under this condition, a violet color is produced. The sample is then dissolved in ethanol, and the rich blue solution obtained can be used as the basis for a test for RDX.

PROCEDURE (29). Add about 200 mg thymol and 6 drops of conc. sulfuric acid to a few milligrams of the sample in a test tube. Warm the tube for 5 min at 100°C and then add 5–10 ml ethanol. The presence of RDX is indicated by the appearance of a rich blue solution. The presence of sugars and aldehydes leads to a brown color and HMX yields a pale plue-green tint.

If the same test is repeated at 150°C, RDX will still produce a blue color, whereas HMX results in an olive color. This test is considered to be specific for RDX.

4.3.8. Dichromates

Alkali dichromates are used as oxidizing agents in homemade explosives. In postexplosion residues, the chromium will be found in a variety of oxidation states, but hexavalent chromium will still be detected.

Spot Tests for Chromium with Diphenylcarbazide (30)

Trace quantities of chromate ions form a characteristic violet coloration with diphenylcarbazide in strongly acidic solution. The chromate ions first oxidize diphenylcarbazide to diphenylcarbazone, and then, in the second stage of the reaction, bivalent chromium ions form a red-violet inner complex salt with the enol form of the diphenylcarbazone.

$$O{=}C\begin{matrix}\diagup NH-NH-C_6H_5\\ \diagdown NH-NH-C_6H_5\end{matrix} \xrightarrow{[O]} O{=}C\begin{matrix}\diagup NH-NH-C_6H_5\\ \diagdown N=N-C_6H_5\end{matrix} \rightleftarrows$$

diphenylcarbazide — diphenylcarbazone (keto form)

$$\rightleftarrows \begin{matrix}\diagup\!\!\diagup N-NH-C_6H_5\\ C-OH\\ \diagdown N=N-C_6H_5\end{matrix} + Cr^{+2} \longrightarrow \begin{matrix}\diagup\!\!\diagup N-N-C_6H_5\\ C-O-Cr(H_2O)_4\\ \diagdown N=N-C_6H_5\end{matrix}$$

diphenylcarbazone (enol form) — inner complex salt of chromous ions

PROCEDURE A [detects only chromium (VI) ions]. Place two drops of the test solution in the depression of a spot plate. Add a drop of conc. sulfuric acid and a drop of the reagent solution. The presence of chromate ions is indicated by the appearance of a red-violet color.

Because during the explosion most of the hexavalent chromium is reduced to lower oxidation states, an alternative spot test procedure describes the detection of all chromium ions irrespective of their valence. Chromium (III) ions are easily oxidized in alkaline solution to chromate ions. Before testing for chromate with diphenylcarbazide, the solution should be acidified to decompose the excess hypobromite, as free bromine destroys the diphenylcarbazide reagent.

$$Br^- + BrO^- + 2H^+ \rightarrow H_2O + Br_2^0$$

In the following test, bromine is removed with sulfosalicylic acid prior to addition of the reagent.

PROCEDURE B (detects all chromium ions). Place a drop of the neutral or weakly acidic test solution in the depression of a spot plate and add a drop of oxidizing solution. After 1 min, add a drop of conc. sulfuric acid and two drops of a 20% aqueous solution of 5-sulfosalicylic acid; then add a drop of the reagent solution. The presence of chromium is indicated by the appearance of a violet color.

PREPARATION OF THE OXIDIZING SOLUTION. Mix 8 ml bromine water with 2 ml 5 *N* sodium hydroxide and 2 g each of potassium and bromine.

PREPARATION OF THE REAGENT SOLUTION. Prepare a saturated solution of diphenylcarbazide in ethanol.

4.3.9. Dinitrotoluene (DNT)

PROCEDURE (19). Place the solid test material in the depression of a spot plate. Add one drop of acetone–ethanol mixture (1 : 1, v/v) and a drop of 20% aqueous tetramethylammonium hydroxide. The presence of DNT is indicated by a blue color; the color becomes dark green when nitroglycerine is also present.

4.3.10. Magnesium

Military and commercial pyrotechnics and homemade explosive devices often use magnesium filings or powder. The filings look brightly metallic, whereas magnesium powder has a dull gray appearance.

Spot Test for Magnesium Using Quinalizarine (31)

Magnesium salts give a blue precipitate or a cornflower blue tint in alkaline solutions of quinalizarine (1,2,5,8-tetrahydroxyanthraquinone).

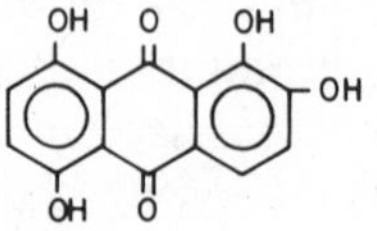

OH O OH
OH
OH O

POSTEXPLOSION TEST PROCEDURE. Place a drop of the test solution in the depression of a spot plate and add two drops of the reagent solution. In an acidic medium, the dye is orange. Add 2 *N* sodium hydroxide solution until the color changes to violet. The presence of magnesium is indicated by the appearance of a blue precipitate or tint that intensifies on standing.

PREEXPLOSION TEST PROCEDURE. Dissolve the unknown metal in 2 *N* hydrochloric acid prior to performing the procedure above.

PREPARATION OF THE REAGENT SOLUTION. Make an 0.02% (w/v) solution of quinalizarine in ethanol.

4.3.11. Nitrates

Nitrate is a constituent of most dynamites as ammonium nitrate, sodium/potassium nitrate, or both; of black powder as potassium nitrate; and of a few flash powders. The facts that nitrates are present in a great number of explosives and occur quite often in the enviornment decrease their significance in explosion investigations. Certain dynamites, however, scatter small particles of ammonium nitrate after exploding. In this case, a nitrate test in conjunction with an ammonium ion test will provide a quick, decisive identification of the material.

Spot Test for Nitrate with the Diphenylamine Reagent

The classical diphenylamine reaction (31) of nitric acid proceeds via the oxidation of diphenylamine to colorless *N,N'*-diphenylbenzidine and results in the formation of a blue quinoid imonium ion.

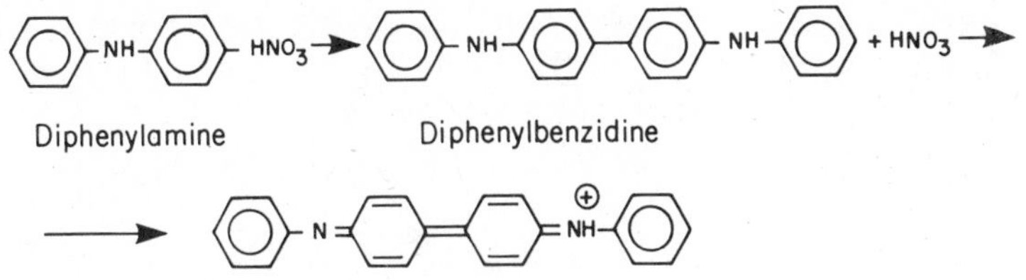

Diphenylamine $\xrightarrow{HNO_3}$ Diphenylbenzidine $\xrightarrow{+HNO_3}$ blue quinoid imonium ion

The color reaction is not specific for nitric acid. Nitrites, chlorates, bromates, iodates, chromates, permanganates, selenites vanadates, molybdates, and peroxides also oxidize diphenylamine. Interferences by alkali halogenates, perchlorates, periodates, permanganates, and persulfates can be avoided if a drop of the weakly alkaline test solution is taken to dryness and the residue heated to 400–500°C. The foregoing compounds are thermally decomposed, whereas the nitrates and nitrites remain unchanged.

PROCEDURE 1. Place ~0.5 ml of the reagent solution on a spot plate and add a drop of the test solution. The appearance of a blue ring indicates the presence of any of the above-mentioned oxidizing compounds, including the nitrates.

Since chlorates and perchlorates are regular constituents of flash powders and react similarly with the diphenylamine reagent, a selective test for nitrates and nitrites in explosives and explosive residues is achieved by Procedure 2.

PROCEDURE 2. Dry a drop of weakly alkaline test solution in a microcrucible and heat the residue to 400–500°C. After cooling, add a drop of the reagent solution to the ignition residue. The formation of a blue ring is a positive result for nitrates and nitrites.

To remove nitrites from the sample and make the test more indicative for nitrates, the test solution is treated with hydrazoic acid (32):

$$HN_3 + HNO_2 \rightarrow H_2O + N_2 + N_2O$$

PROCEDURE 3. Place a drop of the test solution in the depression of a white spot plate. Add a few milligrams of sodium azide, followed by two drops of 12 *N* sulfuric acid. Once the evolution of gas has ceased, add a few milligrams of sodium sulfite and two more drops of 12 *N* sulfuric acid. Stir the mixture with a thin glass rod. Add a drop of diphenylamine reagent. The formation of a blue color indicates the presence of nitrate.

PREPARATION OF THE REAGENT SOLUTION. Dissolve 1 mg diphenylamine in 10 ml conc. sulfuric acid.

Spot Test for Nitrate with Szechrome Nas® Reagent

Szechrome NAS® (*Advanced Products Beer Sheva Ltd., Beer Sheva, Israel*) is a combination of diphenylamine-4-sulfonic acid with *p*-diaminodiphenylsulfone ($H_2N–C_6H_4–SO_2–C_6H_4–NH_2$). The latter compound makes the reagent nitrate-specific.

PROCEDURE. Add one to four drops of the sample solution to a test tube containing 2 ml reagent solution. Compare the intensity of the developed violet color 10–15 min after mixing with the color scale provided with the reagent. The concentration of nitrate can easily be estimated in the 1–100 mg nitrate/liter range.

PREPARATION OF THE REAGENT. Dissolve 5 g Szechrome NAS in 1 liter of a mixture of equal volumes of nitrate-free conc. phosphoric and sulfuric acids.

The influence of other ions is extremely small. Estimations are not affected by the presence of considerable amounts of persulfate, chromate, chlorine, and so forth and of reducing agents such as sulfite and hydroxylamine.

4.3.12. Nitrocellulose

Nitrocellulose is a constituent of smokeless powder and of some dynamites. The $–ONO_2$ group of the nitrocellulose can be detected by fusion with benzoin: At around 150°C, pyrohydrogenolysis occurs and nitrous acid is formed (33).

PROCEDURE A. Heat a minute quantity of the sample together with several centigrams of benzoin in a micro test tube in a glycerol bath preheated to 130°C. The mouth of the test tube should be covered with a piece of filter paper moistened with the Griess reagent. A red stain develops on the reagent paper at 150–160°C.

PREPARATION OF THE GRIESS REAGENT. Mix equal volumes of a 1% solution of sulfanilic acid in 30% acetic acid and a 0.1% solution of 1-naphthylamine in 30% acetic acid just before using.

An alternative test is based on the warming of the sample with syrupy phosphoric acid, which hydrolyzes the nitrocellulose and converts the cellulose to methylfurfuraldehyde. The latter condenses with thiobarbituric acid to form orange products.

PROCEDURE B (34). Treat a minute quantity of the test material in a micro test tube with several centigrams of thiobarbituric acid and one or two drops of syrupy phosphoric acid. Immerse the tube in a preheated 120–140°C glycerol bath. The presence of nitrocellulose is indicated by the appearance of an orange color.

4.3.13. Nitroglycerine

A constituent of most dynamites, nitroglycerin and other nitrate esters are relatively easily hydrolyzed to nitric acid, which oxidizes colorless diphenylamine or diphenylbenzidine in sulfuric acid solution to a deep-violet semiquinoidal compound. In testing for nitroglycerin with this method, interference by ionic nitrates, which may also be present in explosives, must be avoided.

Selective detection of nitroglycerin is achieved by exploiting its relatively high vapor pressure. The suspect material is sampled in the vapor phase, and inorganic nitrates do not have sufficient vapor pressure to enter into the reaction.

Detection of Nitroglycerin with a Diphenylbenzidine-Containing Absorption Tube (19)

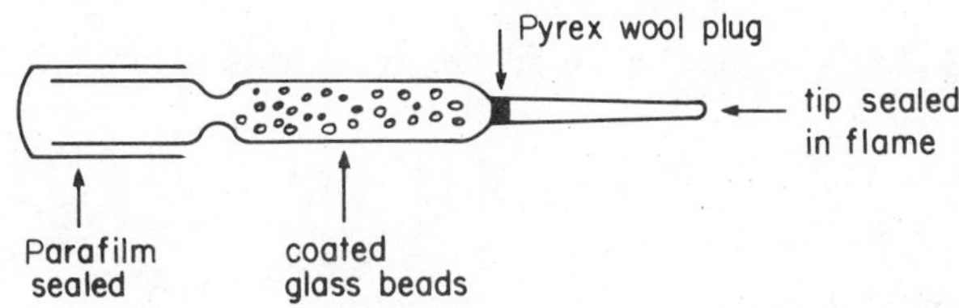

PREPARATION OF THE ABSORPTION TUBE. Add 1.0 ml diphenylbenzidine solution (2.0 mg diphenylbenzidine in 10 ml sulfuric acid) to approximately 30 g of glass beads and mix well. Insert the beads through the open end of the absorption tube and seal the end with polyethylene film. Plug the Pasteur pipet attached to the tube with Pyrex wool and seal its tip in a flame.

PROCEDURE. Remove the Parafilm coating from the tube, and attach the tube to the suction end of a rubber-bulb pump. Break off the sealed tip and aspirate the vapor sample through this broken end. The presence of nitroglycerin is indicated by the appearance of a pale to medium violet color.

The test area searched for nitroglycerin should be warmed to increase the concentration of nitroglycerin in the vapor phase. In practice, a suspect's hand and wrist can be enclosed in a plastic bag for several minutes to enable the nitrate esters to volatilize. The broken tip of a freshly opened tube is then pushed through the wall of the bag and the vapors are aspirated through the tube. The area close to the suspect's palms and fingers is the richest in nitroglycerin vapor. Traces of dynamites can be detected even after the hands have been wiped with a paper towel. The voilet color of a positive test is stable for about a week.

4.3.14. Perchlorates

Perchlorates are constituents of most flash powders. They are used as oxidizing agents in commercial firecrackers, military pyrotechnics, and homemade explosives.

Potassium perchlorate is not very soluble in cold water. When its presence is suspected, the test material should be dissolved in warm water. There is no possibility of detecting perchlorate after an explosion, because it is converted to chloride and oxide salts.

Detection of Perchlorates Through Fusion with Chloride (35)

Nascent oxygen is formed when alkali perchlorate is dry-heated:

$$KClO_4 \rightarrow KCl + 4O$$

In the presence of excess chloride, the chloride ions are oxidized by the atomic oxygen formed in this thermal decomposition. If cadmium chloride, which has a relatively low melting point (568°C without decomposition), is heated to this temperature with alkali perchlorate, free chlorine will form. The chlorine gas can easily be detected in the gas phase on contact with Thio-Michler's ketone paper.

Free halogens in contact with filter paper impregnated with a benzene solution of 4,4′-*bis*-dimethylaminothiobenzophenone (Thio-Michler's ketone) oxidize the reagent to a water-soluble, deep-blue *p*. quinoidal cation.

PROCEDURE. Mix the sample in a microtube with several centigrams of cadmium chloride and heat it over a microburner until the cadmium chloride fuses. Close the mouth of the test tube with a disc of reagent paper. The presence of perchlorate is indicated by the appearance of a blue circular stain on the yellow reagent paper.

PREPARATION OF THE REAGENT PAPER. Soak quantitative filter paper in a 0.1% benzene solution of Thio-Michler's ketone and dry in air. Store the reagent paper in the dark.

Since this spot test cannot be used to detect perchlorates selectively in the presence of halogenates and nitrates, these interferences should be completely eliminated by evaporating the sample with concentrated hydrochloric acid. This treatment completely decomposes halogenates and nitrates; the procedure for perchlorate detection should then be carried out on the evaporation residue.

Detection of Perchlorates with Methylene Blue (36)

Another rather selective test for perchlorates is based on the formation of violet precipitates with these ions in methylene blue solution. The "weighting effect" precipitates the large perchlorate anion with the huge methylene blue cation.

PROCEDURE. Apply a drop of aqueous extract of the explosive residue on a piece of filter paper impregnated with 1 N zinc sulfate–1 N potassium nitrate solution, and spray the paper with 0.05% aqueous methylene blue solution. The presence of perchlorates is indicated by the appearance of a violet spot. Persulfates give an analogous reaction. When they are present, the test is carried out on a thin-layer chromatography plate, which is heated for ~1 hour at 110°C to decompose the persulfates. A violet spot obtained following such treatment indicates the presence of perchlorates.

4.3.15. Permanganate

Potassium permanganate is a common oxidizer in homemade explosives, although never used in commercial products. In the preexploded state, it is a purple-black powder, easily recognized by the characteristic purple color in aqueous solution. After an explosion, the permanganate is generally reduced to manganous ions.

Spot Test for Mn (II) Ions by Catalytic Oxidation to Permanganate

Bivalent manganese is catalytically oxidized in the presence of small quantities of soluble silver salt to the violet permanganate ion (37):

$$2Mn^{2+} + 5S_2O_8^{2-} + 8H_2O = 2MnO_4^- + 10SO_4^{2-} + 16H^+$$

Post Explosion Test Procedure (19). Mix a drop of the test solution with a drop of conc. sulfuric acid in a microcrucible. Add a drop of 4% aqueous silver nitrate solution and a few milligrams of ammonium persulfate, and heat gently. The appearance of a violet color indicates the presence of manganese.

The purple color of the permanganate ion in aqueous solution without any treatment is an adequate indication for the presence of permanganate. However, if the presence of another highly colored ion such as dichromate interferes with the visual reading of the violet color, the reactivity of cellulose in filter paper can be used for the selective reduction of permanganate to insoluble manganese dioxide.

Preexplosion Test Procedure (19). Drop a neutral or slightly acidic test solution onto thick filter paper. In the presence of permanganate, a brown manganese dioxide stain forms at the center of the spot within 3 min. Other water-soluble colored compounds can be washed out easily.

4.3.16. Phosphorus

Red phosphorus, in admixture with oxidizing salts, forms percussion mixtures after detonation that are spread over a large area and in which elemental phosphorus can be detected. If the percussion mixture is confined, the residue may be tested for phosphate, which is the major byproduct of an explosion.

Postexplosion Test for Phosphate (38)

This test is based on the formation of a complex heteropolyacid, phosphomolybdic acid $(H_7(Mo_2O_7]_6$ when phosphates react with molybdates in mineral acid solution. The complexed molybdic acid in phosphomolybdate has an enhanced oxidizing power toward many inorganic and organic compounds. Benzidine, which is not oxidized by free molybdic acid, is instantly transformed to molybdenum blue and benzidine blue by phosphomolybdic acid. This reaction is extremely sensitive, since both the oxidation and the reduction products are blue.

PROCEDURE (19). Place one drop of the test solution in the depression of a spot plate. Add one drop of reagent A, wait 30 sec, add one drop of reagent B, wait another 30 sec, and add three drops of a saturated solution of sodium acetate. A blue-gray color indicates the presence of phosphates.

PREPARATION OF REAGENT A. Dissolve 0.5 g ammonium molybdate in 10 ml water. Add 3 ml conc. nitric acid.

PREPARATION OF REAGENT B. Dissolve 6.5 g benzidine in 10 ml glacial acetic acid and dilute to 35 ml with water.

Preexplosion Test for Elementary Phosphorus (19)

PROCEDURE. Wash out the oxidizing salt from the putative phosphorus-containing mixture by washing it through filter paper with warm water. Phosphorus will remain on top of the filter paper. Place the suspect material in a depression of a spot plate and oxidize any phosphorus to phosphate by adding several drops of a nitric acid–hydrochloric acid mixture (1 : 1). Wait 5 min. Place one drop of the acidic solution on another spot plate and add one drop of a 4 *N* solution of sodium hydroxide, one drop of reagent A (recipe above), and one drop of reagent B (recipe above). Wait ~30 sec; then add three to five drops of a saturated solution of sodium acetate. A blue or blue-gray color indicates the presence of phosphorus.

Soluble silicates and arsenates interfere with this reaction, giving false positive results.

4.3.17. Potassium

Potassium is present in most flash powders as potassium perchlorate; in most black powders as potassium nitrates; and, in small quantities and various formulations, in some dynamites.

Spot Test for Potassium Ions with Dipicrylamine (39)

Dipicrylamine (hexanitrodiphenylamine) dissolves in alkaline solutions and produces an orange-red species. The following equilibrium exists between the yellow baso-form and the orange aci-form:

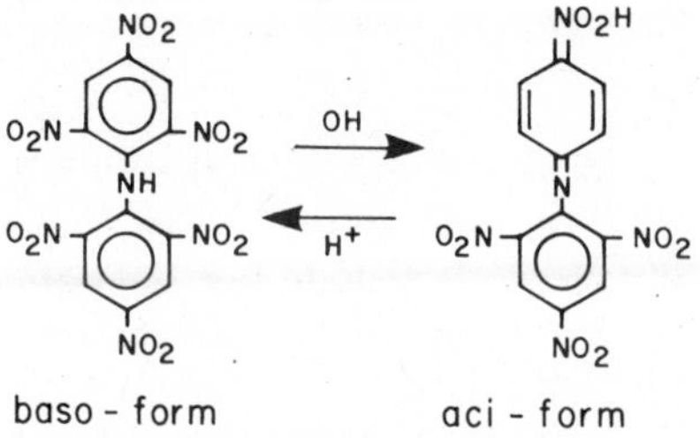

In the presence of potassium ions, red crystalline potassium dipicrylaminate is precipitated, which is distinctly resistant to dilute acids. The only interfering ion that might be present in explosive materials is ammonium. Ammonium salts must be removed by igniting the suspect material before the dipicrylamine test.

PROCEDURE. First test the suspect material for the presence of ammonium ion. If the sample does contain ammonium, ignite a few centigrams of the material in a microcrucible. After cooling, moisten the ignition residue with several drops of water and spot it onto filter paper impregnated with sodium dipicrylaminates. Dry the reagent paper with a heat gun and place it in a 0.1 *N* nitric acid bath. The presence of potassium ions is indicated by a red stain at the site of the spot. All other parts of the reagent paper turn bright yellow.

PREPARATION OF THE REAGENT PAPER. Dissolve 0.2 g dipicrylamine in 2 ml 2 *N* sodium carbonate and 15 ml water. Soak strips of filter paper in the solution and dry with blasts of heated air. The reagent paper should be freshly prepared before the spot test.

4.3.18. Sugar (Sucrose)

Household granulated sugar is often used in homemade explosive mixtures as a reducing material. Following an explosion, the sugar may still be detectable in the residue.

POSTEXPLOSION TEST PROCEDURE (19). Place two drops of the test solution on a spot plate. Add one drop of a 15% ethanol solution of 1-naphthol and two drops of conc. sulfuric acid. A blue or purple-blue color indicates the presence of sugar.

PREEXPLOSION TEST PROCEDURE (19). The procedure is the same as above, but the positive test result is a purple color.

4.3.19. Free Sulfur

Free sulfur is a constituent of flash powder, black powder, and some dynamites.

Spot Test for Sulfur by Conversion to Mercury Sulfide (40)

Crystalline free sulfur can easily be extracted from the explosive material with carbon disulfide. Slight amounts of sulfur dissolved in carbon disulfide can be detected by shaking with mercury; the mercuric sulfide formed collects on the surface of the metal.

PROCEDURE. Place several milliliters of the sulfur extract in a glass test tube with carbon disulfide. Add a drop of mercury and shake the tube vigorously. In the presence of even slight amounts of free sulfur, the surface of the metal will be stained black.

This mercuric sulfide film can be further confirmed by the catalysis of the iodine–azide reaction by sulfide ions:

$$2NaN_3 + I_2 \rightarrow 2NaI + 3N_2$$

PROCEDURE. Decant the carbon disulfide and heat the contaminated mercury on a watch glass to remove the final traces of carbon disulfide. Cover the mercury with the iodine–azide reagent. The appearance of a foam of nitrogen bubbles indicates the presence of free sulfur.
(See preparation of the reagent in Section 4.7.1.)

Sulfur is detected in the presence of metallic aluminum in suspected explosive materials by a simplified field method (41).

PROCEDURE. Dissolve the suspected powder in pyridine. Saturate a cotton plug placed in a disposable pipet with 2 *M* sodium hydroxide, and pour the sample through it. The presence of sulfur is indicated by the appearance of a color reaction at the top of the plug varying from blue to green for low and from red to brown for high concentrations of sulfur.

4.3.20. 2,4,6-Trinitrophenylmethylnitramine (Tetryl) (42)

Tetryl is a military explosive not normally encountered in civilian cases. It can be detected by the formation of nitrous acid when it is heated to 150°.

$$O_2N\text{-}C_6H_2(NO_2)_2\text{-}N(CH_3)NO_2 \xrightarrow{150°C} O_2N\text{-}C_6H_2(NO_2)_2\text{-}N{=}CH_2 + HNO_2$$

If tetryl is heated to the same temperature with hexamethylenetetramine, ammonia and formaldehyde appear in the vapor phase. The tetryl acts as an oxidant with respect to the hexamethylenetetramine and decomposes it oxidatively:

$$(CH_2)_6N_4 + 6[O] \rightarrow 6CH_2O + 2N_2$$

$$(CH_{26}N_4 + 12[O] \rightarrow 6CO_2 + 4NH_3$$

Procedure A. Place the test material in a micro test tube and immerse the tube in a glycerin bath preheated to 150°C. Cover the mouth of the tube with a filter paper disc moistened with the Griess reagent. (See Section 4.3.12 for preparation of the Griess reagent.) In the presence of tetryl, a red stain appears on the paper.

Procedure B. Add to the test material in a micro test tube several centigrams of hexamethylenetetramine and mix thoroughly. Immerse the tube in a glycerol bath preheated to 150°C. Cover the mouth of the tube with a disc of filter paper moistened with the Nessler reagent. (See Section 4.3.2 for preparation of the Nessler reagent.) A brown color indicates the presence of ammonia, and a black color signals that of formaldehyde.

Neither of these tests gives a positive reaction with RDX.

4.3.21. 2,4-Dinitrotoluene and 2,4,6-Trinitrotoluene (TNT)

2,4-Dinitrotoluene is used as a coating on many smokeless powders, whereas 2,4,6-trinitrotoluene is a military explosive not normally encountered on the civilian market.

Characteristic colors develop when di- and trinitroaromatic compounds are treated in alkaline solutions. According to the classical Janovsky reaction (43), characteristic red to violet colors develop when di- and trinitroaromatic compounds in acetone solutions are treated with concentrated aqueous potassium hydroxide solution.

In a modern version of the alkaline reaction, tetramethylammonium hydroxide is used in a selective spot test for 2,4,6-trinitrotoluene (TNT) (44).

Procedure. Add one drop of a 1 : 1 acetone–alcohol mixture and one drop of 25% aqueous tetramethylammonium hydroxide to 5–10 mg of test material. In the presence of 2,4-dinitrotoluene a blue color appears, and in the presence of TNT, a dark red color. Since the color will change with time, the first observation is the decisive one.

This test is selective in the presence of a wide range of blasting explosives.

4.3.22. Zinc

Powdered zinc metal is a part of many homemade explosives. Various combinations of sulfur, zinc, oxidizing agents, and other metals are widespread.

Spot Test for Zinc with Dithizone

Diphenylthiocarbazone (dithizone) forms insoluble, colored inner complex salts with several metal ions, among them zinc.

$$S{=}C\begin{cases} NH-NH-C_6H_5 \\ N{=}N-C_6H_5 \end{cases} + Zn^{+2} \longrightarrow S{=}C\begin{cases} NH-N(C_6H_5)\diagdown \\ \qquad\qquad\quad Zn \\ N{=}N(C_6H_5)\cdots \diagup \end{cases}$$

If the reaction is carried out in alkaline solution, it is selective for zincates.

POSTEXPLOSION TEST PROCEDURE (19). Place a drop of the test solution on a spot plate. Add a drop of 2 *N* sodium hydroxide and a few drops of the dithizone reagent. In the presence of zinc the green dithizone solution turns raspberry red. Compare the outcome of the reaction with that obtained with solutions of zinc of known concentration.

PREEXPLOSION TEST PROCEDURE (19). Dissolve the test material in 2 *N* hydrochloric acid. Place two drops of the solution in a depression of a spot plate and make the solution basic with 2 *N* sodium hydroxide in the presence of litmus paper. Add a few drops of the dithizone reagent solution. The presence of zinc is indicated by the appearance of a raspberry red color in the solution.

PREPARATION OF THE REAGENT SOLUTION. Make a 0.01% (w/v) solution of dithizone in carbon tetrachloride.

4.4. CHEMICAL SPOT TESTS FOR DRUG AND POISON DETECTION

The current high level of narcotic abuse calls for the rapid and reliable identification of drugs and the tracking down of illicit supply sources. As the problem of drug abuse reaches frightening proportions, the use of simple tests gains considerable importance in these investigations.

Prosecution for the use or possession of narcotics or illegal drugs requires analytical data and identification of the suspect material. The U.S. Federal Law Enforcement Assistance Administration has developed programs for professional progress in this field (45). In recent years, various instrumental methods for identifying drugs by their physical properties have been greatly improved. These are enormously useful, but outside the scope of this book; we are concerned here only with quick chemical methods feasible for field conditions.

These tests are intended primarily for the use of hospital biochemists and law-enforcement agents, who need to obtain information in the minimal amount of time without carrying out detailed analyses.

4.4.1. Systematic Spot Tests for Drugs of Abuse

In spite of the shortcomings of spot tests, such as occurrence of false positives and negatives and lack of specificity, they are still very useful in answering preliminary questions quickly under field conditions. With a limited number of tests it is possible to identify a drug tentatively or to obtain definite information about the absence of a compound or a group of compounds.

In one experiment, more than 40 of the most common street drugs were tested with nine different reagents, and the systematic classification of these drugs according to the results obtained was possible (46). The 43 drugs tested were grouped into five classes according to their chemical natures:

1. Alkaloids: atropine, caffeine, cocaine hydrochloride, codeine phosphate, heroin, ephedrine sulfate, lysergide, mescaline hydrochloride, morphine sulfate, nicotine salicylate, psilocin, psilocybin, quinine sulfate, scopolamine hydrobromide, strychnine, and yohimbine hydrochloride.
2. Compounds giving a positive alkaloidal reaction: lidocaine hydrochloride, meperidine, methadone hydrochloride, methaqualone, methylphenidate hydrochloride, pentazocine hydrochloride, phen-

cyclidine hydrochloride, propoxyphene napsylate, and methapyrilene hydrochloride.

3. Barbiturates: amobarbital, phenobarbital sodium, and secobarbital.
4. Amphetamines: amphetamine sulfate and methamphetamine hydrochloride.
5. Miscellaneous: aspirin, benzocaine, cannabidiol, cannabinol, diphenylhydantion sodium, glutethimide, meprobamate, and tetrahydrocannabinol.

Systematic testing for these drugs should be carried out according to the flowsheet of Table 1.

Preparation of the Reagent Solutions

MAYER'S REAGENT. Dissolve 0.68 g mercury chloride and 2.5 g potassium iodide in water and bring volume to 100 ml.

DRAGENDORFF'S REAGENT. *Solution A:* Dissolve 0.85 g bismuth subnitrate in 50 ml 20% aqueous acetic acid. *Solution B:* Dissolve 8.0 g potassium iodide in 20 ml distilled water. Five volumes of A and two volumes of B make up the concentrate. To prepare the working solution of the reagent, mix 10 ml of the concentrate with 100 ml distilled water.

WAGNER'S REAGENT. Mix 1.27 g iodine with 2 g potassium iodide and dissolve in distilled water. Bring to 1 volume of 100 ml.

MARQUIS REAGENT. Mix 10 ml conc. sulfuric acid with 8–10 drops of 40% formaldehyde solution. Prepare fresh prior to each use.

COBALT THIOCYANATE. Prepare a 2% aqueous solution.

ZWIKKER TEST. Solution A is 1% cobalt acetate in methanol and Solution B, 5% isopropylamine in methanol.

MANDELIN'S TEST. Add 1% ammonium vanadate to concentrated sulfuric acid until the reagent acquires an orange color.

EHRLICH'S REAGENT. Add to 100 ml of a 0.2% solution of *p*-dimethylaminobenzaldehyde in 65% sulfuric acid 0.2 ml 5% ferric chloride solution. The reagent is not stable, and should be prepared weekly.

DUQUENOIS REAGENT. Add to 20 ml USP alcohol 5 drops of acetaldehyde and 0.4 g vanillin. This reagent should be kept in a glass-stoppered bottle, protected from light, and discarded when its color changes to deep yellow.

Table 1. Flowsheet for Drug Testing[a]

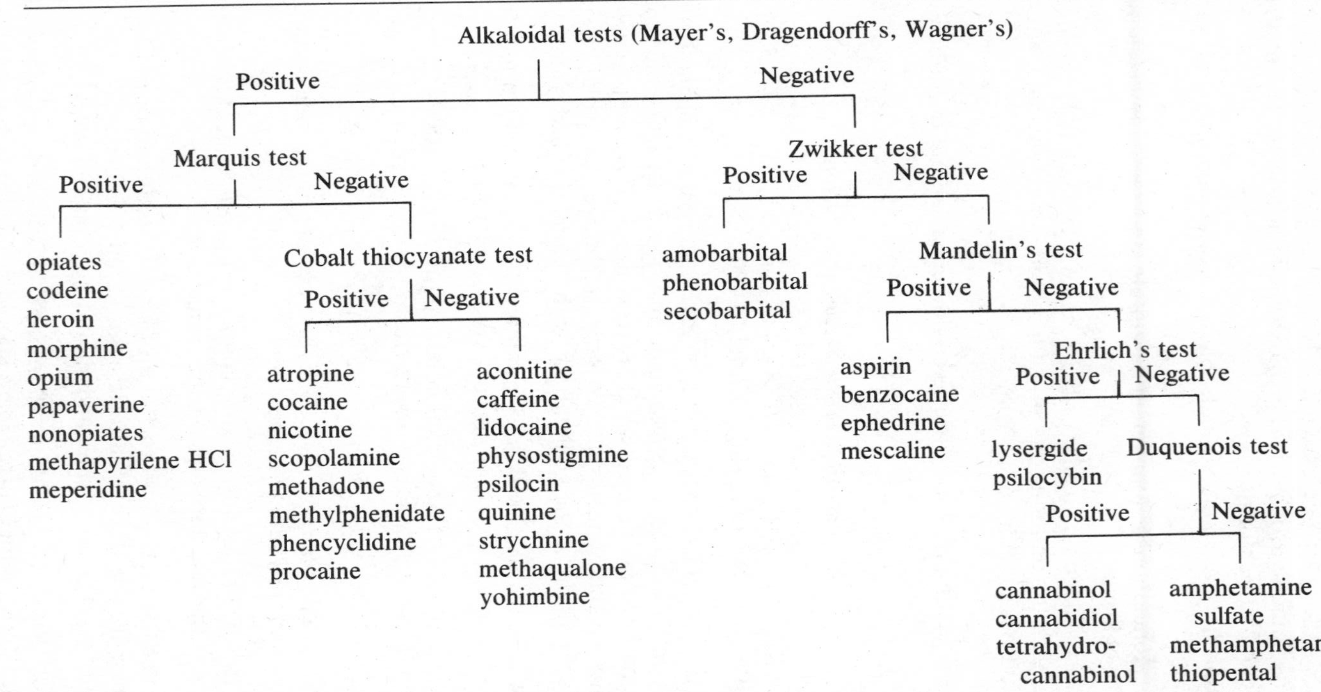

[a] Reproduced with permission of the copyright owner from *J. Pharm. Sci.* **64,** 842 (1975).

Test for Alkaloids

PROCEDURE. Perform spot tests directly on 1–2 mg of powdered drugs in test tubes. Mayer's, Wagner's and Dragendorff's reagents, which are known as the alkaloidal reagents, are used, and precipitate formation was considered a positive reaction. Several alkaloids give negative reactions to one or more of the alkaloidal reagents, whereas a number of nonalkaloids give positive reactions.

Table 2. Alkaloids That Give Negative Tests with One or More Alkaloidal Reagents [a]

	Reagent		
Compound	Mayer's	Dragendorff's	Wagner's
Caffeine	−	+	+
Heroin	+	+	−
Ephedrine	−	−	−
Lysergide	−	−	−
Mescaline	−	−	−
Morphine	+	+	−
Psilocin	−	−	+
Psilocybin	−	−	−

[a] Reproduced with permission of the copyright owner from *J. Pharm. Sci.* **64,** 842 (1975).

Table 3. Nonalkaloids Giving Positive Alkaloidal Tests with One or More Reagents [a]

	Reagent		
Compound	Mayer's	Dragendorff's	Wagner's
Lidocaine HCl	+	+	+
Meperidine	+	+	+
Methadone HCl	+	+	+
Methaqualone	+	+	−
Methylphenidate HCl	−	+	−
Pentazocine HCl	+	+	+
Phencyclidine HCl	+	+	+
Procaine HCl	−	+	−
Propoxyphene napsylate	+	+	+
Methapyrilene HCl	+	+	+

[a] Reproduced with permission of the copyright owner from *J. Pharm. Sci.* **64,** 842 (1975).

Test for Opiates (46)

PROCEDURE. Add several drops of the Marquis reagent to 1–2 mg drug in a test tube. The presence of opiates (codeine phosphate, heroin, morphine sulfate, papaverine, and opium powder) is indicated by the appearance of an intense purple turning to dark blue-violet.

False positives have been recorded with nonopiates such as ephedrine sulfate, methapyrilene hydrochloride, amphetamine sulfate, meperidine, and methamphetamine hydrochloride.

Test for Cocaine, Procaine, and Related Compounds (46)

PROCEDURE. Add several drops of the cobalt thiocyanate reagent to several milligrams of the drug in a test tube. A flaky blue precipitate appears indicates a positive reaction.

Cocaine hydrochloride, codeine phosphate, atropine, heroin, nicotine salicylate, opium, scopolamine hydrobromide, meperidine, methadone hydrochloride, methylphenidate hydrochloride, procaine hydrochloride, methapyrilene hydrochloride, and phenobarbital sodium respond positively to this test.

The cobalt thiocyanate test is not specific for cocaine. A simple, sensitive, and specific field test has been recommended for cocaine (47) that is based on recognition of the odor of methyl benzoate as a test product.

PROCEDURE. Moisten the dry sample with a 5% sodium hydroxide solution in dry methanol. Allow a few minutes to elapse for evaporation of the alcohol; then note whether or not the odor of methyl benzoate is present. If it is, cocaine is present. The reagent and sample must be kept dry, since water interferes with the test.

This test has a very high selectivity. Of over a hundred drugs tested, only cocaine and piperococaine yielded methyl benzoate, although a weak "fishy" odor was noted with amphetamine.

Zwikker-Type Tests for Barbiturates (46)

PROCEDURE. Add to a few milligrams of the drug (residue from evaporation of the aqueous solution if necessary) a few drops of Zwikker A solution, followed by a few drops of Zwikker B. The presence of barbiturates is indicated by the appearance of a blue-violet color.

Amobarbital, phenobarbital sodium, and secobarbital react positively to this test.

In several modifications of the Zwikker alkaline test the alkalizing agent was changed to improve the selectivity. An improved spot test (48) for barbiturates and hydantoin uses 2,6-dimethylmorpholine instead of the isopropylamine of the classical Zwikker test.

PROCEDURE. Place 1–2 mg of the drug on filter paper and treat with a drop of the reagent solution. The appearance of a violet to purple color indicates the presence of barbiturates, whereas hydantoin causes a blue color to form. Submicrogram amounts are detectable, and the color can be preserved indefinitely.

The European Pharmacopeia has accepted a spot test for the identification of barbiturates in which the Zwikker test is modified by the inclusion of cyclohexylamine as the alkalizing agent (49, 50).

PROCEDURE. Dissolve 2–3 mg of the test substance in 2 ml of a 0.1% methanolic solution of cobalt acetate or cobalt nitrate, and mix with 0.1 ml cyclohexamine. A violet color is produced in the presence of barbiturates, but sulfathiazole and theophylline give false positive reactions.

Isobutylamine serves as the alkalizing material in the Zwikker alkaline cobalt reaction for barbiturate detection (51).

PROCEDURE. Place a drop of methanolic sample solution into a depression of a spot plate. Add one drop of methanolic 0.01 *M* cobalt acetate and one drop of 1 *M* isobutylamine in chloroform. The violet color formed by barbiturates is also given by phenytoin, ethosuximide, pyrithyldione, and theophylline. More than a hundred other drugs tested gave negative responses.

Severe poisoning by an overdose of barbiturates is very common, and detection of these substances in powders, tablets, capsule residues, and urine by a screening test procedure should normally precede the estimation of their level in the blood.

Alternative Spot Test Procedures for Detection of Barbiturates

PROCEDURE A. Add 1–2 ml of a 1% ethanolic solution of cobalt nitrate to 1–2 ml of the test material dissolved in ethanol. Add one or two pellets of sodium hydroxide. The presence of barbiturates is indicated by the appearance of a blue color. Bemegride, glutethimide, phenytoin, and primidone also give a positive reaction with the test. Use of lithium hydroxide instead of sodium hydroxide makes the test specific for barbiturates.

PROCEDURE B. Extract the acidic test solution with several drops of ethyl ether and drop the extract onto filter paper. Examine the spot under ultraviolet light before and after exposing it to ammonia vapor. An increasingly dark spot after exposure to ammonia indicates the presence of barbiturates.

Test for Barbiturates in Urine

PROCEDURE. Acidify 50 ml urine to pH 4–5 by adding a few drops of a 10% sulfuric acid solution. Extract twice with 50 ml ether, combine the ether extracts, and wash with water. After evaporation to dryness, dissolve the residue in 1 ml chloroform. Add two drops of a freshly prepared 1% solution of lithium hydroxide in methanol. The appearance of a blue ring at the interface shows the presence of barbiturates.

A commercial test device for the identification of barbituric acid derivates and glutethimide in biological fluids (52) consists of a strip of porous material such as filter paper divided into three areas impregnated with 0.4% potassium dihydrogen phosphate, 0.2% mercuric acetate dissolved in 0.6% methanolic Tris buffer, and 0.02% diphenylcarbazone dissolved in chloroform. The three zones are separated by intermediate reagent-free zones.

PROCEDURE. Absorb the sample into the phosphate zone, wet the middle (acetate) zone with water, and place the strip in a tube containing a solvent suitable for the barbiturates and their mercury complexes. The appearance of a blue color when the solvent reaches the uppermost zone indicates the presence of barbiturates or glutethimide. The detection limit is 20–50 ppm.

The Characterization of Several Drugs by Mandelin's Reagent

The qualitative detection of about 150 different drugs with this reagent has been described (53, 54). The great variety of colors developed with the different drugs, the low rate of color formation, and the interchanges from one to another color make most of these described tests uncertain and extremely difficult to interpret. Easily read colors were produced, however, by four drugs when the tests were carried out in the sequence of Table 1.

PROCEDURE (46). Add several drops of Mandelin's solution to the test material. A stable brick-red color indicates the presence of ephedrine sulfate, while an orange color changing to yellow or green is characteristic of mescaline hydrochloride. Aspirin produces instantly a blue-green color changing to red-violet, and a stable red-violet color shows the presence of benzocaine.

Test for Lysergide (LSD)

Ehrlich's reagent (also referred to as Van Urk's reagent) produces a characteristic purple to deep-blue color with lysergide. Other hallucinogens such as psilocybin and psilocin are also detected by this test.

PROCEDURE. Introduce the sample to be tested into a depression of a spot plate and add two drops of the reagent solution. In the presence of LSD, a violet color will appear.

Tests for Marijuana

The most widely used illegal drug of the present time is marijuana, which actually consists of dried parts of the *Cannabis sativa* plant. Its concentrated, more potent form is hashish, the resin form of the drug, which contains about 30 derivates.

The spot test for marijuana involves both the Duquenois and the Fast Blue B test. Although several herbal materials respond to the Duquenois test, henna is the only substance besides cannabis that responds to both tests.

PROCEDURE I (*Duquenois test*) (55, 56). Place 1–2 mg of the suspected material in a test tube. Add a few drops of the reagent, followed by a few drops of hydrochloric acid, (1:1) and shake the mixture. Wait 1 min; then add a few drops of chloroform and shake again. The presence of cannabis is indicated by the appearance of a blue-violet color in the chloroform layer.

PROCEDURE II (*Fast Blue B test*). Place ~1 mg of the suspected material on the uppermost edge of two sheets of filter paper. Add light petroleum ether drop by drop until the lower paper is just moist. Remove the upper paper and add to the lower one 0.1 mg of a 1:100 (w/w) mixture of Fast Blue B salt–anhydrous sodium sulfate followed by a drop of water. A red-violet color indicates the presence of cannabis.

Marijuana and coca leaf found in illicit cigarette samples can also be revealed by a similar simple test (56).

Procedure. Shake the suspect material with a chloroform solution of Fast Bordeaux Gp salt (C.I. Azoic Diazo Component 1) and with 0.5 *M* sodium hydroxide solution. Cannabinoids are indicated by the appearance of a red color in the chloroform layer. The method can also be applied to urine samples.

Cannabis constituents in the urine of cannabis smokers can be detected by the thermally induced fluorescence of cannabinoids (57).

Procedure. Adjust 75 ml of the sample to pH 3.4 and extract with 75 ml light petroleum ether. Evaporate the extract to dryness under reduced pressure and heat the residue at 230°C for 15 min. Dissolve the cooled residue in 1 ml heptane, filter the solution, and examine the filtrate under an ultraviolet lamp. The wavelengths of maximum emission of the products are 420 nm for Δ^9-tetrahydrocannabinol and 440 nm for cannabidiol and cannabinol.

Spot Test for the Detection of 1-Piperidinocyclohexanecarbonitrile (PCC) in Illicit Phencyclidine (58)

PCC is a toxic precursor of phencyclidine (59) that thermally generates hydrogen cyanide according to the following scheme (60):

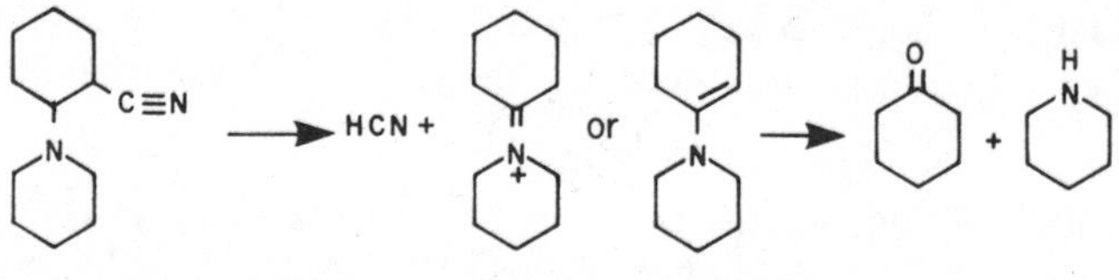

C≡N → HCN + N+ or N → O + H N

A simple and rapid spot test based on this reaction is used in forensic laboratories to detect PCC-contaminated phencyclidine.

Procedure: Place the sample to be used for analysis in a test tube, add 0.5 ml 1.0 *M* sodium acetate–acetic acid buffer (pH 4.7) and suspend the reagent strip above the buffer by wedging it tightly between the cap and the test tube. Agitate gently for 10 sec, taking care not to wet the test strip. Place the tube in a 75 ± 5°C water bath. A positive test is indicated by the appearance of a dark brick-red color.

Preparation of the Reagent Strips. Make a 1 : 1 mixture of 3% sodium hydroxide solution and saturated picric acid solution (61). Impregnate Whatman no. 1 qualitative filter paper with this solution and allow it to dry. Keep the test strips in a closed vial stored in a freezer.

False positive responses were obtained in this test with mercuric cyanide, sodium cyanide, and mandelonitrile glucoside. The limit of detection is 27 μg PCC.

4.4.2. SCREENING TESTS FOR OTHER SPECIFIC DRUGS AND POISONS

Alcohols

The synergistic action of ethanol with other drugs makes the detection of alcohol important and urgent. Poisoning by methanol may be accompanied by marked acidosis. The alcohol is detected by the reduction of potassium dichromate to trivalent chromic ions, indicated by a color change from orange to green:

$$3C_2H_5OH + 2K_2Cr_2O_7 + 8H_2SO_4 \rightarrow 3CH_3COOH + 2Cr_2(SO_4)_3 + 11H_2O + 2K_2SO_4$$

PROCEDURE FOR DETECTION OF ALCOHOL IN URINE (62). Place 1 ml of the urine sample in a test tube. Insert a strip of Whatman glass-fiber filter paper moistened with a drop of sulfuric acid in the neck of the test tube and heat the tube in a boiling water bath for 2 min. The presence of alcohol is indicated by a color change from orange to green. The limit of sensitivity of the reaction is 40 mg%.

In acidic solutions, methanol is readily oxidized to formaldehyde by oxidizing materials. Ethanol is oxidized to acetaldehyde under similar conditions. Since formaldehyde can be detected specifically by a chromotropic acid reaction, the following test for methanol in urine is quite selective.

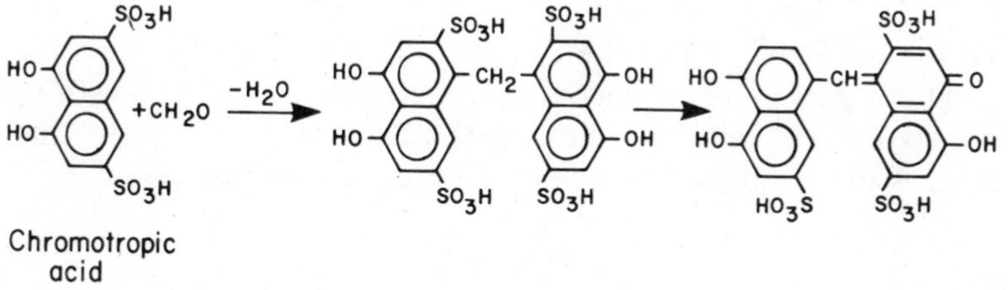

Chromotropic acid

In this test, Chromotropic acid condenses with formaldehyde and the product is then oxidized by concentrated sulfuric acid to a violet semiquinoidal compound.

PROCEDURE FOR THE SELECTIVE DETECTION OF METHANOL (62, 63). To 1 ml of urine add one drop of 25% potassium dichromate solution to 50% sulfuric acid and allow to stand at room temperature for 5 min. Add one drop of ethanol and a few milligrams of chromotropic acid; then add conc. sulfuric acid so that it forms a layer in the bottom of the tube. The presence of methanol is indicated by the appearance of a violet color at the interface.

Salicylates

Overdoses of aspirin and other mild analgesics are extremely common. Poisoning by aspirin is often encountered and may be lethal. The salicylates are detected by formation of the violet chelate iron (III) salicylate.

Test for Salicylates in Urine.

PROCEDURE. Add three drops of 5% ferric chloride solution to 1 ml urine. A violet color indicates the presence of salicylates.

Test for Salicylates, Salicylamide, Paraacetamol, and *p*-Aminosalicylic Acid in Blood.

PROCEDURE. Add 5 ml Trinder reagent (64) to 1 ml blood, shake, and centrifuge. In the presence of salicylates, the supernatant fluid becomes violet.

PREPARATION OF TRINDER REAGENT. Dissolve 4 g mercuric chloride in 85 ml hot water. After cooling, add 12 ml 1 *N* hydrochloric acid and 4 g ferric nitrate.

Test for Salicylates in Gastric Contents. Aspirin recovered from the stomach must be hydrolyzed to salicylic acid before the ferric chloride test can be applied.

PROCEDURE. Boil 3 ml of the sample with an equal volume of 0.1 *N* hydrochloric acid, filter if necessary, neutralize the filtrate with 0.1 *N* sodium hydroxide solution, and add three drops of 5% ferric chloride solution. A violet color develops in the presence of salicylates.

Cyanides

The alkali salts of hydrocyanic acid are extremely toxic, even in minute quantities. Detection of cyanide may be difficult because the cyanide salts are readily decomposed by stomach acid and the hydrocyanic acid formed is volatile and may be lost. The following spot test (65) detects the presence of cyanides in stomach contents, blood, and food slurry.

PROCEDURE. Place several grams of the test material in the outer chamber of a Conway microdiffusion cell and place a square of test paper in the inner chamber. Place the Conway cell in a 50°C incubator for ~30 min. Moisten the test paper with 10% (v/v) hydrochloric acid. The production of Prussian Blue indicates the presence of cyanides. The sensitivity of the test approaches 10 ppm.

PREPARATION OF THE TEST PAPER. Soak filter paper in a 10% aqueous ferrous sulfate solution, remove, and dry. Dip paper in a 10% sodium hydroxide bath, blot, and dry.

Carbon Monoxide

Carbon monoxide is present in automobile exhaust gases and the fumes of some gas heating appliances, and is the cause of accidental deaths from these sources. It is also the means of many suicides. Carbon monoxide, which is a major cause of death, is also a constituent of manufactured water-gas.

Carbon monoxide reduces palladium chloride to metallic palladium:

$$Pd^{2+} + CO + H_2O \rightarrow Pd^0 + CO_2 + 2H^+$$

When treated with sulfuric acid, the carboxyhemoglobin produced in carbon monoxide poisoning releases carbon monoxide, which can then be identified in the gas phase.

PROCEDURE FOR DETECTION OF CARBOXYHEMOGLOBIN IN BLOOD (66). Mix 1 ml blood with 1 ml 10% (v/v) sulfuric acid in the outer chamber of a Conway microdiffusion cell. Place 1 ml 0.1% (w/v) palladium chloride solution in 0.01 *N* hydrochloric acid in the inner chamber of the cell. Leave at room temperature for 30 min. The presence of carboxyhemoglobin is indicated by a silvery film on the surface of the inner chamber. The test detects approximately 10% carboxyhemoglobin.

Arsenic, Antimony, Bismuth, and Mercury

A simple test screens the stomach contents, blood, or urine of a putatively poisoned individual for arsenic, antimony, bismuth, and mercury simultaneously.

PROCEDURE (67). If the test material is urine, it must first be concentrated to one-tenth of its original volume by evaporation. Place about 10 g of the test material (stomach contents, tissue slurry, food, blood, urine concentrate) in a flask and add hydrochloric acid to bring the final acid concentration to 10% (v/v). Clean some copper foil with conc. nitric acid and then wash it carefully. Place the foil in the flask and heat the flask in a steam bath for 15 min. Examine the copper foil for an eventual staining reaction. If no stain forms, repeat the boiling for an additional 15 min and reexamine. The presence of >5 ppm mercury in the sample leads to a formation of a silvery stain on the copper foil. A content of >2 ppm arsenic results in a dull black coating, and antimony (>2 ppm) forms a purple-black stain.

Fluoride

The most common form of fluoride is sodium fluoride, a constituent of ant poisons. The industrial applications of fluorine are increasing rapidly and a series of new fluorine compounds have appeared on the market, possibly leading to an increased significance for fluorides in the field of toxicology.

Toxicologically significant fluoride concentrations in tissue, blood, urine, and stomach content can be selectively detected in special fluoride-resistant Conway dishes. Fluoride ions change the red color of the chelate of alizarin complexan (1,2-dihydroxyanthraquinone-3-methylamino-*N*,*N*-diacetic acid) with trivalent lanthanum to lilac blue, the color of the alizarin complexan itself (68). A test for forensic application uses a plastic Conway microdiffusion cell (69).

PROCEDURE. Place several grams of the blood, urine, or stomach contents in the outer compartment of the microdiffusion cell along with a mixture of iodic and sulfochromic acids. Add 1 ml 0.001 *M* lanthanum nitrate and 1 ml 0.001 *M* alizarin complexan buffered to pH 5.4 with an acetate buffer to the inner compartment. Seal the cell with a cover glass and leave the cell at room temperature for an hour. The presence of fluorides is indicated by a color change from red to lilac blue.

4.4.3. Systematic Screening for Drugs in Urine

The following systematic qualitative search for drugs in urine was described by Clarke (70).

The Detection of Basic Drugs in Urine

PROCEDURE. Place 75 g of anhydrous sodium sulfate and 1 g sodium tetraborate in a screw-capped bottle. Add ~30 ml chloroform and 25 ml of the urine sample and shake vigorously for 2–3 min. Allow the mixture to separate and decant the chloroform extract. If less than 15 ml chloroform extract is obtained, repeat the procedure and combine the chloroform extracts. Dry the chloroform extract by shaking with ~1 g of anhydrous sodium sulfate. Almost all basic drugs are extracted into the chloroform phase by this procedure.

Spot a strip of filter paper with six separate small drops of 2 *N* sulfuric acid. Place the paper strip into the bottle containing the chloroform extract, close the bottle, shake it for 15 min, and allow it to stand. Remove the paper and allow it to dry.

In the presence of a phenothiazine, the spots turn purple.

Cut the filter paper into five strips, cutting between the spots. Treat the strips as follows:

Strip 1. Add to the strip a drop of iodoplatinate reagent (0.25 g platinic chloride and 5 g potassium iodide dissolved in 100 ml water). A dark purple spot is given by almost all basic drugs. Amphetamine may give a white spot. If a dark spot forms, add a drop of freshly prepared 10% sodium sulfite solution. The dark spots of different drugs disappear at different rates. Morphine and nicotine spots fade almost at once. Methadone spots turn reddish brown and fade slowly, as do those of imipramine, chlorpromazine, and diphenhydramine.

Strip 2. Treat the paper strip with Marquis reagent (see Section 4.4.1). A change in color to violet indicates codeine or morphine; to bright yellow, diphenhydramine; to orange-brown, amitriptyline, amphetamine, or methamphetamine; to red, chlorpromazine; and to brownish purple, thioridazine.

Strip 3. This test is carried out when a color is obtained with the Marquis reagent (strip 2). Treat the test strip with a drop of concentrated sulfuric acid. A blue color indicates thioridazine; bright yellow, diphenhydramine or orphenadrine; orange, amitriptyline; and red, chlorpromazine.

Strip 4. Treat the strip with FPN reagent [5 ml 5% (w/v) solution of ferric chloride in water, 45 ml 20% (w/w) perchloric acid, and 50 ml 50% (w/w) nitric acid mixed together]. Shades of red are given by phenothiazines. Thioridazine and imipramine produce shades of blue.

Strip 5. Nitrazepam (a benzodiazepine) can be detected by the Bratton Marshall test. Place the strip in a small porcelain basin. Add two drops of 10% sulfuric acid and a drop of freshly prepared 0.1% sodium nitrite solution, allow to stand for half a minute, add one drop of 0.5% sulfamic acid solution, allow to stand for a further half minute, and then add a drop of 0.1% *N*-1-naphthyl ethylenediamine dihydrochloride solution. The slow formation of a magenta color indicates the presence of nitrazepam.

The Detection of Acid Drugs in Urine

PROCEDURE. Add 0.5 ml 10% sulfuric acid to 25 ml urine and pour into a screw-capped bottle containing 75 g anhydrous sodium sulfate. Extract with chloroform, shake, and decant. It is not necessary to dry the chloroform. Spot a strip of filter paper with a drop of 5% barium hydroxide solution, immerse it in the chloroform extract, and shake for 20 min. Remove paper when dry and test with 5% ferric chloride. A violet color appearing after addition of a drop of the Mandelin reagent (see Section 4.4.1) indicates the presence of salicylates. The appearance of a dull purple color indicates the presence of phenylbutazone.

A test kit for the field detection of some narcotics and psychotropic substances such as LSD, hashish, opium, morphine derivatives, morphine, and cocaine is marketed by E. Merck (Darmstadt, F. R. G.) (71).

4.5. SPOT TESTING FOR THE PRESENCE OF BLOOD

Chemical testing for blood or traces of blood is often needed to provide physical evidence in cases in which direct visual inspection is not a decisive proof. Small blood spots on a dark background may not be visible; on the other hand, a stain may resemble a bloodstain. In such cases, chemical spot tests can verify the presence or absence of bloodstains.

All microchemical tests for blood actually detect heme in the blood, and this fact makes possible interference by plant or animal tissue that also contains the heme molecule. Mashed insects and fly specks may give false positive reactions for human blood, although in general it is relatively easy to distinguish morphologically between blood carried by insects and human blood stains resulting from a crime. Plant peroxidases found in onions, cabbages and potatoes do not interfere if samples are preheated to 100°C before testing.

Blood is identified in most cases on the basis of the peroxidase-like activity of the heme group of hemoglobin. The enzyme peroxidase decomposes hydrogen peroxide or organic peroxides to form free hydroxyl radicals:

$$H_2O_2 \rightarrow 2(OH^{\cdot})$$

The reaction between hydrogen peroxide or sodium perborate and a reducing agent is extremely slow unless a catalyst such as peroxidase or hemic protein is present. Minute quantities of hemoglobin catalyze the reaction; however, this catalysis is effected by other heme proteins as well.

The reducing agents generally applied in this catalytical redox reaction are benzidine, phenolphthalein, the leuco base of malachite green, and luminol. In the classical blood test, benzidine was oxidized to the quinoidal benzidine blue by peroxide in the presence of hemoglobin.

$$H_2N-C_6H_4-C_6H_4-NH_2 \xrightarrow[Hb]{H_2O_2} HN=C_6H_4=C_6H_4=NH$$

Since benzidine and its salts are now considered to be carcinogenic (72), its use has been met with very strong reservations by the Occupational Safety and Health Administration. Forensic laboratories began to apply *o*-tolidine (3,3′-dimethylbenzidine) as a blood test reagent until unfortunately this reagent too became classified as a highly hazardous carcinogen (73). In 1974, the synthesis 3,5,3′,5′-tetramethylbenzidine was described (74). According to limited animal studies, this compound is noncarcinogenic, and it has since been used for spot testing of blood (75).

4.5.1. Spot Test for Traces of Blood by the Hydrogen Peroxide–3,5,3′,5′-Tetramethylbenzidine Reaction

PROCEDURE. Rub the tip of a folded piece of filter paper several times over the dried suspected stain. Put two drops of the test reagent on the paper tip. Wait ~30 sec to ensure that no color develops and then add one drop of 3% hydrogen peroxide. Alternatively, place a piece of the stained object in a test tube and cover with the test reagent. After agitating the tube for 1 min, add a drop of 3% hydrogen peroxide. A positive reaction is indicated by the appearance of a deep blue coloration.

If the stained pieces in the tube are treated in a steam bath for 5–10 min prior to testing, the labile plant peroxidases are destroyed but a true bloodstain will still give a positive test.

PREPARATION OF THE REAGENT SOLUTION. Add 10 ml 1*M* acetate buffer (7.9 ml acetic acid and 9.2 g sodium acetate diluted to 250 ml with water), pH 4.55, to a solution of 10 mg 3,5,3′,5′-tetramethylbenzidine in 0.5 ml glacial acetic acid.

4.5.2. Spot Test for Traces of Blood by Phenolphthalein

Free hydroxyl ions liberated by the peroxidase-like action of heme oxidize colorless phenolphthalin (reduced phenolphthalein) to pink phenolphthalein:

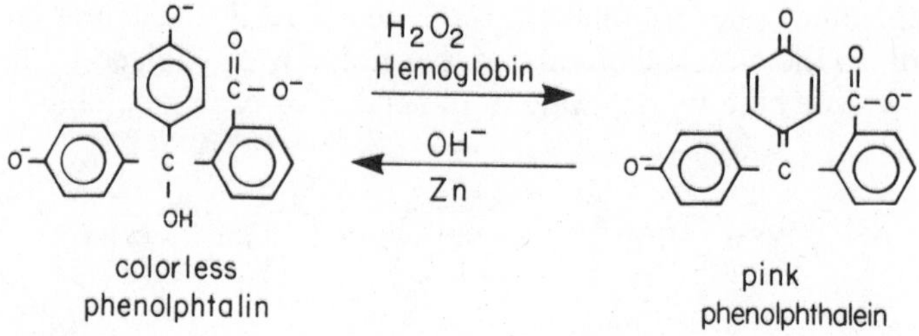

H_2O_2, Hemoglobin →
← OH^-, Zn

colorless phenolphtalin

pink phenolphthalein

Phenolphthalein is believed to be noncarcinogenic (76). Test results also indicate that the enzyme peroxidase, which is widely distributed in plants, does not contribute to false positive results in this test.

PROCEDURE. Rub the stain with a moistened cotton swab. Add a drop of phenolphthalin reagent solution (reduced phenolphthalein with metallic zinc in basic solution). Add a drop of 3% hydrogen peroxide solution. The formation of a pink color indicates the presence of blood traces.

4.5.3. Leucomalachite Green

Leuco malachite green (tetramethyldiaminotriphenylmethane) structurally resembles phenolphthalin. The leuco form is prepared by the reaction of malachite green with zinc dust in basic solution:

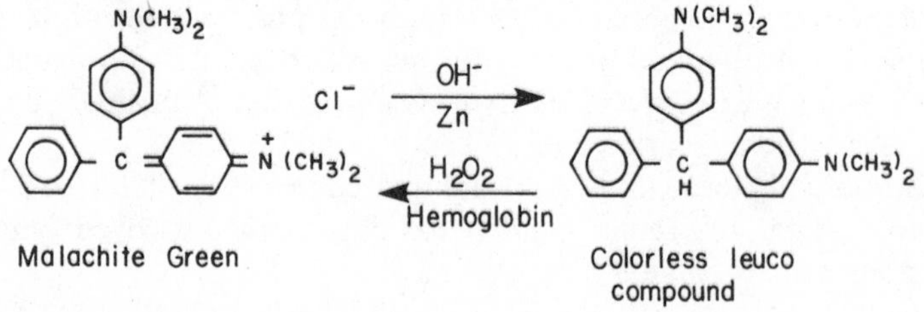

OH^-, Zn →
← H_2O_2, Hemoglobin

Malachite Green

Colorless leuco compound

The reduced leuco compound is reoxidized by hydrogen peroxide in the presence of traces of blood.

4.5.4. Test for Invisible Bloodstains

The luminol reagent is used to visualize invisible bloodstains (77). This chemoluminescence reaction works especially well on decomposed blood. Preliminary spraying with hydrochloric acid decomposes hemoglobin quickly and raises the sensitivity of the test.

Blood testing kits employing luminol are on the market for ready use for field screening (Microchemical Specialities Co., Berkeley, California). Dry, premixed reagents are in the solid state, and reagent solutions are prepared instantly with prepared solvents before use. Hydrogen peroxide, which is rather unstable, is generated on the spot using solid peroxides.

4.5.5. Microcrystal Tests for the Detection of Blood

The crystal tests for blood are far less sensitive than the chemical spot tests, but are nonetheless used as confirmatory tests because of their higher selectivity.

Teichman Test

Hemoglobin treated with acid hydrolyzes into its protein and prosthetic group components. The globin component is denatured. The oxidation of ferrous iron to ferric iron is greatly enhanced in acid solution. In the presence of chloride, insoluble brown crystals of hemin (ferriprotoporphyrin chloride) form.

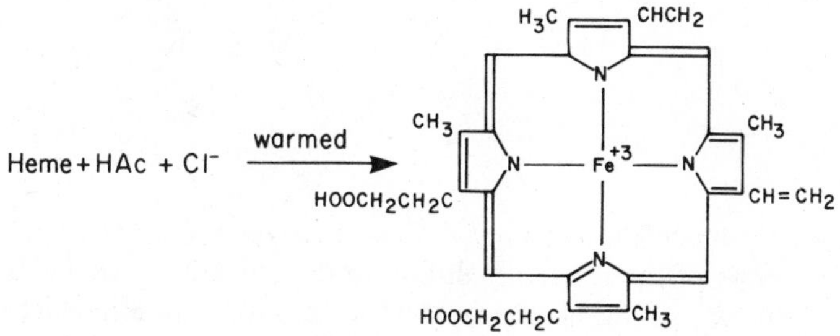

$$\text{Heme} + \text{HAc} + \text{Cl}^- \xrightarrow{\text{warmed}} \text{Hemin}$$

Hemin

PROCEDURE. Place the sample on a microscope slide and cover it. Add one drop of the reagent solution under the cover slip, and warm the slide on a slide warmer. The appearance of brownish rhomboid crystals observed under the microscope indicates a positive reaction.

PREPARATION OF THE REAGENT SOLUTION. Dissolve 0.1 g potassium chloride, 0.1 g potassium bromide, and 0.1 g potassium iodide in 100 ml glacial acetic acid.

Takayama Test

The formation of pink pyridine hemochromogen crystals is the basis of this widely used microcrystal test. The prosthetic group of the hemoglobin (whether ferroprotoporphyrin or ferriprotoporphyrin) is able to add to further nitrogen-containing molecules, such as ammonia, cyanide, nicotine, and pyridine. The new complexes, called hemochromogens, are characteristic in their colors and crystal forms.

The alkaline hydrolysis in this test separates into the prosthetic group and the globin. On heating, ferric iron is reduced to ferrous iron, which attaches to pyridine molecules, forming an insoluble pyridine ferroprotoporphyrin.

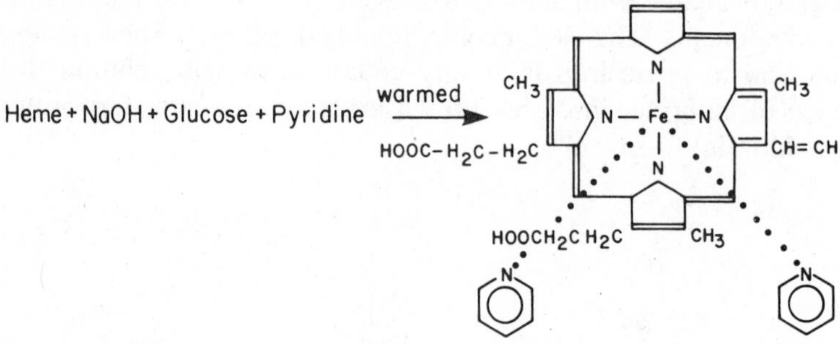

Heme + NaOH + Glucose + Pyridine $\xrightarrow{\text{warmed}}$

CH_3 CH_3
N
N — Fe — N
N
$HOOC\text{-}H_2C\text{-}H_2C$ $CH{=}CH_2$
$HOOCH_2CH_2C$ CH_3
N N

PROCEDURE. Place the sample on a slide, cover with a cover slip, and allow two drops of the reagent solution to flow under the slip. Warm the slide for ~30 sec. In the presence of traces of blood, pink needle-shaped or rhomboid crystals are observed under the microscope.

REAGENT SOLUTION. Mix 3 ml 10% sodium hydroxide solution with 3 ml pyridine, 3 ml saturated aquous glucose solution and 7 ml deionized water. The solution keeps one or two months.

4.6. DETECTION OF SEMEN

The high incidence of sex crimes necessitates the rapid identification of semen. Seminal fluid contains vast amounts of spermatozoa, except in cases of sterility, when spermatozoa may be lacking. The male germ cells are very rich in choline, acid phosphatase, citric acid, and fructose. They also contain acriflavine, the presence of which may have direct diagnostic value. At high dilutions, acriflavine fluoresces under ultraviolet light, and this bluish-white fluorescence is used as a very simple test for the presence of semen. This method today is useful only in the examination of objects that have had no contact with laundry agents, since the modern "brighteners" added to these agents also fluoresce very intensively.

The concentration of acid phosphatase in semen is sometimes several orders of magnitude higher than in other bodily fluids. The detection of this enzyme is an appropriate method of screening for seminal fluid, especially when the test applied is not sensitive enough to detect acid phosphatase at the levels found in other fluids.

4.6.1. Spot Test for Semen by the Identification of Acid Phosphatase

Phosphatase, particularly acid phosphatase, hydrolyzes aliphatic and aromatic orthophosphoric acid esters. Although acid phosphatase is present in many animal and plant tissues, there are greater concentrations of it in seminal fluid, so that an insensitive test will give a strong indication for the presence of this enzyme in semen, but not in other fluids. Disodium phenylphosphate is hydrolyzed by acid phosphatase, and the resulting phenol can be detected with the Folin-Ciocalteu reagent.

O⁻ | OP−O⁻ | O — phenyl ring —(Acid phosphatase)→ OH — phenyl ring + PO_4^{-3}

Phenol forms molybdenum blue in reaction with with the phosphomolybdates and phosphotungstates incorporated in this reagent.

PROCEDURE. Cut or scrape the suspected stain from its location, and incubate it for 30 min at 37°C with two drops of a 0.1% aqueous solution of disodium phenylphosphate. Add two drops of the Folin-Ciocalteu reagent and a few centigrams of sodium carbonate. The presence of acid phosphatase is indicated by the appearance of a dark blue color.

Preparation of Stock Solution of Folin-Ciocalteu Reagent. Dissolve 100 g sodium tungstate and 25 g sodium molybdate in 800 ml water in a 1500-ml flask. Add 50 ml phosphoric acid and 100 ml hydrochloric acid and reflux for 10 hours. Cool; add 150 g lithium sulfate, 50 ml water, and four to six drops of bromine; and allow to stand for 2 hours. Boil for 15 min to remove the excess of bromine, cool, filter, and dilute to 100 ml with water.

This stock solution should be stored at a temperature not exceeding 4°C and used within 4 months of its preparation. For use, dilute one volume of the stock solution with two volumes of water.

4.6.2. Detection of Semen Using the Florence Microcrystalline Test for Choline

Freshly ejaculated semen contains phosphorylcholine, which breaks down on standing into phosphate and choline by the action of acid phosphatase:

$$H_2PO_3O{-}CH_2{-}CH_2{-}\overset{+}{N}(CH_3)_3 \xrightarrow[+H_2O]{\text{Acid phosphatase}} H_3PO_4 + OH{-}CH_2CH_2\overset{+}{N}(CH_3)_3$$

Phosphorylcholine Choline

Choline can be identified by the microscopic crystal test. The aqueous extract is treated with the Florence iodine reagent, and the characteristic rhombic crystals are observed under a microscope. The advantage of this test is that it has two functions: (a) as a test for spermatozoa in a stain; and (b) as a test for detection of choline. The formula of the choline periodide formed is $C_5H_{14}NOI \cdot I_8$.

The test is not selective, since considerable concentrations of choline are found in cells of various organs (e.g., brain, liver, pancreas, stomach, uterus, and blood) and in various food substances (e.g., egg yolk and albumin, cow and human milk, and brewer's yeast). Therefore, the test has only an orientation value, and visual observation of spermatozoa is needed for exact identification.

Procedure. Place the saline extract of the suspected substance on a slide. Add a drop of the Florence reagent. Cover the slide with a cover slip. The appearance of rhombic microcrystals indicates the presence of choline.

PREPARATION OF THE FLORENCE REAGENT. Dissolve 6.6 g potassium iodide and 10.16 g iodine in 120 ml distilled water.

4.6.3. Detection of Semen Using the Microcrystalline Test for Spermine

Spermine (diaminopropyltetramethylenediamine) is found in high concentrations in semen. It can be detected by the Barberio test (78), in which this potent basic polyamine compound is precipitated by picric acid as needle-like spermine picrate crystals.

Although spermine is also found in muscle and nervous tissue, its concentration is almost twice as high in semen.

PROCEDURE. Place a drop of the saline extract of the suspect stain on a slide. Add a drop of a saturated aqueous solution of picric acid. In the presence of spermine, the insoluble needle-like crystals of spermine picrate will be observed under the microscope.

4.7. SPOT TESTS FOR THE DETECTION OF SALIVA

Saliva sometimes requires forensic identification. Its occurrence may be related to spittle, or on objects in contact with the mouth. It contains thiocyanate and, in fresh samples, nitrate. Two thirds (0.6%) of the solid content of saliva are amylase mucin. Salivary amylase, also called ptyalin or diastase, hydrolyzes starch to sugars, and this hydrolysis is used as a basis for saliva identification. Amylase is also found in blood and in some other bodily fluids, but at much lower levels than in saliva.

4.7.1. TEST FOR SALIVA USING THE CATALYSIS OF THE IODINE–AZIDE REACTION BY THIOCYANATES

Thiocyanates accelerate the very slow reaction

$$2NaN_3 + I_2 \rightarrow 2NaI + 3N_2$$

The evolution of nitrogen gas bubbles and loss of the brown color of iodine identify the presence of thiocyanates. Sulfides and thiosulfates catalyze the reaction also, but the absence of these salts in saliva makes the following test quite reliable (79).

PROCEDURE. Mix a drop of the test solution in a depression of a spot plate with a drop of the iodine–azide reagent. In the presence of thiocyanate, and thus saliva, gas bubbles appear and the brown color disappears.

PREPARATION OF IODINE–AZIDE REAGENT. Dissolve 3 g sodium azide in 100 ml of 0.1 *N* iodine. The sodium solution is stable.

4.7.2. Spot Test for Saliva Through Identification of the Enzyme Amylase

Salivary amylase (ptyalin) catalyzes the hydrolysis of starch in the digestive system. Because ptyalin is unique to saliva, the ptyalin test for saliva is quite reliable.

Starch–Iodine Test (80)

Mixing starch and the iodine reagent (iodine dissolved in potassium iodide) results in formation of the familiar blue-black starch–iodine complexes by trapping of the iodine molecule in the helical part of the starch molecule. This reaction can serve as a basis for amylase detection. A saliva stain incubated with an appropriate starch solution and tested with the iodine reagent will give a colorless reaction picture, whereas in the absence of amylase, a dark blue color appears.

PROCEDURE I. Cut out the suspected saliva stain and place it in a test tube. Add five drops of the starch working solution and incubate for 30 min at 37°C. Add one drop of iodine reagent to the tube. The appearance of a blue coloration indicates a negative test; a colorless solution shows the presence of the amylase enzyme. Amylase samples of known concentrations may be used to obtain a standard curve for quantitation.

PROCEDURE II (81). Extract an approximately 0.125-square-inch stained cloth with about seven drops of barbitone-buffered saline (pH 7.2) for 30 min. Dilute the extract serially with the buffer solution, and incubate the dilutions for 1 hour at 37°C with a starch solution. Add iodine solution and observe the resulting colors. The highest dilution at which hydrolysis of the starch is still complete is a measure of amylase activity, and can be used in court as evidence for the presence of saliva.

PROCEDURE III (82). Dampen a piece of reagent paper impregnated with the dyed amylopectin Lyosine Red (83), dampen the stained fabric, and press them together for 15–30 min at room temperature. If the stain is rich in amylase, a pale area appears on the pink paper. Washing this zone with water will make it colorless, because the products of the enzymatic reaction are water soluble. The zone can be "fixed" by immersing the paper in 0.5 *N* sodium hydroxide solution for ~5 min and drying it at room temperature.

Contact between the reagent paper and the specimen for 15 min produces a positive result with saliva stains, but not with stains due to urine, blood, sweat, semen, or nasal secretions. Positive results have been obtained with 20-day-old stains in many fabrics.

Spot Test for Saliva Through the Solubilization Effect of Amylase (80)

The cross-linked blue starch polymer is completely insoluble in water. Amylase enzymatically solubilizes the polymer by hydrolyzing it into smaller soluble fragments that are still (max. abs. 620 nm).

PROCEDURE. Place a cutting from the suspected saliva stain in a tube with 1 ml distilled water. An unstained control, a known sample, and a water blank should be run in parallel with the sample. Add three drops of a slurry made by suspending one reagent tablet in 5 ml distilled water. Incubate for 30 min at 37°C. Stop the enzyme action by adding 1 ml 0.5 *M* sodium hydroxide solution, and centrifuge. The appearance of a blue color in the supernatant fluid indicates the presence of amylase. Although semen and blood contain the enzyme as well, this procedure will not detect amylase derived from either of these two matrixes, and thus is useful for saliva detection even in their presence.

REAGENT TABLET. The cross-linked blue starch polymer is marketed in tablet form by Pharmacia Laboratories, Inc., Piscataway, New Jersey 08854.

4.8. TESTS FOR THE PRESENCE OF URINE

The detection of urine stains on a garment for forensic purposes is possible because it fluoresces on ultraviolet illumination. This method, however, is rather doubtful, because most of the detergents now used contain highly fluorescent "brighteners." The simplest method of confirming whether a fluorescent spot contains urine is to brush the piece of fabric over a small flame until fumes develop (84); the very characteristic odor of urine is unlike that of any other physiological fluid.

Urea, creatine, and creatinine are characteristic constituents of urine, but qualitative chemical testing for these compounds alone is not sufficient for absolute identification of urine. These compounds appear, although in smaller quantities, in perspiration, blood, and other bodily fluids as well. Finding urea and creatinine together, however, is strongly indicative of the presence of urine in a stain.

The detection of urea for forensic work (80) is easily accomplished by enzymatic decomposition by urease and detection of ammonia in the gas phase:

$$C(=O)(NH_2)_2 \xrightarrow[+H_2O]{\text{urease}} CO_2 + 2NH_3$$

PROCEDURE. Place cuttings from a suspected urine stain in a test tube. Add three drops of distilled water to cover the sample. Appropriate known and blank controls should be run in parallel with the samples. Add one drop of 1% aqueous urease solution. Affix a strip of litmus paper to a cork that will fit the test tubes used by cutting a slit in the bottom of the cork and inserting the end of the paper. The paper strip should hang down into the tube without touching the walls of the tube. Incubate at 37°C for 30 min. The liberation of ammonia is indicated by a change in the color of the litmus paper from red to blue.

As a confirmatory test, repeat the above procedure with the Nessler reagent (for preparation of the reagent, see Section 4.3.2). Strips wetted with the Nessler reagent should be placed in corks in the same manner as described above. A brown color indicates the presence of ammonia.

Test for Creatinine

Adults normally exrete 1.2–1.3 g creatinine/day. Selective testing for creatinine is based on the fact that in basic solution, creatinine forms a red-orange addition compound, creatinine picrate, with picric acid.

PROCEDURE. Place a cutting containing the suspected urine stain on a watch glass. Add a drop of a saturated aqueous solution of picric acid at the border between the stained and unstained part. Add a drop of 10% sodium hydroxide solution. In the presence of creatinine, a dark red-orange coloration appears on the stained portion.

REFERENCES

1. V. P. Guinn, *Methods Forensic Sci.* **3,** 64 (1964).
2. Feigl and H. A. Suter, *Ind. Eng. Chem. Anal. Ed.* **14,** 843 (1942).
3. I. Whitehead, unpublished research. University of Rhode Island, Kingston (1954).

4. Y. Leisst, *Characterization of Firearm Discharge Residence,* Master's Thesis, Hebrew University, Jerusalem (1981).
5. G. Price, *J. Forensic Sci. Soc.* **5,** 199 (1965).
6. F. Feigl, *Mikrochemie* **2,** 88 (1924).
7. D. S. Vaitsman, A. Caldas, and D. P. Miranda, *Rev. Quim. Ind. (Rio de Janeiro)* **48,** 29 (1979).
8. H. C. Harrison and R. Gilroy, *J. Forensic Sci.* **4,** 2 (1959).
9. S. F. Bosen and D. R. Scheuing, *J. Forensic Sci.* **21,** 163 (1976).
10. I. Castellanos, *J. Crim. Law* **33,** 482 (1943).
11. I. M. Kolthoff and G. E. Naponen, *J. Am. Chem. Soc.* **55,** 448 (1933).
12. "Diphenylamine Paraffin Test for Gunpowder Residues—Admissibility in Evidence," *J. Crim. Law* **27,** 908 (1937) Police Science Note edited by F. E. Inban and M. E. O'Neill.
13. J. T. Walker, *N. Engl. J. Med.* **216,** 024 (1937).
14. N. K. Nag and M. Mazumdar, *Forensic Sci.* **5,** 69 (1975).
15. S. S. Krishnan, *J. Forensic Sci.* **12,** 112 (1967).
16. F. Feigl, *Spot Tests in Organic Analysis,* 7th ed., Elsevier, Amsterdam, London, New York (1966).
17. R. G. Parker, M. O. Stephenson, J. M. McOwen, and J. M. Cherolis, *J. Forensic Sci.* **20,** 133 (1975).
18. E. Eegriwe, *Z. Anal. Chem.* **76,** 440 (1929).
19. *General Information Bulletin,* Vol. 74-8, Federal Bureau of Investigation, Washington, D.C. (1974), p. 4.
20. N. A. Tananaeff and A. A. Budkewitsch, *Chem. Abstr.* **30,** 5095 (1936).
21. E. Eegriwe, *Z. Anal. Chem.* **70,** 400 (1927).
22. H. Weisz, *Mikrochim. Acta,* 1225 (1956).
23. F. Feigl, *Rec. Trav. Chim.* **58,** 479 (1939).
24. D. G. Sanger, *J. Forensic Sci. Soc.* **13,** 177 (1973).
25. F. Feigl and D. Hagenauer-Castro, *Chemist–Analyst* **51,** 5 (1962).
26. E. Sawicki, T. W. Stanley, and J. Pfaff, *Chemist–Analyst* **51,** 9 (1962).
27. S. Semel, *Chemist–Analyst* **51,** 6 (1962).
28. S. A. Amas and H. J. Yallop, *J. Forensic Sci. Soc.* **6,** 185 (1966).
29. S. A. Amas and H. J. Yallop, *The Analyst* **94,** 828 (1969).
30. F. Feigl and V. Anger, *Spot Tests in Inorganic Analysis,* 6th ed., Elsevier, Amsterdam, London, New York (1972), p. 188.
31. F. L. Hahn, H. Wolf, and G. Jaeger, Berichte *Ber.* **57,** 1394 (1924).
32. H. A. Suter and P. H. Suter, *Mikrochim. Acta,* 1136 (1956).
33. F. Feigl and E. Liebergott, *Chemist–Analyst* **52,** 47 (1963).
34. F. Feigl and E. Liebergott, *Anal. Chem.* **36,** 132 (1964).
35. F. Feigl, *Spot Tests in Inorganic Analysis,* 5th ed., Elsevier, New York (1958), p. 300.
36. J. E. Chrostovsky, W. O. Thurman, and J. J. Javorsky, *Arson Anal. Newslett.* **5,** 14 (1981).
37. H. Marshall, *Z. Anal. Chem.* **43,** 418 (1904).
38. See ref. 35, p. 333.
39. N. S. Poluektoff, *Mikrochemie* **14,** 265 (1934).
40. See ref. 35, p. 374.

41. R. E. Meyers, *Arson Anal. Newslett.* **4,** 1 (1980).
42. F. Feigl and D. Hagenauer-Castro, *Chemist-Analyst* **51,** 16 (1962).
43. J. V. Janovsky and L. Erb, *Ber.* **19,** 2156 (1886).
44. A. H. Amas and H. J. Yallop, *The Analyst* **91,** 336 (1966).
45. *Chemical Spot Test Kits for Preliminary Identification of Drugs of Abuse,* U.S. Dept. of Justice, Law Enforcement Assistance Administration, Washington, D.C. (1978).
46. A. N. Masoud, *J. Pharm. Sci.* **64,** 841 (1975).
47. F. W. Grant, W. C. Martin, and R. W. Quackenbush, *Bull. Narc.* **27,** 33 (1975).
48. M. J. de Faubert, *Analyst* **100,** 878 (1975).
49. A. Bult and H. B. Klasen, *Pharm. Weekbl. Ned.* **109,** 389 (1974).
50. A. Bult, *Pharm. Weekbl. Ned.* **110,** 1161 (1975).
51. F. Schmidt, *Dtsch. Apoth. Ztg.* **113,** 443 (1978).
52. Orion-Yhtima Oy, British Patent 1,361.964 (1973).
53. E. G. C. Clarke, *Isolation and Identification of Drugs,* The Pharmaceutical Press, London (1969), pp. 130, 132.
54. C. H. Thienes and T. J. Haley, *Clinical Toxicology,* 5th ed., Lea & Febiger, Philadelphia (1972), pp. 341, 347, 434.
55. D. J. Harvey, *Trends Anal. Chem.* **1,** 66 (1981).
56. P. V. Win, *Forensic Sci.* **10,** 261 (1977).
57. A. Dionyssion-Asterion and C. J. Miras, *J. Pharm. Pharmacol.* **27,** 135 (1975).
58. D. M. Dulik and W. H. Soine, *Clin. Toxicol.* **18,** 737 (1981).
59. K. Bailey, A. Y. K. Chow, R. H. Downie, and R. K. Pike, *J. Pharm. Pharmacol.* **28,** 713 (1976).
60. W. H. Soine, W. C. Vincek, and D. T. Agee, *N. Engl. J. Med.* **301,** 438 (1979).
61. J. V. Kostir and V. Rabek, *Biochim. Biophys. Acta* **5,** 210 (1950).
62. E. G. C. Clarke, *Isolation and Identification of Drugs,* vol. 2, The Pharm. Press, London (1975), p. 893.
63. E. Eegrive, *Mikrochim. Acta* **2,** 329 (1937).
64. P. Trinder, *Biochem. J.* **57,** 301 (1954).
65. A. O. Gettle and L. R. Goldbaum, *Anal. Chem.* **19,** 270 (1947).
66. A. S. Curry, *Poison Detection in Human Organs,* Charles C. Thomas, Springfield, Illinois (1963).
67. S. Kaye, *Handbook of Emergency Toxicology,* 2nd ed., Charles C. Thomas, Springfield, Illinois (1961).
68. R. Belcher, M. A. Leonard, and T. S. West, *Talanta* **2,** 92 (1959).
69. F. J. Frere and M. Rieders, *Am. Acad. Forensic Sci. Toxicol.* **5,** 615 (1961).
70. See ref. 62, p. 908.
71. Merck Patent Gesellschaft, British Patent Application, February 12, 1972.
72. Department of Labor, Occupational Safety and Health Administration, *Federal Register* **39,** 3754 (1974).
73. H. E. Christensen and T. T. Luginbyhl, editors, *Suspected Carcinogens, A*

subfile of the NIOSH Toxic Substance List, U.S. Government Printing Office, Washington, D.C. (1975), p. 55.

74. V. R. Holland, B. C. Saunders, F. L. Rose, and A. L. Walpole, *Tetrahedron* **30,** 3299 (1974).
75. J. L. Pinkus and L. S. Goldman, *J. Chem. Ed.* **54,** 380 (1977).
76. R. S. Higaki and W. M. S. Philip, *Can. Soc. Forensic Sci. J.* **9,** 97 (1976).
77. F. Proescher and A. M. Moody, *J. Lab. Clin. Med.* **24,** 1183 (1939).
78. T. Mann, *The Biochemistry of Semen and of the Male Reproductive Tract,* Barnes & Noble, New York (1964).
79. See ref. 30, p. 442.
80. R. P. Spalding and W. F. Cronin, *Technical and Legal Aspects of Forensic Serology,* Federal Bureau of Investigation, Washington, D.C. (1976).
81. S. J. Baxter and B. Rees, *Med. Sci. Law* **15,** 37 (1975).
82. P. H. Whitehead and A. Kipps, *Ann. J. Forensic Sci. Soc.* **15,** 39 (1975).
83. S. M. Sax, A. B. Bridgewater, and J. J. Moore, *Clin. Chem.* **17,** 311 (1971).
84. P. L. Kirk, *Crime Investigation,* 2nd ed., John Wiley & Sons, New York (1974), p. 211.

CHAPTER

5

APPLICATION OF SPOT TEST ANALYSIS IN GEOCHEMISTRY

5.1. GENERAL

The specific aim of exploration geochemistry (also called geochemical prospecting) is to find new deposits of metals and nonmetals and to detect crude oil and natural gas accumulations. Chemical tests may also indicate extensions of existing deposits. Most of the analytical methods used in geochemistry today are instrumental; some are nondestructive, like x-ray fluorescence, and others destructive, like emission spectrographical, ICP–AES, and AAS methods. These techniques require the use of many standards, measure elements through one of their physical characteristics, and are subject to rather significant errors due to variation of the sample matrix. But because a precision of $\pm 10\%$ at the 95% confidence level is good enough for geochemical purposes, at least in the initial stages of a project, this shortcoming does not hinder their wide application. For the same reason, simple colorimetric field tests are still applied for the semiquantitative determination of trace elements in soil and sediment samples.

These techniques can be carried out in the field by relatively unskilled operators. The methods are very versatile, produce moderately precise results, and are very inexpensive. There is a very low capital cost associated with these methods, and the operators can be trained in a relatively short time. Approximately 30 elements of interest in exploration geochemistry can be analyzed by rapid tests not requiring sophisticated or expensive instruments. Stanton (1), Ward et al. (2), and Hawkes (3) have published a series of methods of geochemical interest that have proven to be very functional as semiquantitative tests.

Qualitative tests have also orientation value in geochemical explorations and in mineralogy. Selective detection of heavy metals, and rapid chemical differentiation between parent minerals have definite significance in preliminary prospecting work, even when not providing quantitative results. Several selective tests applicable to geochemical exploration were described by Feigl and Anger (4). Some characteristic examples are discussed in the following section.

5.2. SELECTIVE SPOT TESTS USED IN GEOCHEMISTRY

5.2.1. Specific Spot Test for Antimony in Minerals (5)

Antimony (III) salts are oxidized by alkaline solutions of mercury cyanide. The visible change is the deposition of black metallic mercury:

$$Sb_2O_3 + 2Hg(CN)_2 + 4KOH \rightarrow 2Hg^0 + Sb_2O_5 + 4KCN + 2H_2O$$

This reaction may serve for the specific detection of antimony in minerals. Fusion of a powdered sample with potassium persulfate will result in trivalent antimony irrespective of the oxidation number of antimony in the mineral sample [metallic antimony, antimony (III), or antimony (V)]. This means that in the fusion persulfate may act as both oxidant and reductant.

PROCEDURE. Place a solid sample in a micro test tube, add several centigrams of potassium persulfate, and heat the mixture over a microburner for several minutes after melting has occurred. After cooling, dissolve the melt in several drops of water. Neutralize the solution with 10% sodium hydroxide. Add a drop of basic 5% mercury cyanide solution. A gray or black precipitate appears in the presence of antimony.

A positive response to the antimony test was given by the following antimony minerals: bournonite, leevisite, pyrargyrite, stibnite, valentinite, stibiconite, and bindheimite.

5.2.2. Spot Test for Arsenic in Minerals

Arsenic compounds heated in the presence of sodium formate produce highly reducing arsine gas. The release of atomic hydrogen in the thermal decomposition of sodium formate causes this reduction (6). The gaseous arsine reduces silver nitrate to metallic silver.

PROCEDURE (7). Place a minute quantity of the solid mineral in a test tube and mix with several grains of ignited lime. Heat the tube over a free flame for several minutes. After cooling, add an excess of solid sodium formate, cover the mouth of the tube with a filter paper disc moistened with 10% silver nitrate solution, and heat the tube again. The presence of arsenic is indicated by the appearance of a dark circle on the paper.

Naturally occurring arsenic minerals such as mispickel, scorodite, and nicolite all give positive reactions to this test.

5.2.3. Spot Test for Differentiation Between Barium- and Strontium-Containing Minerals (4)

Barite ($BaSO_4$) and celestite ($SrSO_4$) can easily be differentiated by their color reactions with sodium rhodizonate. Both barium and strontium ions form a red-violet insoluble complex compound with sodium rhodizonate, but the extremely low-solubility product of barite (pS = 10.01) does not liberate enough ions for the reaction, whereas the more soluble celestite (pS = 9.03) is tinted red-violet readily by contact with a freshly prepared reagent solution.

PROCEDURE. Place several milligrams of the pulverized test material to be differentiated into a depression of a spot plate. Add a drop of a 0.2% solution of sodium rhodizonate. Celestite will be tinted red-violet; barite will not.

In the differentiation of witherite ($BaCO_3$) and strontianite ($SrCO_3$), the minerals have to be transformed to sulfates prior to the test.

PROCEDURE. Heat in a microcrucible a minute quantity of the mineral along with ~0.5 g ammonium sulfate. Spot the cold residue with a 0.2% solution of sodium rhodizonate. Strontianite will be tinted red-violet; witherite will not.

5.2.4. Spot Test for Beryllium in Minerals and Ores (4)

Beryllium-containing minerals, even when in the silicate complex form, are easily converted to alkali beryllium fluoride by fusion with alkali bifluoride. The alkali bifluoride decomposes thermally to hydrofluoric acid, which attacks the beryllium-containing mineral, and ultimately alkali beryllium fluoride is formed. The complex alkali beryllium fluoride reacts with ammoniacal quinalizarine solution to form a blue-violet, stochiometrically well-defined, insoluble complex.

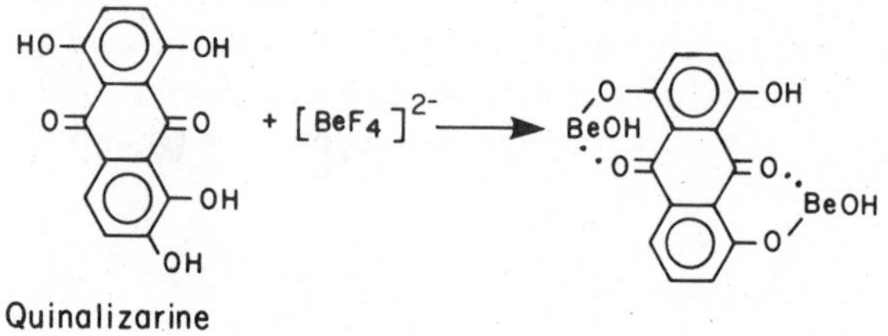

$$\text{Quinalizarine} + [BeF_4]^{2-} \longrightarrow \text{(BeOH complex)}$$

Quinalizarine

This test enables the specific and sensitive detection of beryllium in ores. The fluoride complexes formed by fusion of magnesium, calcium, ferric, aluminum, titanium, and zirconium compounds with potassium bifluoride do not interfere.

PROCEDURE. Mix a few milligrams of the sample in a platinum spoon with an excess of potassium-bi-fluoride and fuse for several minutes. Treat the cold residue with 2 ml cold water and centrifuge the suspension after 5 min. Place a drop of the supernatant fluid into a depression of a spot plate and add a drop of 0.05% quinalizarine solution in 10% ammonia. In the presence of beryllium a blue tint or precipitate appears. Add a few drops of bromine water. The violet color of the reagent disappears, but the blue beryllium quinalizarinate should remain unchanged.

5.2.5. Test for Differentiation Between Calcite and Dolomite

In geochemical prospecting an urgent need to differentiate between calcite and dolomite rocks sometimes arises. A spot test for this differentiation was elaborated based on the different behaviors of the two carbonate rocks toward weak acids. Calcite is attacked by dilute acids at room temperature, whereas the reaction rate between dolomite and acids is very low.

PROCEDURE (8). Place some crystals of the sample material in a depression of a spot plate. Add a drop 0.1% alizarin S solution dissolved in saturated tartaric acid solution. In the presence of calcite a purple-red precipitate appears. Dolomite does not react.

5.2.6. Test for Differentiation Between Calcite and Aragonite (4)

Hexagonal calcite and rhombic aragonite may be differentiated without microscopic examination. A spot test for this is based on a slight solubility difference between the two modifications.

Suspension of calcium carbonate in water produces a low concentration of hydroxyl ions in the solution due to the hydrolysis of carbonate ions. Because the solubility of aragonite is somewhat higher than that of calcite, a sensitive test for hydroxyl ions indicates the presence of aragonite, but barely reacts with calcite. The extremely sensitive indicator for hydroxyl ions is the redox reaction between silver (I) and manganese (II) salts:

$$Mn^{2+} + 2Ag^{+} + 4OH^{-} \rightarrow MnO_2 + 2Ag^{0} + 2H_2O$$

Because both the oxidation product (manganese dioxide) and the reduction product (metallic silver) are black, this reaction allows the detection of minute quantities of hydroxyl ions, and thus of the trace quantities resulting from suspending aragonite in water.

PROCEDURE. Place a small quantity of the test material in a depression of a spot plate and add a drop of the reagent solution. Examine the plate after 2 min. In the presence of aragonite a black deposit is observed. Calcite does not react within this time period.

PREPARATION OF THE REAGENT SOLUTION. Dissolve 1 g solid silver sulfate in water; add 11.8 g $MnSO_4 \cdot 7H_2O$. Bring to 100 ml and boil. After cooling, filter the suspension and then add some drops of dilute sodium hydroxide solution. Filter again after 1–2 hours. Keep the solution in a dark bottle.

5.2.7. Test for Boron in Geochemical Screening (9)

Boron can be extracted from silicate rocks by fusion with potassium hydroxide. Borates change the blue-violet color of *p*-nitrobenzeneazochromotropic acid (chromotrope 2B) to greenish blue. The color change indicates the formation of an inner complex phenol ester of ortho- or metaboric acid.

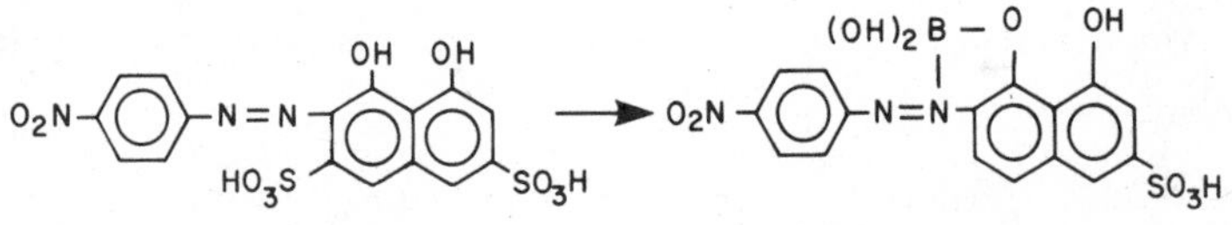

Chromotrope 2B (blue)

Phenol ester of boric acid (greenish blue)

PROCEDURE. Fuse several centigrams of the finely powdered rock samples with potassium hydroxide in a silver crucible. Extract the cold melt with a few drops of water, and evaporate to dryness in a porcelain disc. Add two drops of a 0.005% solution of chromotrope 2B in conc. sulfuric acid. Observe the color after cooling. The presence of borate is indicated by a color change from blue-violet to greenish blue.

Fluorides interfere with this test; extremely stable boron fluorides form, preventing the esterification. Fluorides can be removed by evaporation with silica and sulfuric acid. Oxidizing agents interfere as well. They oxidize chromotrope 2B to a red product. Evaporation with hydrazine sulfate can easily eliminate this interference.

5.2.8. Test for Calcium Oxide in Magnesite (10)

The presence of lime lowers the quality of magnesite; it is an undesirable constituent. Its rapid detection in the field by an unsophisticated screening test performed on the rock sample itself has significant value.

Rapid differentiation between lime and magnesite is based on the selective dissolution of lime in dilute acids and subsequent formation of calcium alizarinate.

PROCEDURE. Polish an area of the rock sample and drop it in a boiling 0.1% solution of sodium alizarinsulfonate in $\frac{1}{15}$ M or 0,066 M hydrochloric acid. The lime-containing regions turn violet; the magnesite remains yellow.

5.2.9. Test for Chromium in Rocks (4)

The formation of the red-violet inner complex salt of Cr (VI) ions and diphenylcarbazide (see reaction mechanism, section 4.3.8) is utilized for the quick detection of chromium-containing rocks. The test is feasible even when the rock contains chromium in concentrations as low as 0.04%.

PROCEDURE. Grind a tiny particle of the rock to a fine powder in an agate mortar and mix with a fourfold excess of a 1 : 1 mixture of potassium carbonate and sodium peroxide. Dip a red-hot platinum wire into the mixture twice. Beads are formed. Place the cold bead in a depression of a spot plate, dissolve in two drops of 1 : 1 sulfuric acid, and add a drop of 1% alcoholic diphenylcarbazide solution. The presence of chromium in the rock is indicated by the appearance of a violet color.

An alternative method involves fusion of the mineral or rock sample with potassium fluoride in a platinum spoon and subsequent reaction with diphenylcarbazide. The latter procedure is especially recommended in the presence of interfering molybdate ions, because these will be masked with fluoride as $[MoO_3F_2]^{2-}$.

5.2.10. Test for Chlorine-Containing Rocks (11)

Chlorine is a structural element in only a few minerals and rocks, so its selective identification may be an important preliminary test. Its selective estimation (in the presence of other halogens) is based on oxidative decomposition, producing the free halogens, and subsequent oxidation of diphenylamine dissolved in ethyl-acetate-containing trichloroacetic acid. This reagent solution is a selective indicator for chlorine, and is not affected by bromine and iodine.

PROCEDURE. Place a small amount of the rock or mineral in a micro test tube and mix with a few milligrams of potassium permanganate and two drops of 6 *N* sulfuric acid. Cover the mouth of the tube with a filter paper disc moistened with a drop of freshly prepared reagent solution. Immerse the tube in a boiling water bath. The appearance of a blue-green stain within 2–3 min indicates the presence of chlorine in the rock sample.

Preparation of the reagent solution: Add 0.5 g trichloroacetic acid to 10 ml of saturated solution of diphenylamine in ethyl acetate.

5.2.11. Test for Fluorite (4)

Fluorite in rock may be estimated by selective identification of fluorine and calcium. The basis of one selective test for fluoride is its corrosive effect on glass, which is demonstrated by a decrease in transparency or in wettability of the glass surface.

For the selective detection of fluoride, the acid-insoluble mineral is decomposed by heating with solid ammonium chloride. The sublimed ammonium chloride acts as a water-free hydrochloric acid and dissolves the fluorite.

PROCEDURE A. Place a few grains of potassium dichromate and 1–2 ml conc. sulfuric acid in a micro test tube. Introduce a few particles of the sample and warm the mixture. The presence of fluoride is indicated by an uneven flow along the unmoistened areas of the tube when it is rotated.

PROCEDURE B. Place a minute quantity of the solid sample in a microcrucible, mix with finely ground ammonium chloride, and heat on an open flame. After cooling, dissolve in a few drops of water, add two drops of sodium selenosulfate, and warm slightly. In the presence of calcium salts, red elemental selenium is precipitated (12).

$$CaCl_2 + Na_2SSeO_3 \rightarrow CaSO_3\downarrow + 2NaCl + Se^0$$

PREPARATION OF THE REAGENT SOLUTION. Warm several milliliters of 1% sodium sulfite solution with an excess of finely powdered selenium for 5–10 min. Cool and centrifuge. The colorless reagent solution is stable for several days.

5.2.12. Test for Lead in Ores and Minerals (13)

The very high sensitivity of the rhodizonate reaction with lead (see reaction mechanism, section 4.2.1) allows the direct estimation of lead in ores and minerals even if they are insoluble in water. Because barium and strontium also form colored products with sodium rhodizonate, in the presence of these two possibly interfering metals the sample must be fumed with sulfuric acid. The resulting barium and strontium sulfate are unaffected by the reagent, whereas lead sulfate reacts readily.

PROCEDURE. Place a small quantity (~1 mg) of the finely powdered sample in a depression of a spot plate. Add a drop of pH 2.8 buffer solution (1.9 g sodium bitartrate and 1.5 g tartaric acid dissolved in 100 ml water) and a drop of freshly prepared 0.2% aqueous sodium rhodizonate solution. In the presence of lead the surface of the sample turns red-violet.

Galena, cerussite, stolzite, anglesite, crocoite, and all water-insoluble lead ores give positive results with the test.

5.2.13. Rough Discrimination of Different Lead-Containing Ore Bodies (14)

The following simple geochemical prospecting field tests were developed to discriminate between:

1. Ores containing less than 1% lead and those containing more than 1% lead; and
2. Ores containing less than 0.5% lead and those containing more than 0.5% lead.

PROCEDURE A (1% lead cutoff). Crush the sample to −80 mesh or finer (percussion drill dust is of sufficiently fine grain size). Measure in a test tube calibrated at 10 ml one leveled sampling scoop of ground material.

Add 1 : 1 hydrochloric acid to the mark, and let react overnight. Put a drop of the supernatant fluid on a Whatman GF/B glass-fiber paper strip. Suspend the glass-fiber strip over a hot plate until all the hydrochloric acid is driven off, using porcelain boats on both sides as support. Treat sample spots with one drop of 0.2% aqueous sodium rhodizonate solution. Wash the reagent strip on both sides with a direct stream of tap water. A reddish-blue spot on one side of the strip indicates the presence of more than 1% lead in the ore sample. A few drops of 0.1 *N* hydrochloric acid turns the reddish-blue precipitate to a violet color and dissolves the brown iron oxide background.

PROCEDURE B (0.5% lead cutoff). This method is exactly the same as Procedure A except that two sampling scoops are measured into the graduated test tube instead of one.

SAMPLING SCOOP. Calibration showed a mean value of sampled material of 91 ± 7.8 mg (S.D.). The sampling scoop consists of an aluminum bar into which a hemispherical indentation 6.3 mm in diameter is drilled to a depth of 3.2 mm (volume ~0.065 cm^3).

Table 1 compares results of sampling tests using these two tests with those obtained with AAS. Inherent inaccuracies and approximations of these screening methods result in an uncertain region, a "gray area" within which sharp distinctions are no longer possible (compare samples 14 and 15).

5.2.14. Test for Manganese in Minerals and Rocks (4)

Insoluble manganese dioxide and manganese silicate deposits are all transformed to soluble manganese (II) ions by sintering them with potassium persulfate:

$$MnO \cdot SiO_2 + K_2S_2O_8 \rightarrow MnSO_4 + K_2SO_4 + SiO_2 + \tfrac{1}{2}O_2$$

$$MnO_2 + K_2S_2O_8 \rightarrow MnSO_4 + K_2SO_4 + O_2$$

The catalytic oxidation to permanganate by persulfate in the presence of small amounts of soluble silver salts reveals relatively small amounts of manganese in minerals and rocks.

PROCEDURE. Heat a minute quantity of the pulverized ore sample with an excess of potassium persulfate in a small test tube over a free flame.

Table 1. Calibration of the Screening Tests Against Actual Analyzed Ore Samples[a]

Sample No.	Lead Content Analyzed by AAS (%)	Screening Test[b] Proc. A	Screening Test[b] Proc. B
1	1.721	++	+++
2	0.0257	−	−
3	0.254	−	−
4	0.141	−	−
5	0.1365	−	−
6	0.062	−	−
7	0.091	−	−
8	0.135	−	−
9	0.1785	−	−
10	0.712	−	+
11	1.532	++	+++
12	1.050	+	++
13	1.62	++	++
14	0.421	−	+
15	0.569	−	−
16	1.823	++	+++
17	0.784	−	+
18	0.851	−	+
19	0.884	−	+
20	1.135	++	++
21	0.217	−	−
22	5.708	+++	+++
23	2.31	++	+++
24	0.044	−	−
25	0.134	−	−
26	0.286	−	−
27	0.049	−	−

[a] Reproduced with permission from Anglo American Research Laboratories, Johannesburg.

[b] -,

After cooling, add two to three drops of 2 *N* sulfuric acid and some tiny crystals of silver sulfate and boil. Add several centigrams of persulfate to the hot solution. The appearance of a violet color indicates the presence of manganese of any oxidation number.

5.2.15. Test for the Presence of Pyrolusite (4)

Pyrolusite (MnO_2) oxidizes chromium (III) salts in acidic solutions to hexavalent chromium, which in turn can be easily identified with diphenylcarbazide (see reaction mechanism, section 4.3.8).

PROCEDURE. Heat a small quantity of the pulverized ore in a micro test tube with several centigrams of chromic sulfate dissolved in 1 : 1 sulfuric acid for several minutes. After cooling, add a drop of 1% diphenylcarbazide solution in ethanol. The appearance of a violet color indicates the presence of pyrolusite in the sample.

5.2.16. Test for Molybdenum in Ores (15)

Trace quantities of molybdenum in ores are identified by applying the extremely sensitive and specific color reaction with potassium xanthate. The identification is based on the formation of a violet complex compound, $MoO_3 \cdot 2[SC(SH)(OC_2H_5)]$.

PROCEDURE. Digest a few grains of the ore in a small crucible with a minimal amount of conc. sodium hydroxide and acidify with phosphoric acid. Add a few crystals of potassium xanthate. In the presence of molybdenum a dark violet color forms that fades on standing.

5.2.17. Test for Nickel in Garnierite or Other Nickel-Containing Silicate Rocks (16)

Water-insoluble complex silicate can be solubilized by fusion with zinc chloride. Zinc chloride reacts with its own water of hydration, releasing anhydrous hydrochloric acid far above its boiling point. This acid partially decomposes insoluble complex silicates, allowing identification of the participating nickel.

$$ZnCl_2 + H_2O \rightarrow Zn(OH)Cl + HCl\uparrow$$

$$Zn(OH)Cl \rightarrow ZnO + HCl\uparrow$$

PROCEDURE. Pulverize the sample rock and fuse a minute quantity of it with zinc chloride in a platinum loop. With a drop of hot water, wash the cold mass from the loop into a depression of a spot plate. Add a drop of conc. ammonium hydroxide and a drop of a 1% alcoholic solution of dimethylglyoxime. The presence of nickel is indicated by the appearance of a red precipitate.

Rocks containing bivalent iron interfere with the test by reacting similarly to nickel. In their presence, warm the ammoniacal suspension with hydrogen peroxide, centrifuge, and discard the ferric hydroxide precipitate.

5.2.18. Differentiation Between the Sodium and Potassium Felspars (16)

The detection of potassium in silicate rocks is of definite value in preliminary geochemical work, since only few of the ortho-, meta-, and anhydrosilicates contain potassium.

Potassium is not ionogenically bound; the ion is released only after zinc chloride sintering (15). It is detected by precipitation with sodium tetraphenylborate, with which it forms insoluble $K[B(C_6H_5)_4]$.

PROCEDURE. Pulverize the sample and mix it in a quartz crucible with excess zinc chloride. Heat the crucible for 5 min over a bare flame. Take up the cooled mass in several drops of hot water, and centrifuge. Treat a drop of the supernatant fluid with a 1% aqueous solution of sodium tetraphenylborate. A white precipitation signals the presence of potassium.

5.2.19. Detection of Tin-Bearing Ores (4)

Tin-containing minerals dissolved in concentrated sulfuric acid produce $Sn(SO_4)_2$. Addition of alkali iodide forms, and extraction with benzene separates, SnI_4, which can be selectively identified with the morin reagent.

Many water-insoluble metal hydroxides show a yellow-green fluorescence when in contact with an acetone solution of morin.

HO OH OH OH O

Although the benzene extraction eliminates most of the potentially interfering metals, the presence of antimony in the ore still presents a problem, since $Sb(OH)_3$ reacts similarly to tin (IV) with morin. This interference can be eliminated by the oxidation of $Sb(OH_3)$ to the noninterfering $Sb(OH)_5$. This is accomplished by including iodine in the benzene. The excess of iodine is reduced with sodium sulfite.

PROCEDURE. Place a small quantity of the pulverized ore in a micro test tube, add three drops of conc. sulfuric acid, and heat for several minutes. Cool, dilute with water, and extract with 10 drops of a 5% solution of iodine in benzene. Place a drop of the benzene solution on filter paper, and then hold the paper over ammonia. Treat the spot with one drop of 0.5% aqueous sodium sulfite solution and one drop of 0.05% morin solution in acetone. Wash the paper briefly with 50% acetic acid. In the presence of tin a bluish-green fluorescence appears under ultraviolet light. The test is specific for tin.

5.2.20. Detection of Cassiterite in Granite Rocks (17)

Geochemical prospecting sometimes has the objective of the quick recognition of favorable host rocks. A preliminary rock survey determines some geochemically diagnostic feature that permits the recognition of rocks likely to be associated with mineralization and the differentiation of such rocks from barren rock. The detection of stannic oxide may be important in this preliminary exploration, since certain granites have above-average contents of tin.

Stannic-oxide-containing rock samples heated with solid ammonium chloride produce sublimed tin (IV) chloride. This reaction is feasible at Bunsen-burner flame temperatures and effectively separates the tin compound from the rock residue. If the mixture of the sample and ammonium chloride contains metallic magnesium as well, stannous chloride, which can be easily identified by the familiar phosphomolybdic redox reaction, forms during dry heating.

PROCEDURE. Place a small quantity of the pulverized sample in a small test tube and mix it with several centigrams of solid ammonium chloride and some fine granules of metallic magnesium. Heat the tube over a low, free flame until all the ammonium chloride has sublimed. After cooling, press some cotton wool moistened with several drops of 2% phosphomolybdic acid solution against the sublimate ring with a glass rod. In the presence of tin, molybdenum blue appears immediately.

This test may also be applied for the detection of sulfidic tin minerals. In this case, a prior roasting procedure is needed to convert the tin sulfides to dioxides.

5.2.21. Test for Titanium Minerals (4)

The naturally occurring titanium minerals partially dissolve when heated with concentrated sulfuric acid, and the titanic ions can be detected with the chromotropic acid reaction.

PROCEDURE. Mix some finely powdered sample with several drops of conc. sulfuric acid in a microcrucible and heat to fuming. Cool, add a small crystal of solid chromotropic acid, and warm gently. The presence of titanium is indicated by the appearance of a violet-colored complex.

5.2.22. Tests for Uranium in Minerals

Uranium extracted selectively from rocks and ores can be easily identified with the familiar potassium ferrocyanide reaction.

PROCEDURE (18). Boil a 5-g sample with conc. nitric acid; then add an equal volume of deionized water to the cooled solution. Extract with a few milliliters of triisobutylphosphate cyclohexane mixture (1 : 4). Place a drop of the extract on a piece of filter paper impregnated with 5% potassium ferrocyanide solution. A brown precipitate indicates the presence of uranium. The detection limit is 10–20 ppm.

Neutral solutions of uranyl ferric or bismuth salts turn the colorless benzene solution of Rhodamine B red when shaken with this solution. This dye also shows an intense orange fluorescence. Because iron compounds are almost omnipresent in uranium-containing ores, and bismuth may be present too, these two potentially interfering ions have to be precipitated with sodium carbonate.

Uranium is soluble in the form of the complex anion $[UO_2(CO_3)_2]^{2-}$. When this is taken to dryness in the presence of nitric acid, neutral uranyl nitrate forms, which is easily detected with the Rhodamine B method.

PROCEDURE (4). Fuse a minute quantity of the pulverized sample with sodium carbonate on a platinum wire. Dissolve the melt in water in a small centrifuge tube, and centrifuge. Add to a drop of the supernatant fluid in a microcrucible two drops of conc. nitric acid and evaporate to dryness. Treat the aqueous solution of the residue in a micro test tube with five drops of the reagent solution and shake. In the presence of uranium a red or pink benzene layer appears that shows an orange fluorescence under ultraviolet illumination.

PREPARATION OF THE REAGENT SOLUTION. Add an excess of solid Rhodamine B to a 0.5% solution of benzoic acid in benzene, shake, and filter.

5.2.23. Field Test for Rare Earth Elements (19)

Rare earth metals in rocks and minerals can be detected using spot tests on filter paper impregnated with a 0.25% solution of arsenazo III.

PROCEDURE. Leach a powdered sample with hydrochloric or nitric acid (sometimes with the addition of hydrogen peroxide). Place a drop of the clarified solution on the arsenazo III test paper. The rapid appearance of a green color indicates the presence of rare earth metals.

5.3. SEMIQUANTITATIVE FIELD TESTS USED IN GEOCHEMICAL EXPLORATION

Methods designed for field use in geochemical prospecting that are characterized by simplicity and commonly produce results within 30–50% of correct values have been elaborated by the U. S. Geological Survey Laboratories (2), R. E. Stanton (1), H. E. Hawkes (3), A. A. Levinson (20), and others. These methods are generally based on colorimetric procedures carried out under field conditions.

Under such conditions, use of plastic standards, which are prepared by adding suitable dyes to a transparent thermosetting polyester resin (21), is preferred. These standards are far more stable than the conventionally used liquid standards. The plastic standards are prepared in tubes of the same dimensions as those containing the liquid standards. The advantages of plastic standards over liquid ones are quite obvious: No time-consuming preparation of a standard series in the field is necessary, and the hardened plastic endures rough handling.

PROCEDURE FOR PREPARATION OF PLASTIC STANDARDS. Prepare a series of liquid standards. Pour about 3 tablespoons of transparent thermosetting polyester resin, known commercially as Castolite, into a cup, and add small amounts of dye until the color matches that of the most intensely colored liquid standard. Dilute the colored resin with clear Castolite to produce the proper series. For standards differing in color with different concentrations of analyte, each standard must be prepared separately. Pour alternating clear and colored Castolite layers until the whole standard series is matched. Cure the plastic standard by heating the tube for 30 min in a water bath at 70°C. Remove the tube from the water bath and remove the casting from the glass by adding 5 ml acetone and inverting the tube.

The preparation of a plastic rod takes about 5 hours. Castolite is a clear, syrupy liquid that pours easiy and forms a crystal-clear solid extremely resistant to heat and chemicals. The addition of a hardener and cold-setting promoter leads to solidification.

5.3.1. Rapid Screening for Copper, Lead, and Zinc in Soils and Rocks

Many rapid methods have been developed for the estimation of copper, lead, and zinc for geochemical prospecting purposes. In most cases the total amount of these metals is not determined; rather, varying amounts of the metals enter the solution depending on the dissolution procedure. The determination of only the readily soluble copper, lead, and zinc is sometimes of greater diagnostic value for prospecting than are the total values.

Various extraction modes have been used successfully in geochemical prospecting. Samples of hard rock were leached, for example, with dilute sulfuric acid or dilute hydrochloric acid (22) for semiquantitative screening for copper, lead, and zinc. Sediments and soils were screened for the same metals by partially dissolving the samples in dilute nitric acid (23, 24).

In contrast to the above-mentioned leaching procedures, the pyrosulfate fusion screening technique dissolves almost totally the copper, lead, and zinc in almost all naturally occurring soils and rocks.

The chromogen for the colorimetric estimation of zinc and lead is dithizone. Carbon tetrachloride solutions of dithizone form red zinc dithizonate when shaken with a buffered sample. Dithizone is also used as a specific reagent for the determination of lead when the masking agent cyanide is used to sequester the interfering ions. Copper is determined on the basis of the formation of the complex of monovalent copper with 2,2′-biquinoline.

PROCEDURE FOR PREPARATION OF THE SAMPLE SOLUTION. Crush the sample to minus 80 mesh. Scoop a 0.1-g sample into a 16 × 150 mm test tube, add by scooping 0.5 g potassium persulfate powder, mix intimately, and heat. Fuse the mixure for about 2 min after the flux melts. After cooling, add 3 ml 50% HCl to the tube and place the tube in a hot water bath until the melt disintegrates completely. Crushing with a glass rod helps the disintegration process. After removal and cooling, dilute the sample to 10 ml with deionized water. Take aliquots from the sample for the lead, zinc, and copper determinations.

PROCEDURE FOR THE ESTIMATION OF LEAD. Put a 2-ml aliquot of the sample solution into a 125-ml separation funnel containing 10 ml lead buffer solution. Add conc. ammonia dropwise in the presence of thymol blue indicator until the pH range 8.5–9.0 is reached. This point is indicated by a color change from yellow to blue. Add 5 ml 0.001% dithizone solution in carbon tetrachloride and shake the separating funnel for ~15 sec. Drain the carbon tetrachloride layer into a 25-ml glass-stoppered graduated cylinder containing 10 ml 0.1% potassium cyanide solution. Shake the cylinder briefly. Compare the color of the carbon tetrachloride layer with those of similarly prepared standard solutions.

Lead standards (0, 1, 2, and 3 μg) are prepared by pipetting aliquots from a 10-ppm standard lead solution. Calculation of the metal concentration is according to the general formula:

$$\text{ppm} = \frac{\text{volume of sample solution in ml}}{\text{sample weight taken for analysis in g}} \times \frac{\mu\text{g of trace element found}}{\text{aliquot of sample solution in ml}}$$

Preparation of Lead Buffer Solution. Put 50 g ammonium citrate, 10 g potassium cyanide, and 8 g hydroxylamine hydrochloride into a large separation funnel. Add 800 ml deionized water and dissolve the materials by shaking. Add 2 ml 0.2% aqueous thymol blue indicator; then add conc. ammonia until the color turns blue (pH 8.5).

Procedure for the Estimation of Zinc. Transfer a 2-ml aliquot of the sample solution to a 22 × 175 mm test tube containing 8 ml zinc buffer solution. Add 5 ml 0.001% dithizone solution in carbon tetrachloride, cap the tube, and shake for 30 sec. Compare the color of the carbon tetrachloride layer with those of similarly prepared standard zinc solutions (0, 1, 2, 3, and 4 μg).

Preparation of Zinc Buffer Solution. Dissolve 125 g sodium thiosulfate in ~400 ml deionized water in a large separation funnel. Remove heavy metals by extracting with 0.01% dithizone solution and discarding the colored extract. Dissolve 300 g sodium acetate in 400 ml deionized water, add 60 ml glacial acetic acid, and remove heavy metals as before. Combine these two solutions and dilute to 2 liters.

In the presence of sodium thiosulfate, elements potentially interfering with the zinc determination are masked. Only palladium and bivalent tin react under similar conditions, but palladium is unlikely to occur in significant concentrations, and tin occurs almost always in the stannic form.

PROCEDURE FOR THE ESTIMATION OF COPPER. Transfer a 2-ml aliquot of the sample solution to a 22 × 175 mm test tube containing 10 ml copper buffer solution. Add 2 ml 0.02% 2,2′-biquinoline solution in isoamyl alcohol. Cap the tube and shake vigorously for 45 sec. After separation of the layers, compare the color of the isoamyl alcohol layer with those of similarly prepared standards of known copper content.

PREPARATION OF COPPER BUFFER SOLUTION. Dissolve 400 g sodium acetate, 100 g sodium tartrate, and 20 g hydroxylamine hydrochloride in 1 liter deionized water. Check for possible copper contamination with biquinoline reagent, and if a pink color appears in the organic layer, remove copper by extraction with biquinoline.

PREPARATION OF PLASTIC STANDARDS FOR ESTIMATING COPPER. No commercially available dye duplicates satisfactorily in plastic the color of the copper biquinoline complex. The complex formed by the reaction of tin (IV) and 4,5-dihydroxyfluorescein forms a red-colored complex whose color closely resembles that of the copper biquinoline complex. This complex is prepared in acetate-buffered medium at pH 2.0–2.5, and is extracted with a binary mixture of *o*-dichlorobenzene and cyclohexanone.

The rapid determination of zinc and copper in soils, sediments, and rocks for exploration purposes can be done on several levels. Hot-acid-extractable metal can be differentiated from cold-acid-extractable metal by performing the extractions in that order. Such procedures roughly differentiate strictly residual metal in the soil from nonresidual metal (25).

Extraction with cold hydrochloric acid from a soil or sediment yields only about 10–30% of the total copper in most samples, but for geochemical exploration these results are more significant than measurements of total copper. Moreover, the analytical method is rapid and the equipment used suits rough field conditions.

The semiquantitative screening tests can be carried out very quickly, and no single determination requires more than 2 min. Interfering materials in soil are organic compounds (e.g., humic acid). The yellow color they produce in the organic solvent can interfere with the pink color of the copper–biquinoline complex.

5.3.2. Screening for Acid-Extractable Copper in Soils and Sediments (26)

PROCEDURE. Crush the sample to −80 mesh or finer. Scoop 0.2 g of the sample into a 22 × 175 mm test tube. Add ~1 ml 6 *M* hydrochloric acid and mix for ~30 sec. Add 10 ml buffer solution (see Section 5.2.1) and

2 ml 0.02% biquinoline reagent in isoamyl alcohol, cap the tube, and shake for ~30 sec. After separation of the layers, compare the color of the organic layer with those of either biquinoline standards or a plastic standard rod.

Spot Test for Copper in Stream Sediments (27)

A simple method for determining extractable copper in acetic acid–sodium acetate leaching solution is used at varius types of geochemical exploration site. The buffer attacks, at least superficially, the alteration products of sulfides, oxides, or carbonates of copper. The reagent applied is dithiooxamide (rubeanic acid), which forms a dark green precipitate with copper ions:

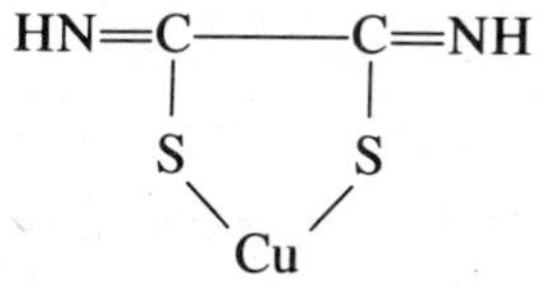

```
HN=C———C=NH
   |     |
   S     S
    \   /
     Cu
```

Although this method does not claim better precision than other field methods, it is far simpler: It does not use two liquids for extraction, and it combines filtration and reaction on the same strip of filter paper.

PROCEDURE. Scoop a 1-g soil or sediment sample into a 15- to 18-mm-wide, 30-mm-long test tube. Add 1 ml buffer solution, mix, and add some more buffer to yield some supernatant fluid. Cork the tube with two halves of a cork stopper, placing the reagent strip between the two halves in such a fashion that the thin end of the paper dips into the buffer. The buffer should not touch the paper strip anywhere else. After a few minutes, wet ~40 mm^2 of the strip above the reagent spot with the buffer. For contents of a few micrograms per milliliter, the spot obtained is a thin dark line. Its intensity of coloration is much more related to copper concentration of the solution than to the total amount that has passed through the reagent spot.

Standard spots are prepared in the same way, only using standard solutions of 30, 10, 3, 1, and 0.3 mg copper instead of the sample.

Compare the test with the standard samples.

PREPARATION OF THE REAGENT PAPER STRIPS. Use 55-mm-long acid-washed paper strips. The lowest 20 mm of the strip should be 2 mm wide; the middle 10 mm, 4 mm wide; and the upper 25 mm, ~12 mm wide. Apply a droplet of a 1% acetone solution of dithiooxamide at the tapering part of the strip.

PREPARATION OF THE BUFFER SOLUTION. Dissolve 250 g hydrated sodium acetate in water, add 250 ml glacial acetic acid, and dilute to 1 liter.

The copper content of water samples can be very indicative, and field examination can be carried out on the spot (28).

PROCEDURE. Add to 5 ml sample solution 5 ml buffer solution and 10 ml quinolyl reagent solution (0.2 g/liter in amyl alcohol). Shake the mixture vigorously for 10 sec and compare the color of the complex against a set of standards.

PREPARATION OF BUFFER SOLUTION. Dissolve 20 g sodium acetate, 10 g potassium sodium tartrate, and 3 g hydroxylammonium chloride in 100 ml water.

The detection limit is ~0.01 ppm copper.

5.3.3. Rapid Screening for Dispersion of Heavy Metals

Geochemical dispersion of base metal deposits is mapped quite efficiently by measuring total heavy metal content. The following simple method suits field conditions very well and is widely used in geochemical prospecting.

This method determines only the heavy metals soluble in aqueous ammonium citrate solution at ambient temperature. The screening is extremely rapid; samples can be analyzed as rapidly as they are collected.

PROCEDURE (29). Crush the sample to minus 80 mesh or finer, or use percussion drill dust. Scoop ~0.1 g of the powdered soil or sediment sample into a 25-ml glass-stoppered cylinder and add 5 ml ammonium citrate solution and 1 ml 0.001% dithizone solution in xylene. After shaking the cylinder vigorously for several seconds, allow the phases to separate, and observe the color of the upper layer. If this layer is green, green-blue, or blue, record as 0, 0.5, or 1 ml, respectively.

If the color of the upper layer ranges from blue-purple to red, add successively 1-ml portions of the 0.001% dithizone solution in xylene. Shake vigorously after each addition until a blue color is obtained.

The heavy metal content is indicated by the volume of dithizone solution used in this titrationlike procedure.

Make a calibration curve by titrating a series of solutions that contain known amounts of zinc. From the calibration curve the zinc-equivalent metal content can be obtained in parts per million.

Preparation of Ammonium Citrate Solution. Dissolve 25 g ammonium citrate and 4 g hydroxylamine hydrochloride in ~300 ml deionized water. Neutralize with conc. ammonia until pH 7 is reached, and dilute to 500 ml with water. Extract the traces of heavy metals present with 0.01% dithizone in chloroform in a separation funnel until the organic layer remains green. Extract the dissolved dithizone by shaking with 25-ml portions of chloroform until the organic layer is colorless. Adjust the pH to 8.5 with ammonia.

5.3.4. Field Determination of Cobalt in Soils

A number of rapid screening tests are applied for the estimation of cobalt in soil and sediment samples.

The spot test for cobalt with 2-nitroso-1-naphthol, which is extractable in carbon tetrachloride, was the basis of the test of Almond (30). Because this reagent forms colored complexes with nickel and copper as well, these ions constitute potential interferences in the estimation of cobalt. Such interference is eliminated by the complex-binding action of dilute cyanides, in the presence of which the 2-nitroso-1-naphthol complexes of nickel and copper are unstable.

Procedure. Crush the sample to −80 mesh or finer. Scoop 0.1 g of the sample into a 16 × 150 mm test tube. Scoop in 0.5 g fused potassium pyrosulfate and mix well. Fuse the mixture over a burner for 3–5 min. After cooling, dissolve the melt with 5 ml 1 *M* hydrochloric acid with intermittent shaking, and dilute the solution to 10 ml with deionized water. Transfer a 2-ml aliquot to a separation funnel. Add 1 ml 10% sodium citrate solution, 4 ml 2.2% Na_2HPO_4 solution, and two drops of bromothymol blue indicator, and shake the funnel. Add conc. ammonia dropwise until the solution turns blue, then back-titrate with 1 *M* hydrochloric acid to a faint yellow-green (pH ~6.5). Add 3 ml 2-nitroso-1-naphthol solution and 1 ml carbon tetrachloride to the funnel and shake for 1 min. After the separation of the layers, drain the organic layer into a 16 × 150 mm test tube containing 10 ml 0.05% potassium cyanide solution. Compare the color in the carbon tetrachloride layers with the colors of similarly prepared standard solutions of known cobalt content. A recommended set of cobalt standards is 0, 0.2, 0.4, 0.8, 1.5, 3.0, and 6.0 μg cobalt. Calculate the cobalt concentration according to section 5.2.1.

Preparation of 2-Nitroso-1-Naphthol Reagent Solution. To 0.05 g reagent add 10 drops 2 *M* sodium hydroxide. Add deionized water dropwise until complete dissolution is achieved. Dilute to 500 ml.

PREPARATION OF BROMOTHYMOL BLUE INDICATOR SOLUTION. Grind 0.1 g bromothymol blue in a mortar with 16 ml 0.01 *M* sodium hydroxide solution, and dilute to 250 ml with deionized water.

An alternative field method used for the semiquantitative estimation of trace quantities of cobalt in soil and sediment is based on formation of the familiar thiocyanate complex with the cobalt ion and its extraction with isoamyl alcohol (31).

The soil sample is decomposed by potassium bisulfate fusion, in which the metals are converted to sulfates, which are easily leached out with dilute hydrochloric acid. In tri-*n*-butylamine, medium blue tri-*n*-butylammonium tetrathiocyanato cobaltiate is formed. Potentially interfering cupric ions must be reduced with hydroxylamine hydrochloride to monovalent copper. Sodium pyrophosphate is used as a masking agent to avoid interference by many other elements, but the insoluble pyrophosphate precipitate may carry coprecipitated cobalt, thus lowering the results, especially in the presence of $\geq$5% calcium. The presence of up to 10% iron, aluminum, and manganese is tolerated in this screening test.

PROCEDURE. Crush the sample to −80 mesh or finer. Scoop 0.025 g of the sample powder into a 16 × 150 mm test tube. Scoop in 1 g of fused and powdered potassium bisulfate, and fuse until frothing has ceased. Leach the cold melt with 5 ml 0.5 *M* hydrochloric acid in a hot water bath. Take an aliquot into a test tube calibrated at 10 and 12 ml and containing 10 ml buffer solution. Add 0.5 ml 10% tri-*n*-butylamine solution in isoamyl alcohol, cork the tube, and shake vigorously for 1 min. Compare the intensity of the blue in the solvent phase with the intensities in a set of standards.

PREPARATION OF THE BUFFER SOLUTION. Dissolve 170 g sodium acetate trihydrate, 40 g sodium pyrophosphate decahydrate, and 5 g hydroxylamine hydrochloride in water; add 140 ml 6 *M* hydrochloric acid, and dilute to 1 liter with deionized water (solution 1). Dissolve 360 g potassium thiocyanate in water and dilute to 1 liter (solution 2). Mix solutions 1 and 2 in the ratio 9 : 1.

5.3.5. Field Determination of Nickel in Soil and Sediment Samples

The nickel content of soil and rock samples can be estimated in the concentration range 1–15,000 ppm by visual comparison of the amount of the yellow product of benzene extraction of the nickel α-furildioxime inner complex formed from the sample with the amounts in a series of standards. Extensive field tests carried out by the Anglo American Corporation of South Africa Ltd. show that this method can be used for geochemical prospecting purposes (32).

Nickel reacts with the furildioxime reagent to form an insoluble red complex. Extraction with benzene produces a yellow solvent phase. The optimum pH range for the extraction procedure is 9.0–10.3. The presence of ammonium citrate in the buffer solution has the dual function of controlling the pH and masking heavy metals from hydroxide precipitation. The sample is disintegrated by fusion with potassium bisulfate. This oxidative decomposition must be complete to avoid the presence of ferrous iron, which reacts similarly to nickel with furildioxime.

PROCEDURE. Crush the sample to −80 mesh or finer. Scoop 0.2 g of the powder into a 16 × 150 mm test tube. Scoop in 1 g of fused and powdered potassium bisulfate and fuse until 2 min after frothing has ceased. After cooling the melt, leach with 5 ml 1 *N* hydrochloric acid in a boiling water bath. Place a 2-ml aliquot in an 18 × 180 mm test tube containing 5 ml buffer solution and calibrated at 5, 7, 8, 10, ..., 20, and 22 ml. Add 1 ml α-furildioxime solution using an automatic pipet, cork the tube, and shake vigorously for 2 min. Compare the intensity of the yellow color of the solvent phase against a series of standards. The nickel content in ppm is obtained using the expression bv/wa, where b = matching standard (g), v = volume of solvent phase (ml), w = weight of sample (g), and a = aliquot taken (ml).

PREPARATION OF STANDARDS. Add to 10 test tubes, each containing 5 ml buffer solution, 0, 0.05, 0.1, 0.2, 0.3, ..., 0.7, and 0.8 ml, respectively, of a standard solution containing 5 μg nickel/ml; then add 1 ml α-furildioxime solution to each.

PREPARATION OF THE BUFFER SOLUTION. Dissolve 20 g triammonium citrate in water, add 130 ml of ammonia (sp. gravity 0.880), and dilute to 1 liter with water.

PREPARATION OF THE α-FURILDIOXIME REAGENT. Dissolve 0.3 g reagent in 90 ml benzene and 10 ml absolute ethyl alcohol.

When the copper/nickel ratio in the soil exceeds 30 : 1, the copper will cause interference by turning brown. If the zero standard is yellow, the buffer solution should be purified with respect to nickel. This may be accomplished by an extraction procedure at a pH greater than 8.9.

Ferrous ion forms a purple complex with α-furildioxime, but only when the sample is inadequately fused with potassium bisulfate.

5.3.6. Screening for Antimony in Soils and Rocks (1)

This field method for the estimation of antimony in rocks is based on the xantho dyestuffs reaction with pentavalent antimony in strong hydrochloric acid solution. The red color of Rhodamine B changes to violet or blue in the presence of the $[SbCl_6]^-$ anion.

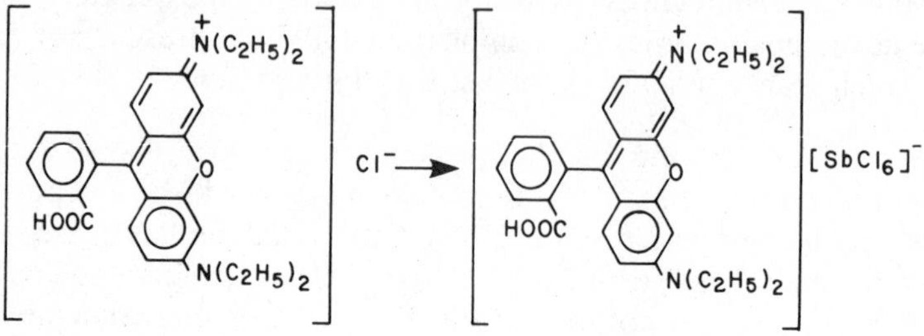

$$\left[\text{Rhodamine B}\right] \xrightarrow{Cl^-} \left[\text{Rhodamine B}\right][SbCl_6]^-$$

Rhodamine B (red)

Rhodamine B-chloroantimonate (blue)

In this field test, the powdered sample is fused with bisulfate and leached with hydrochloric acid. The antimony is oxidized to the pentavalent state and subsequently extracted into isopropyl ether. This technique separates antimony from bismuth, chromium, cobalt, lead, and tungsten, but not from iron. The ether layer is shaken successively with mild reducing agents and dilute acids to isolate the antimony from interferences. Shaking with Rhodamine B produces a fine red-violet dispersion of Rhodamine B chloroantimonate. This fine suspension behaves like a true solution.

Procedure (33). Crush the sample to −80 mesh or finer. Scoop a 0.2-g sample into an 18 × 150 mm test tube, add ~1 g sodium bisulfate, and fuse. After cooling, add 6 ml of a hydrochloric acid–glycerin mixture (100 parts 6 *M* HCl, 20 parts glycerin) and heat the tube in a hot water bath until disintegration is complete. Add 1 ml 1% sodium sulfite solution to the cooled solution and 3 ml hydrochloric acid–glycerin mixture. Filter the solution through Whatman no. 42 filter paper and collect the filtrate in a 125-ml separating funnel. Wash the residue from the test tube into the filter paper with two separate 3-ml portions of hot hydrochloric acid–glycerin solution; then pour 2 ml hot water through the filter paper. Cool the filtrate to room temperature, add 3 ml 5.2% ceric sulfate solution, and mix. Add 10 drops of 1% hydroxylamine hydrochloride solution, and shake. Add 40 ml water and 5 ml isopropyl ether and shake the funnel.

After separation of the phases, discard the aqueous phase and wash the ether phase with 2 ml 1 *M* hydrochloric acid. Add 2 ml 0.02% Rhodamine B solution dissolved in 1 *M* hydrochloric acid, and shake the funnel for 10 sec. Discard the Rhodamine B reagent phase and pour 3 ml of the ether phase into a flat-bottomed tube. Compare the color of the ether phase with the color of similarly prepared standards of known antimony content. This color is stable for at least 6 hours.

The extraction of antimony pentachloride with isopropyl ether ensures a preliminary separation of antimony from iron. It is somewhat cumbersome in the field, since one operator cannot deal with more than 20 samples a day. The low boiling point of isopropyl ether seriously restricts use of the method in tropical climates.

It was shown (34) that heating of soil samples with ammonium chloride decomposes the sample as well as fusion with bisulfate does. The anhydrous hydrogen chloride formed from the thermal dissociation of ammonium chloride attacks the sample, and excess ammonium chloride is easily removed by sublimation. Antimony trichloride deposits on the cooler parts of the tube, and then is oxidized by sodium nitrite to pentavalent antimony. The qualitative test for chloroantimonate recommended by Lapin and Gein (35) involving its complexation with brilliant green is used for the screening. Because ferric iron forms a colored complex with brilliant green too, sodium hexametaphosphate is used as a sequestering agent for iron. The blue complex between chloroantimonate and brilliant green in toluene is stable for more than 8 hours.

PROCEDURE (36). Crush the sample to −80 mesh or finer. Scoop 1 g of the sample into an 18 × 180 mm test tube. Scoop 1 g of ammonium chloride into the tube and heat until all the ammonium chloride has sublimed. After cooling, add 5 ml 6 *M* hydrochloric acid and react on a sand tray or in a boiling water bath until the sublimate has dissolved. After cooling, pipet a 2-ml aliquot of the clear solution into an 18 × 180 mm test tube calibrated at 3, 13, 17, and 21 ml, and dilute it to 3 ml with 6 *M* hydrochloric acid. Add 0.1 ml of 5% aqueous sodium nitrite solution and mix. Dilute to 10 ml with the reagent solution and add 1 ml toluene. Compare results with standards. If indicated concentration is greater than that of the highest-concentration standard, dilute with toluene until it is within the standard range. To calculate the antimony concentration in ppm, use the formula 5 × (μg in matching standard) × (ml of solvent phase).

PREPARATION OF STANDARDS. To eleven 18 × 180 mm test tubes add, respectively, 0, 0.2, 0.4, 0.6, 0.8, 1.0, 1.2, 1.4, 1.6, 1.8, and 2.0 μg antimony. Dilute to 3 ml with 1 *M* hydrochloric acid and add 0.1 ml 5% sodium nitrite solution. Mix; then dilute to 10 ml with the reagent solution. Add 1 ml toluene and shake for 30 sec.

PREPARATION OF THE REAGENT SOLUTION. Dissolve 100 g sodium hexametaphosphate in 1 liter deionized water (solution A). Dissolve 10 mg brilliant green in 200 ml water (solution B). Solution B must be prepared freshly each day. Mix solutions A and B in the ratio 5 : 1 daily as required.

The range of antimony concentration determinable by the above method is 0.2–110 ppm, but may be extended to 2200 ppm by using an aliquot of 0.1 ml. One hundred samples can be analyzed by one operator during an 8-hour working day.

5.3.7. Field Determination of Arsenic in Soil

The following spot procedure for semiquantitative estimation of arsenic in soil was first presented by Almond (37). The basis of the test is the reduction of arsenic to arsine and arsine's color reaction with mercuric chloride. The test is carried out in a modified Gutzeit apparatus. The yellow to orange-color of the product formed between mercuric chloride and arsine slowly fades; consequently, artificial standards are recommended.

The dissolution of the soil sample is carried out in a hot potassium hydroxide melt. The duration of the heating has to be limited because of the significant solubility of glass in hot potassium hydroxide. An alternative preparation of the sample consists of oxidative decomposition of all reducing sulfides in the soil. Subsequent reactions in the procedure reduce arsenic to arsine (38).

Both field tests use mercuric-chloride-saturated filter paper discs; the intensity of the yellow spot formed quantitates the arsenic concentration. Hydrogen sulfide evolved along with the arsine has to be removed before it reaches the indicator paper, since the formation of highly insoluble mercuric sulfide would quickly deplete the reagent. A lead-acetate-impregnated glass-wool plug located between the indicator and the reaction removes the hydrogen sulfide effectively.

Heavy metals in high concentration (50 mg/0.1 g or more) interfere with the test, but the presence of such high concentrations is most unlikely.

PROCEDURE A (37). Scoop 0.1 g of a crushed sample into a 25 × 150 mm test tube, add eight pellets of potassium hydroxide, and heat over a free flame for ~1 min. After melting, further heating should be avoided in order not to dissolve the glass. Add 3 ml water and shake until the melt dissolves. Add 3 ml conc. hydrochloric acid and 4 ml water and mix. Add 0.5 ml 10% stannous chloride in conc. hydrochloric acid and 2–4 g mossy zinc. Attach the test tube immediately to the rubber stopper of a modified Gutzeit apparatus. After 1 hour, read the arsenic concentration from the reagent paper disc by comparing it with a standard series.

The artificial standard series is prepared using chrome yellow and deep chrome yellow dyes by visual comparison with the colors obtained with a standard series of 1, 2, 4, 8, and 16 μg arsenic. The modified Gutzeit apparatus consists of a reaction tube coupled through a rubber stopper with a pipe plugged with lead-acetate-coated glass wool. The mercuric chloride paper is prepared by impregnation of filter paper with a 25% alcoholic mercuric chloride solution and subsequent drying.

PROCEDURE B (38). Crush the sample to minus 80 mesh or finer. Scoop 0.25 g of the powdered sample into a 16 × 150 mm test tube. Fuse with ~1 g potassium bisulfate. Leach the melt with 5 ml 0.5 *M* hydrochloric acid on a sand tray or in a boiling water bath. Pipet a 2-ml aliquot of the cold solution into a 19 × 150 mm test tube calibrated at 2 and 10 ml and containing 2 ml 2.5% potassium iodide solution. Dilute to the 10-ml mark with 0.75% stannous chloride solution. Add 2–4 g of zinc pellets and immediately connect the reaction tube to the Gutzeit tube containing the mercuric chloride paper. After 1 hour, compare the results with artificial standards (see Procedure A).

The range of this test is 1–50 ppm, which may be extended to 1000 ppm by using an aliquot of 0.1 ml. One operator can handle 100–110 samples during an 8-hour working day.

Rapid Geochemical Survey Analysis of Arsenic in Geochemical Samples with the Silver Diethyldithiocarbamate Method (39)

Arsenic can be determined in a convenient system for rapid geochemical survey analysis utilizing the reaction of arsine with silver diethyldithiocarbamate (40). The powdered sample is leached with either a perchloric acid–nitric acid or a hydrofluoric acid–aqua regia mixture. Arsenic in aliquots is first reduced with potassium iodide and stannous chloride to the trivalent state, and arsine is finally liberated by zinc pellets. Arsine changes the color of silver diethyldithiocarbamate in chloroform–quinoline solution from yellow to red. The method is apparently free from matrix interferences usually encountered in soils, rocks, and ores.

To eliminate possible interferences, EDTA is incorporated into the reaction as a masking agent for cobalt and nickel cations.

PROCEDURE. Heat 200 mg of a finely pulverized sample with 4 ml of a mixture of 2 parts 40% hydrofluoric acid and 1 part aqua regia. After putting the sample into solution, evaporate slowly to dryness at 110°C on a hot plate. Dissolve the residue in 10 ml 10% hydrochloric acid solution while warming. Place a 5-ml aliquot into the arsine-generator tube and add 2 ml Fe/KI solution and 2 ml 5% stannous chloride dissolved in 50% hydrochloric acid. Bring the arsenic to the trivalent oxidation state. Warm the solution to complete the reaction. After cooling, add 5 ml 10 *M* hydrochloric acid–1.5% EDTA solution and four arsenic-free zinc pellets. The use of glass wool impregnated with lead acetate as in the alternative estimation method (38) is not necessary, since under the described working conditions all sulfides are oxidized to sulfates. Bubble the arsine generated into an absorption tube containing 3 ml silver diethyldithiocarbamate solution. The same absorption tube can be used as an absorption cell for rapid geochemical analysis.

PREPARATION OF Fe/KI SOLUTION. Dissolve 14 g ferrous ammonium sulfate hexahydrate and 20 g potassium iodide in 400 ml of approximately 3% sulfuric acid. Add a trace of ascorbic acid to prevent oxidation.

PREPARATION OF THE REAGENT SOLUTION. Dissolve, with warming, 1000 g silver diethyldithiocarbamate in 300 ml chloroform and 100 ml quinoline. Store under refrigeration and away from light.

5.3.8. Screening for Barite Content in Prospecting Samples (41)

Geochemical prospecting samples containing in excess of 0.5% barium can be screened for barite by the application of a relatively simple solubilization process followed by the use of a new selective spot test reagent, Chlorindazone-DS.

Because the barium-bearing mineral is usually barite, the preparation of the sample under field conditions presents a formidable problem. Barite dissolves only in hot concentrated sulfuric acid, a procedure that is neither attractive nor feasible under field conditions.

The thermal decomposition of sulfamic acid is used in a relatively simple procedure for solubilization of barium sulfate. Sulfamic acid is a harmless solid that thermally decomposes at 260°C to sulfur dioxide, sulfur trioxide, nitrogen, and water. The boiling sulfuric acid formed *in situ* dissolves barite rapidly.

Chlorindazone-DS [(1,6-chloroindazol-3-ylazo)-2-hydroxynaphthalene, 3,6-disulfonic acid], synthesized by Schweppe (42), turned out to be a selective precipitation agent for barium ions in the pH range 1–5. Under these conditions dissociation of the acidic group is sufficiently suppressed and barium ion is precipitated as a violet water-insoluble salt. The presence of relatively large quantities (1 mg) of sodium, potassium, lithium, ammonium, calcium, strontium, magnesium, beryllium, lead, copper, mercury, cobalt, nickel, aluminum, zinc, chromium, manganese, ferric, ferrous, silver, cadmium, and uranyl ions does not interfere with the test.

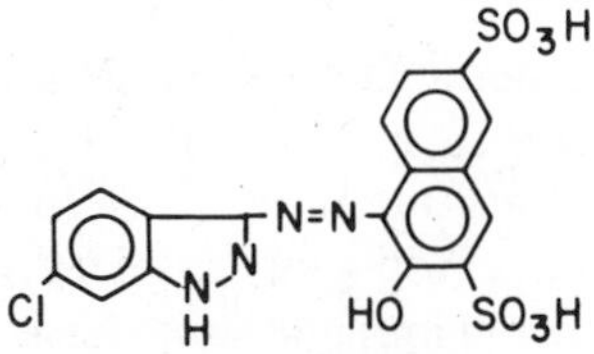

Chlorindazone-DS

PROCEDURE. Crush the sample to −80 mesh or finer. Scoop dust up to the 10-ml mark in an 18 × 180 mm graduated test tube. (On average, this is 9.1 ± 7.5 mg.) Mix a few centigrams of solid sulfamic acid intimately with the sample, and heat the tube on a direct flame on a butane stove until it melts and boils. After air-cooling (but while the tube is still hot) fill the test tube up to the 10-ml mark with pH 4 buffer solution and mix with a glass rod. After the precipitate settles, dispense one drop of the supernatant liquid onto a strip of Whatman GF/B glass-fiber filter paper. Treat the sample spot with a drop of 0.1% aqueous Chlorindazone-DS solution on both sides of the glass-fiber paper, and expose the paper to conc. ammonia vapor for a few minutes. Wash the colored stains on both sides with a direct stream of tap water. Sample solutions prepared from an ore sample containing more than 0.5% barium give a positive response in the form of a violet stain remaining on one side of the glass-fiber paper after washing; those samples containing less barium generally fail to give an observable reaction.

Barium is a useful guide in regions of deep weathering and oxidation for differentiation between base-metal-sulfide-related gossans and pseudogossans. For rapid field discrimination this sample test can be used.

5.3.9. Rapid Determination of Molybdenum in Soil and Sediment Samples

The molybdenum content in soil can be determined using a spot test (43) involving the formation of red $H_3[Mo(SCN)_6]$ (44) after addition of a reducing agent and thiocyanate. The extraction of molybdenum is achieved by fusing the rock sample with a sodium carbonate–potassium nitrate flux. The analytical method itself consists of the reduction of the hexavalent molybdenum to the trivalent state, formation of the red molybdenothiocyanate complex, and extraction of this complex in isopropyl ether.

The fusion with carbonate practically separates insoluble-hydroxide-forming heavy metals from molybdenum. The addition of tartrate prevents possible interference by tungsten.

PROCEDURE. Grind the sample to −80 mesh size and scoop a 0.1-g sample into a 16 × 150 mm test tube. Mix with ~0.5 g of a pulverized sodium carbonate–potassium nitrate mixture. Heat the tube over a free flame until the mixture is sintered. Add 4 ml deionized water after cooling. Pipet a 1-ml aliquot into a 16 × 150 mm test tube marked at 5 ml. Add one drop of 1% phenolphthalein indicator; then add 1 *M* hydrochloric acid until the pink color disappears. Add 0.5 ml 10% potassium nitrate solution and ~0.2 g powdered sodium tartrate. Shake to dissolve, and bring the volume to the 5-ml mark with water. Add successively 0.5 ml conc. hydrochloric acid, 0.3 ml 5% potassium thiocyanate solution, and 0.5 ml 0.5% stannous chloride solution. Allow the tube to stand for 1 min and then add 0.5 ml isopropyl ether and shake the tube vigorously. Compare the red color of the organic layer with that obtained from standard solutions.

The digestion of most molybdenum minerals, such as molybdenite and powellite, is done with a mixture of concentrated sulfuric and nitric acids over a low flame (2). Molybdenum (VI) is readily extracted from such a medium into isoamyloacetate, and forms the upper layer. An aliquot from the organic layer is taken to be reduced, and forms an amber-colored complex with thiocyanate. The omnipresent iron (III) should be reduced with stannous chloride; otherwise, the formation of the familiar red $(FeSCN)^{2+}$ complex interferes with the test.

Field Determination of Molybdenum in Soils and Sediments with Zinc/Dithiol

The colorimetric method for the estimation of molybdenum with zinc toluene, 3,4-dithiol (toluene, 3,4-dithiol-"dithiol") in geochemical samples has been used by a number of authors (45–50). Fusion of the powdered sample with the carbonate–chloride–nitrate flux (45) converts molybdenum, tungsten, aluminum, and chromium into water-soluble molybdate, tungstate, aluminate, and chromate, respectively, whereas other heavy metals are precipitated as carbonates. Potassium nitrate in the fusion mixture oxidizes organic matter and sulfides, but the potassium nitrite resulting from this reaction has to be removed carefully with hydroxylamine hydrochloride or it will destroy the reagent. The green complex formed between molybdenum and dithiol is extracted into amyl acetate. A high silica content in the sample can blur the color of the solvent layer through the formation of a silicic acid emulsion in both phases. Washing the amyl acetate phase with concentrated hydrochloric acid resolves this problem.

PROCEDURE. Crush the sample to −80 mesh or finer, and scoop 0.25 g into a nickel crucible containing ~1.25 g of fusion mixture (Na_2CO_3, NaCl, and KNO_3 in the ratio 5 : 4 : 1). Fuse until frothing has ceased. Add 2 ml deionized water and leave overnight. Transfer to a 16 × 150 mm test tube marked at 5 ml and adjust to this volume with water washings from the crucible. Complete the disintegration of the melt on a sand tray. Take a 2-ml aliquot of the clear, cold solution into a 16 × 150 mm test tube calibrated at 2 and 4 ml and containing 2 ml 2.5% hydroxylamine hydrochloride solution in hydrochloric acid. Shake the tube to liberate the carbon dioxide and leave it to cool to room temperature. Add 0.5 ml 1% dithiol reagent solution. Shake gently at frequent intervals over a period of 20 min. Compare results with the standards.

PREPARATION OF STANDARDS. To 12 test tubes containing 2 ml 2.5% hydroxylamine hydrochloride each, add, respectively, 0, 0.2, 0.4, 0.6, 0.8, 1.0, 2.0, 3.0, 4.0, 5.0, 7.0, and 10.0 μg molybdenum. Dilute to 4 ml with water and add 0.5 ml 1% dithiol reagent solution. Shake gently at frequent intervals over a period of 20 min.

PREPARATION OF 1% DITHIOL REAGENT SOLUTION. Weigh 0.5 g zinc "dithiol" into a 50-ml stoppered cylinder, add 1 ml conc. hydrochloric acid, and shake for 10 sec. Dilute to 50 ml with amyl acetate and mix well.

The main shortcomings of the above procedure are inadequate leaching of the fused sample and the incomplete formation and extraction of the molybdenum complex. Decomposition of the soil and sediment samples by fusion with potassium hydrogen sulfate and leaching with hydrochloric acid (46) correct these imprecision-causing defects.

PROCEDURE. Scoop 0.25 g of the crushed sample into a borosilicate test tube and fuse with 1 g potassium hydrogen sulfate until a quiescent melt is obtained. Continue heating for a further 2 min. Leach on a sand tray with 5 ml 6 *M* hydrochloric acid, mix, and let settle. Take 5 ml of the clear sample solution into a 16 × 150 mm test tube, add 2 ml reducing solution, mix, and let stand for 2 min. Add 2 ml 50% (w/w) potassium iodide solution and 1 ml dithiol solution and let stand for 2 min. Add 0.5 ml petroleum spirits, stopper the tube with silicone greased cork, shake vigorously for 90 sec, and compare the intensity of the color in the solvent phase with that of a standard series.

PREPARATION OF THE STANDARD SERIES. To each of thirteen 16 × 150 mm test tubes add 2 ml 1% iron (III) salt solution, followed by 0, 0.2, 0.4, 0.6, 0.8, 1.0, 1.5, 2.0, 3.0, 4.0, 5.0, 7.0, and 10.0 μg molybdenum, respectively. Dilute to 5 ml with 6 *M* hydrochloric acid and proceed according to the procedure above.

PREPARATION OF REDUCING SOLUTION. Dissolve 75 g citric acid and 150 g ascorbic acid in water and dilute to 1 liter.

PREPARATION OF DITHIOL SOLUTION. To 0.3 g zinc "dithiol" add 2 ml ethanol, followed by 4 ml water, 2 g sodium hydroxide, and 1 ml thioglycolic acid. Dilute the solution to 50 ml with 50% (w/w) potassium iodide solution and store in the refrigerator.

The zinc dithiol method of molybdenum determination in soil has been modified (49) to avoid formation of an emulsified organic layer, which may interfere with the colorimetric determination. Clouding of the organic layer appears in the nitric acid–perchloric acid digestion method when soluble perchlorates precipitate on addition of potassium iodide and emulsify the organic layer. Addition of small quantities of acetone is immediately effective in clarifying the emulsion and cleaning the precipitate from the test tube walls.

PROCEDURE (45). Take a 1.0-g sample, dissolve in 20 ml 4 : 1 nitric acid–perchloric acid mixture, evaporate to dryness, and extract in 10 ml 6 *M* hydrochloric acid. Add 0.2–0.5 ml acetone to the organic phase and gently transfer it to the aqueous phase after clarification of the emulsion.

After the acetone was transferred to the aqueous phase, no difference in the color of the green molybdenum dithiol complex was visually apparent between calibrated standards treated with acetone and a second set of untreated standards.

In the last decade, the dithiol method for the colorimetric estimation was still used very extensively (50) in geochemical prospecting in spite of the tendency for classical methods to be replaced by instrumental techniques such as AAS and ICP. The following is a representative instrumental test.

PROCEDURE. Crush the sample to −100 mesh or finer and scoop 0.2 g into a borosilicate test tube. Add ~1 g potassium hydrogen sulfate, mix, fuse over a free flame until a quiescent melt is obtained, and continue heating for a further 2 min. Leach the melt in a boiling water bath with 10 ml 4 *M* hydrochloric acid until the melt can be broken with a glass rod. Let the precipitate settle. Transfer 5 ml of the supernatant liquid into a 150 × 16 mm test tube. Add 1 ml of the reducing solution and let stand for 5 min. Add 3 ml potassium iodide solution, mix, and then add 0.02 ml thioglycolic acid and mix again. Allow to stand for 30 sec; then add 1 ml dithiol solution, mix, and allow to stand for 1 min. Add 5 ml isoamyl acetate and shake vigorously for 1 min. Measure the absorbance of the organic layer at 680 nm within 2 hours.

PREPARATION OF REDUCING SOLUTION. Prepare a solution containing 15% (w/v) of ascorbic acid and 2% (w/v) citric acid in deionized water.

PREPARATION OF POTASSIUM IODIDE SOLUTION. Prepare a 100% (w/v) solution of potassium iodide in deionized water. Add 0.5% (w/v) ascorbic acid.

PREPARATION OF DITHIOL SOLUTION. Add 6 ml ethanol to 1 g zinc dithiol, followed by 10 ml water, 4 g sodium hydroxide, and 2 ml thioglycolic acid. Mix and dilute to 300 ml with deionized water. Store in the refrigerator.

Up to 150 geochemical samples can be analyzed per man-day by this method.

Sodium tungstate interferes with the method slightly by forming a blue-green dithiol complex. But since sodium tungstate is only slightly soluble in acid, the presence of 0.2 g tungstate gives an absorbance equivalent to only 8 ppm molybdenum. The addition of 20 mg citric acid complexes tungsten and prevents the formation of tungsten dithiol. Iron (III), which interferes in a number of ways, is completely suppressed by ascorbic acid, which reduces iron (III) but not Mo (VI). Copper (II) interferes with the formation of the molybdenum dithiol complex by rapidly forming a very stable purplish-black dithiol complex. This complex does not interfere colorimetrically, since it is insoluble in isoamyl acetate. It does interfere, however, by consumption of the reagent, so it must be reduced to copper (I) and precipitated as copper (I) iodide.

5.3.10. Field Determination of Chromium in Soils and Rocks (51, 52)

Rock samples containing more than 100 ppm chromium can be screened using a test based on the oxidation of chromium to chromate by fusion of the sample with sodium hydroxide and sodium peroxide. Leaching the melt with hot water separates all the insoluble hydrated oxides formed during the fusion from the soluble chromate. All the potentially interfering cations are precipitated and the remaining ions do not impair or interfere with the test. The only colored, soluble ions that might mask the color of chromate are uranium, cerium, gold, and platinum, but these ions are unlikely to be present in significant concentrations. Although during the fusion the manganese is oxidized to manganate, it is quickly reduced to insoluble hydrated manganese dioxide by addition of alcohol to the hot water leach of the melt.

PROCEDURE. Crush the sample to minus 80 mesh or finer and scoop a 0.1-g sample into a nickel crucible. Add five pellets of sodium hydroxide and ~0.25 g sodium peroxide. Heat the crucible over a moderate flame (in order to avoid sizzling) until a quiescent melt is obtained. Add 10 ml 10% aqueous ethyl alcohol to the crucible, and bring the solution to a boil. Use a glass stirring rod to ensure complete disintegration. Filter the solution into a 16 × 150 mm test tube and compare the yellow color of the filtrate with that of a series of standard solutions.

PREPARATION OF CHROMIUM STANDARD SOLUTIONS. Dissolve 0.283 g potassium dichromate in water, and dilute to 100 ml with water. Dilute this to an 100-ppm chromium working solution. Prepare a series containing 0, 10, 20, 40, 80, 150, and 300 μg chromium in 16 × 150 mm test tubes. Add 2 ml 2 *M* NaOH to each tube and dilute the solutions to 10 ml with water.

For a lower soil chromium content (5–250 ppm) the diphenylcarbazide method leads to a better evaluation. Following oxidation of the chromium in the sample to bichromate, the red-violet diphenylcarbazide complex is formed (see reaction mechanism in section 4.3.8). The main interfering ion is molybdate, which forms a similarly colored complex with the reagent. Vanadate, permanganate, and ferric ions interfere by forming brown complexes. Addition of EDTA masks the vanadate, and the leaching procedure with alcohol reduces permanganate and precipitates iron.

The acidity of the solution should be kept in a rather narrow range around 0.4 *N*. The complex does not develop fully below 0.2 *N*, and above 0.6 *N* it decomposes gradually.

PROCEDURE (52). Follow the preceding procedure up to the filtration of the extract. Pipet a 2-ml aliquot into a test tube calibrated at 10 ml and boil for a few seconds. Add 1 *M* hydrochloric acid dropwise until the aluminum hydroxide precipitate present is just redissolved. In the absence of aluminum use a few drops of phenolphthalein indicator to obtain the proper acidity (0.05% in 50% ethyl alcohol). Add 2 ml 0.5% EDTA dissolved in 22% aqueous hydrochloric acid. Add 2 ml of a freshly prepared 0.2% solution of diphenylcarbazide dissolved in 1 : 9 glacial acetic acid–acetone. Dilute to 10 ml with water and mix. Compare with the standard series after 10 min.

PREPARATION OF STANDARDS. Add 0, 0.25, 0.5, 1.0, 1.5, 2.0, 2.5, 3.0, 3.5, 4.0, 4.5, and 5.0 μg chromate to 12 test tubes calibrated at 10 ml. Add 2 ml EDTA solution and 1 ml diphenylcarbazide solution (both as in Procedure, above), dilute to the mark with water, mix, and let stand for 10 min.

5.3.11. Screening for Manganese in Rocks and Soil

Several methods have been described for the sensitive, rapid field determination of manganese for geochemical exploration. The methods (53, 54) are based on the oxidation of manganese to permanganate in acid solution. The sample is fused with an acid flux and the melt leached with acid. Such treatment readily decomposes most of the manganese minerals. A portion of the sample solution is further acidified and the manganese is oxidized to permanganate by potassium periodate. The characteristic purple permanganate color developed is compared visually with standard solutions to estimate the manganese concentration.

The method is relatively free from interferences. Masking with orthophosphoric acid prevents interference by ferric ions. Other interference experienced is caused by the iron color of copper, nickel, cobalt, chromium, and uranium ions, but as much as 20 ppm of these can be tolerated.

PROCEDURE. Crush the sample to minus 80 mesh or finer and scoop a 0.1-g sample into a 16 × 150 mm Pyrex test tube. Mix with ~0.5 g potassium bisulfate and fuse. Leach with 5 ml 0.5 *M* sulfuric acid. Pipet a 1-ml aliquot into an 18 × 180 mm test tube calibrated at 5 ml and 10 ml and containing 5 ml of the acid mixture. Add 0.2 g potassium periodate and boil for 10 min. Dilute to the 10-ml mark and compare the result with standards.

PREPARATION OF THE ACID MIXTURE. Add 125 ml sulfuric acid (s.g. 1.84) to ~500 ml water, mix well, and cool. Add 62.5 ml nitric acid (s.g. 1.42) and 62.5 ml orthophosphoric acid (sp. gravity 1.75), mix, and dilute to 1 liter.

PREPARATION OF STANDARDS. Add to 19 test tubes, each containing 5 ml of the acid mixture, 0, 2, 4, 6, 8, 10, 12, 14, 16, 18, 20, 30, 40, 50, 60, 80, 100, 150, and 200 μg manganese. Add 0.2 g potassium periodate to each and bring to a boil in a water bath for 10 min. Keep in the dark when not in use.

5.3.12. Field Determination of Beryllium in Soil, Sediment, and Rock Samples

In geochemical samples, beryllium is present mostly as beryl, a beryllium aluminosilicate, and since no other beryllium mineral is more resistant to chemical agents than beryl, the procedure for solubilizing beryl decomposes all other possible beryllium minerals. The thermal decomposition of ammonium fluoride leads to the formation of anhydrous hydrogen fluoride, which effectively attacks the insoluble silicates and aluminosilicates by silicon tetrafluoride formation. Excess ammonium fluoride must be removed; otherwise it will complex beryllium. *N,N*-di(2-hydroxyethyl)-glycine and the disodium salt of EDTA serve as masking agents to prevent interference from foreign metals. The estimation is based on the spot test (55) in which beryllium reacts very selectively at high pH with beryllon II (the tetrasodium salt of 1′,8,8′-trihydroxy-1,2′-azonaphthalenetetrasulfonic acid) to form a blue product.

PROCEDURE (56). Crush the sample to minus 80 mesh or finer and scoop 0.5 g into a silica crucible containing 2 g ammonium fluoride. After mixing, heat gently until bubbling ceases. Break up any lumps with a spatula and continue heating until the evolution of fumes has stopped. Leach the cooled material by boiling with 5 ml 0.2 *N* nitric acid. Add a further 5 ml 0.2 *N* nitric acid, mix, and let the crucible stand until the residue settles. Transfer 2 ml of the supernatant fluid into an 18 × 180 mm test tube previously calibrated at 5, 7, and 12 ml and containing 5 ml complexing solution. Add 5 ml 0.25 *N* sodium hydroxide and mix. Add 0.02% aqueous beryllon II solution, mix, and allow to stand for 10 min. Compare with standards.

PREPARATION OF STANDARDS. Add respectively 0, 0.25, 0.5, 0.75, 1.0, 1.25, 1.5, 2.0, 2.5, 3.0, 3.5, and 4.0 μg beryllium to 12 test tubes each containing 5 ml complexing solution. Dilute to 7 ml with 0.2 *N* nitric acid, mix, add 5 ml 0.25 *N* sodium hydroxide, and mix again. Add 1 ml 0.02% aqueous beryllon II, mix, and allow to stand for 10 min. The standard beryllium solution (100 ppm) is prepared by dissolving 196.4 mg $BeSO_4 \cdot 4H_2O$ in 100 ml 0.2 *N* nitric acid.

PREPARATION OF COMPLEXING SOLUTION. Dissolve 10 g EDTA (disodium salt) in water, add 10 ml 25% *N,N*-di(2-hydroxyethyl)-glycine, and dilute to 1 liter.

5.3.13. Rapid Estimation of Tin in Soils and Rocks

The following simple and moderately accurate method for screening geochemical samples for tin (57, 58) is based on the volatilization of stannic iodide by heating the sample with ammonium iodide and the formation of a pink-colored product between quadrivalent tin and 4,5-dihydroxyfluorescein. Although the estimation cannot be described as highly accurate, field tests have been sufficient to show areas of tin mineralization. The procedure has been tried successfully on samples containing 2 μg tin in the presence of 10,000 μg titanium or 100 μg molybdenum or tungsten.

PROCEDURE. Crush the sample to minus 80 mesh or finer and scoop a 0.2-g sample into the bulb of the sublimation vessel along with 0.5 g ammonium iodide. Mix the contents and heat the tube over a moderate flame until the mixture is quiescent; then cool. Add 5 ml of a mixture of equal volumes of ethyl alcohol and 2 *M* hydrochloric acid. Heat the tube gently and swirl the contents until all the sublimate is in solution. Do not boil. Let the tube stand for ~15 min until all the undecomposed sample settles to the bottom. Transfer 2 ml of the supernatant liquid to a 16 × 150 mm test tube. Add 5 ml buffer solution and heat gently without boiling until the solution loses its color. Cool the tube to room temperature, dilute to 10 ml, and add 1 ml 0.02% alcoholic 4,5-dihydroxyfluorescein reagent. Shake the tube several times, and within 2 min add 2 ml of a mixture of 46 ml cyclahexanone and 50 ml *o*-dichlorobenzene. Stopper the tube and shake vigorously for 60 sec, and allow the phases to separate. Compare the pink color formed in the unknown with the standard series.

PREPARATION OF STANDARDS. Add respectively 0, 0.5, 0.1, 0.2, 0.4, 0.8, and 1.5 ml of the 10-ppm tin solution to seven 16 × 150 mm test tubes. Add 5 ml buffer solution to each tube, dilute to 10 ml, and proceed as above by adding the reagents.

PREPARATION OF BUFFER SOLUTION. Dissolve 50 g monochloroacetic acid, 50 g chloroacetic acid (sodium salt), and 25 g hydroxylamine hydrochloride in 500 ml deionized water.

PREPARATION OF STANDARD TIN SOLUTION (1000 ppm). Dissolve 0.1 g tin metal in 50 ml hydrobromic acid, A.C.S. grade. Add 0.5 g potassium bromate and swirl to bring into solution. If the color due to free bromine fades as the tin goes into solution, add another 0.5 g potassium bromate. Dilute to 100 ml with hydrobromic acid.

In an alternative procedure (59) the tin-bearing ore (usually cassiterite) is converted, much like in the preceding method, to stannic iodide, which can be kept in acid solution after hydrolysis. At pH 2.0–2.5, stannic ion forms with gallein a pink lake. This lake is slowly precipitated in the absence of a protective colloid. To avoid the precipitation gelatin is added to the standard series.

Hydroxylamine hydrochloride is present to reduce iodine liberated by the heating of ammonium iodide with the sample. The iodide formed prevents gallein from reacting with copper, iron, lead, and manganese. Massive interference is caused by antimony, which forms a similarly colored complex to that of tin.

PROCEDURE. Crush the sample to minus 80 mesh or finer and scoop 1 g into an 18 × 180 mm Pyrex test tube. Add ~1 g ammonium iodide and heat until ammonium iodide is no longer sublimed and the residue becomes red hot. After cooling, add 5 ml 1 *M* hydrochloric acid, and leach by heating in a boiling water bath. Allow the residue to settle, and transfer 1 ml of the supernatant fluid to an 18 × 180 mm test tube. Add 5 ml buffer solution and warm gently to reduce iodine. After cooling the tube, add 0.1 ml reagent solution, mix, and allow to stand for 10 min. Compare result with the standards.

PREPARATION OF STANDARDS. Add respectively 0, 0.25, 0.5, 1.0, 1.5 2.0, 2.5, 3.0, 3.5, 5.0, and 5.0 μg tin to eleven 18 × 180 mm test tubes and to each add 0.5 ml 1 *M* hydrochloric acid. Add 0.2 ml 0.5% aqueous gelatin solution, 5 ml buffer solution, and 0.2 ml reagent solution. Let stand for 10 min.

PREPARATION OF BUFFER SOLUTION. Dissolve 45 g chloroacetic acid, 75 g sodium chloroacetate, and 20 g hydroxylamine hydrochloride in water and dilute to 1 liter.

PREPARATION OF THE REAGENT SOLUTION. Dissolve 0.1 g gallein in 100 ml ethyl alcohol by heating gently. Dissolve 0.03 g methylene blue in 200 ml water by warming gently and combine these two solutions in equal proportions to form the reagent solution.

PREPARATION OF STANDARD TIN SOLUTION (100 ppm). Dissolve 50 ml tin powder in 50 ml conc. hydrochloric acid and dilute to 500 ml with water.

5.3.14. Semiquantitative Estimation of Iron in Geochemical Samples

The following field method is based on the decomposition of the ore sample by a mixture of nitric and perchloric acid. Insoluble silicates form, whereas the metals are solubilized as perchlorates. The chromogen in this test is thioglycolic acid, which produces a purple anion complex with ferric ion in ammoniacal solution. The precipitation of heavy metal hydroxides is avoided by the use of tartrate as a masking agent.

PROCEDURE (52). Crush the sample to minus 80 mesh or finer, and weigh 0.1 g into a beaker. Add 5 ml nitric acid–perchloric acid mixture (4 : 1, v/v). Evaporate to fumes of perchloric acid. After cooling, transfer the content of the beaker into a 16 × 150 mm test tube calibrated at 10 ml, dilute to the mark with water washings from the beaker, and let settle. Transfer 1 ml of the clear supernatant fluid into an 18 × 180 mm test tube calibrated at 5 and 10 ml and containing 5 ml reagent solution. Dilute to 10 ml with deionized water, and compare result with a standard series.

PREPARATION OF STANDARDS. To 12 test tubes containing 5 ml reagent solution, add, respectively, 0, 5, 10, 20, 30, 40, 50, 60, 80, 100, 150, and 200 μg iron. Dilute to 10 ml with deionized water and mix. Keep in the dark when not in use.

PREPARATION OF STANDARD IRON SOLUTION (100 ppm). Dissolve 432 mg ammonium ferric sulfate crystals (A.R.) in ~50 ml water and 5 ml hydrochloric acid (sp. gravity 1.18) and dilute to 500 ml with water.

PREPARATION OF REAGENT SOLUTION. Dissolve 40 g tartaric acid in ~200 ml water, add 50 g thioglycolic acid and 200 ml ammonia (sp. gravity 0.880), and dilute to 1 liter with water.

Results can be obtained within ±25% at the 95% confidence level for the iron content range 100 ppm to 2%.

5.3.15. Estimation of Bismuth in Rocks

In the following test, the pulverized rock sample is decomposed by fusion with pyrosulfate, and the fused mass is leached with hot 8 *M* nitric acid. EDTA and cyanide are added as masking agents to prevent the formation of iron (III), silver, and copper (II) diethyldithiocarbamates.

PROCEDURE (60). Crush the sample to −80 mesh or finer and scoop 0.5 g into a 60-ml beaker. Add 1.5 g potassium bisulfate and fuse. After cooling, add 2 ml 8 *M* nitric acid and heat until boiling. Add 8 ml deionized water, heat to boiling, and add ~1 g solid EDTA. Add ammonia dropwise until the pH paper shows a pH value of 5–6. Add 5 ml 5% sodium cyanide solution and filter into a separating funnel. Add 1 ml 1% aqueous sodium diethyldithiocarbamate solution and 5 ml chloroform. Shake the funnel for 1 min. Place a plug of cotton into the stem of the funnel and drain the organic layer into a test tube. Compare the resultant color with the standards.

Preparation of Standards. To a series of five separating funnels, each containing 8 ml water and 1 g solid EDTA, add 0, 10, 20, 30, 40, and 80 μg bismuth and proceed with the dropwise addition of ammonia and so forth as above.

Preparation of Bismuth Standard Solution. Dissolve 0.5 g bismuth metal in 500 ml conc. nitric acid. (1000 ppm)

5.3.16. Estimation of Mercury in Soil and Rocks

In the following test, rock samples containing cinnabar heated at low heat with ammonium iodide form a sublimate of mercuric iodide. This settles, together with the excess ammonium iodide and iodine, in the upper, cooler parts of the test tube. The nonvolatile residue is discarded, and the sublimate dissolved in a pH 4 acetate buffer. Dithizone is used as a reagent. Interference by silver is avoided by masking with thiocyanate; other metals are sequestered with EDTA. The mercuric iodide volatilization method is most effective on samples containing cinnabar, and has been tested in geochemical explorations in California, Nevada, and Texas.

Procedure (61). Crush the sample to minus 80 mesh or finer and scoop 0.2 g into an 18 × 180 mm test tube calibrated at 10 ml. Add ~0.1 g ammonium iodide and heat the mixture at a low temperature for 1 min. After cooling, invert the tube and discard the loose solid material. Add 10 ml pH 4 acetate buffer solution [a 2:1 mixture 2 *M* acetic acid–sodium acetate 2 *M*], scoop in 0.5 g complexing agent mixture and 0.1 g solid ammonium thiocyanate, and dissolve. Transfer the content of the tube to a separating funnel and add 1 ml 0.0015% dithizone solution. Stopper the funnel and shake for 1 min. Discard the aqueous phase. Add 5 ml water and shake the funnel for 10 sec. Drain and discard the aqueous phase. Add 5 ml metal-free 0.4 *M* ammonia and shake the funnel for 5 sec. Drain and discard the aqueous phase. Compare the intensity of the color in the organic phase with the colors of the standards.

Preparation of Standards. Mercury standards do not perform well in field use, because mercuric dithizonate is thermally decomposed when exposed to heat and sunlight. An azo dye—orange II—closely resembles the color of mercury dithizonate. This dye is incorporated in a thermosetting resin to provide stable and sturdy artificial plastic standards (see Section 5.2).

PREPARATION OF COMPLEXING AGENT MIXTURE. Mix 5 g hydrazine sulfate and 10 g EDTA.

PREPARATION OF 0.0015% DITHIZONE. Dissolve 0.01 g pure dithizone in 100 ml reagent-grade chloroform. Mix 15 ml 0.01% dithizone solution in chloroform with 85 ml soltrol.

5.3.17. Estimation of Niobium in Rock Samples

The following procedure is applicable to rocks containing as little as 50 ppm niobium.

The sample is fused with potassium pyrosulfate and the fused mass extracted with hot tartaric acid. Strong hydrochloric acid is added to an aliquot of the tartaric acid solution, and thiocyanate is brought into reaction in an aqueous medium. The niobium thiocyanate is then extracted into diethyl ether, which is then treated with stannous chloride dissolved in hydrochloric acid. The color of the niobium thiocyanate in the organic phase is calibrated against standard solutions. Interference is caused by the presence of large amounts of molybdenum and tungsten.

The extraction of the niobium through fusion with pyrosulfate is effective on columbite, bauxite, monazite, and nephelene syenite, but less effective on euxenite and wolframite.

PROCEDURE (62). Crush the sample to −80 mesh or finer, scoop 0.2 g into a test tube, and add 4 g potassium pyrosulfate. Fuse the mixture for 15 min. After cooling, add 10 ml 1 *M* tartaric acid and break up the fused mass with a glass stirring rod. Clear the solution by heating the tube for 2–3 min in a boiling water bath, and transfer a 1-ml clear aliquot into a separating funnel containing 5 ml 15% tartaric acid solution dissolved in 9 *M* hydrochloric acid. Add 5 ml freshly prepared 20% aqueous ammonium thiocyanate solution and mix. Extract with 5 ml diethyl ether. After separation of the phases, drain and discard the aqueous layer. Add 5 ml 10% stannous chloride in conc. hydrochloric acid, shake for 20 sec, and drain off the aqueous phase. Repeat this step. Transfer the organic phase to a 10-ml volumetric flask, add acetone to the mark, and mix. Compare the color of the sample visually with that of a standard niobium solution prepared according to the same procedure.

PREPARATION OF STANDARD NIOBIUM SOLUTION (200 ppm). Fuse 0.0286 g niobium pentoxide with 1.5 g potassium pyrosulfate in a porcelain crucible. Dissolve the mass in 1 *M* tartaric acid, and bring volume to 100 ml with 1 *M* tartaric acid.

5.3.18. Rapid Determination of Sulfur in Rocks (63)

Sulfur content can be estimated in rocks turbidimetrically as the barium sulfate formed after attack with aqua regia. This method does not require complex instrumentation. Reproducibility of the turbidity is achieved by adding barium chloride as a solid rather than as a solution. To a solution made by attacking a 100-mg sample with aqua regia in a test tube, is added calcium chloride. The solution is boiled and made ammoniacal and the precipitate centrifuged down. To an aliquot of the clear supernatant fluid a small spike of sulfate is added, and the resulting sulfate is precipitated by adding powdered barium chloride. The turbidity is measured and compared with that of a reference solution prepared from pure sodium sulfate. The result obtained represents all of the sulfide that has been converted to sulfate plus all of the acid-soluble sulfate. Using hydrochloric acid instead of aqua regia enables the estimation of the acid-soluble sulfate content.

PROCEDURE. Crush the sample to −80 mesh or finer and scoop 0.1 g into an 18 × 150 mm test tube. Prepare a standard by adding 2 ml of the standard sulfate solution to an additional test tube. Another tube serves as a blank. Add 0.2 ml conc. nitric acid to all the tubes. Boil for several seconds, and after cooling, add 0.4 ml conc. hydrochloric acid and heat again. Add 1 ml 5% calcium chloride solution and 5 ml water and bring to a boil over a burner. Keep at or near boiling for 30 sec. Add 2 ml 50% ammonium hydroxide and 5 ml water and agitate briefly. Add 2 ml 50% ammonium hydroxide to the blank and standard. Add 10 ml water to the blank and 8 ml to the standard test tube. Centrifuge all the samples. To each of a second set of dry test tubes add 0.1 ml of the standard sulfate solution to serve as a spike and transfer 5 ml of the supernatant fluid from the previously prepared set to the second set. Set the spectrophotometer to zero absorbance at 650 nm using a test tube containing water. Record the absorbance of each test tube of the second set. These readings provide a means of correcting for any color, cloudiness, or tube variations, and should be subtracted later. Scoop ≈30 mg powdered barium chloride into to each tube without agitation. Note the time and then agitate each tube for a few seconds. Two minutes after the first tube has been agitated start reading the absorbance.

Calculate sulfur concentrations by subtracting the absorbance readings prior to precipitation from those after precipitation for each tube and then subtracting the value for the black from those for the samples and the standard. The standard corresponds to 0.32% sulfur and its absorbance is ~0.400.

PREPARATION OF STANDARD SULFATE SOLUTION. Dissolve 177.5 mg sodium sulfate in 250 ml water. One milliliter contains 0.4 mg of sulfate or 0.16 mg of sulfur.

5.3.19. Estimation of Sulfate in Soil and Sediment (64)

The screening test for those sulfates that are soluble in hot, dilute acid is based on comparison of the turbidity of the samples with those of standards. In this test barium and strontium sulfate in the geochemical sample remain undissolved, whereas lead and calcium sulfate may be partially dissolved.

PROCEDURE. Crush the sample to −80 mesh or finer and scoop 1 g into a 16 × 150 mm test tube. Digest for 1 hour on a sand tray with 5 ml 1 *M* hydrochloric acid. After settling has occurred, take 1 ml of clear supernatant fluid into a 18 × 180 mm test tube calibrated at 10 ml and dilute to the mark with acid–salt solution. Scoop 250 mg barium chloride crystals into the tube and shake for ~30 sec. Compare with standards against a dark background. The sulfate content in ppm is calculated using the formula 5000 × (mg sulfate in matching standard).

PREPARATION OF STANDARDS. Add 0, 0.05, 0.1, 0.2, 0.3, 0.4, 0.5, 0.75, 1.0, 1.25, 1.5, and 2.0 mg sulfate to 12 test tubes. Add 1 ml 5% gum acacia solution and dilute to 10 ml with acid–salt solution. Add 250 mg barium chloride crystals and dissolve by shaking for ~30 sec.

PREPARATION OF STANDARD SULFATE SOLUTION (500 ppm). Dissolve 907 mg potassium sulfate in water and dilute to 1 liter.

PREPARATION OF ACID–SALT SOLUTION. Dissolve 48 g sodium chloride in water, add 4 ml conc. hydrochloric acid (s.g. 1.18) and 125 ml orthophosphoric acid (s.g. 1.75), and dilute to 1 liter with water.

5.3.20. Estimation of Tungsten in Soils

There are two widely known methods for geochemical field testing of tungsten. In the first (65) the sample is sintered with a carbonate. The sinter is leached with hot water. An aliquot is treated with stannous chloride and potassium thiocyanate in hydrochloric acid. The yellow tungsten thiocyanate formed is extracted into a small volume of isopropyl ether. Heavy metals are precipitated as hydroxides during the carbonate fusion procedure. Molybdenum may interfere by forming amber-colored molybdenum thiocyanate, but increasing the hydrochloric acid concentration of the aqueous phase and shaking the tube minimizes this interference. Rhenium interferes, by reacting similarly to tungsten, but because it is extremely rare, can be ignored.

PROCEDURE. Crush the sample to −80 mesh or finer, scoop 0.25 g into a 16 × 150 mm test tube along with 1.25 g flux, and sinter. Cool the tube, add 5 ml water, and place in a boiling water bath. Cool and transfer 1 ml of the supernatant to another tube calibrated at 5 ml. Adjust to 5 ml with deionized water. Add 4 ml 10% stannous chloride solution and 0.5 ml 25% potassium thiocyanate solution, shaking the tube after each addition. Place the tube into a boiling water bath for ~5 min. Cool, add another 0.5 ml 25% potassium thiocyanate solution and 0.5 ml isopropyl ether. Stopper the tube and shake. Let settle. Compare the color in the ether layer with the color in the standards.

PREPARATION OF STANDARDS. Add to six 16 × 150 mm tubes the following amounts of the 10-ppm tungsten standard solution: 0, 0.1, 0.4, 0.8, 1.5, and 3.0 ml, corresponding to 0, 1, 4, 8, 15, and 30 μg tungsten. Dilute the standards to 5 ml and proceed with addition of reagents as above.

PREPARATION OF TUNGSTEN STANDARD SOLUTION (100 ppm). Dissolve 0.09 g sodium tungstate dihydrate in water and dilute to 500 ml.

PREPARATION OF FLUX. Mix by weight 5 parts of sodium carbonate, 4 parts of sodium chloride, and 1 part of potassium nitrate, and grind the mixture to minus 80 mesh.

In the alternative method (45) the sample is decomposed by fusing it with a sodium carbonate–sodium chloride–potassium nitrate mixture, and the quantitation is based on formation of the tungsten dithiol complex. Potentially interfering molybdenum is reduced with stannous chloride before addition of the dithiol.

PROCEDURE. Crush the sample to minus 80 mesh or finer and scoop 0.25 g of the sample into a nickel crucible containing 1.25 g of fusion mixture. Mix and fuse until frothing has ceased. Cool, add 2 ml water, and leave overnight. Transfer to a 16 × 150 mm test tube calibrated at 5 ml, and adjust to the mark with water washings from the crucible. Break the melt by boiling it on a sand tray. After cooling, transfer 0.5 ml of the clear supernatant to a 16 × 150 mm test tube calibrated at 5 ml and containing 5 ml 2% stannous chloride solution. Heat in a boiling water bath for 4 min and add 0.5 ml dithiol solution. Continue to heat for ~30 min, shaking the tube periodically. Remove from the water bath when the ester phase has been reduced to a small globule. Add 0.5 ml white spirits and shake gently to dissolve the globule. Compare with a standard series.

PREPARATION OF STANDARDS. To twelve 16 × 150 mm test tubes calibrated at 5 ml and containing 5 ml 2% stannous chloride solution add 0, 0.2, 0.4, 0.6, 0.8, and 1.0 μg tungsten, using a standard solution of 1 ppm tungsten, and 2.0, 3.0, 4.0, 5.0, 7.0, and 10.0 μg tungsten, using a 10-ppm tungsten solution.

PREPARATION OF STANDARD TUNGSTEN SOLUTION (100 ppm). Dissolve 90 mg sodium tungstate in water and dilute to 500 ml.

PREPARATION OF FUSION MIXTURE. Mix 500 g sodium carbonate, 500 g sodium chloride, and 100 g potassium nitrate.

PREPARATION OF DITHIOL SOLUTION. Weigh 0.5 g zinc dithiol into a 50-ml cylinder, add 1 ml conc. hydrochloric acid, stopper, and shake for 10 sec. Dilute to 50 ml with isoamyl acetate and mix well.

PREPARATION OF WHITE SPIRITS. Purify by shaking 1 liter spirits with 10 ml sulfuric acid in a separating funnel. Discard the acid extract, shake the spirits with 10 g lime, and filter.

5.3.21. Field Determination of Vanadium in Geochemical Samples

Rapid field determinations of vanadium in preliminary prospecting work are based either on the formation of yellow phosphotungstovanadic acid or on the reaction of trivalent vanadium with thiocyanate and extraction of the complex with ether.

In the first method the sample is decomposed by fusion with potassium bisulfate and digestion with hydrochloric or nitric acid. Most of the common reducing ions are oxidized in this procedure, especially in the next step, when a portion of the sample solution is boiled with nitric acid. Phosphoric acid and sodium tungstate are then added to form the yellow heteropolyacid, and the solution is heated in order to reach equilibrium quickly. Molybdenum tends to interfere, but only in relatively high concentrations. Iron interference is masked with orthophosphoric acid by formation of a stable, colorless complex. Ions that are themselves colored (chromium, cobalt, copper, and nickel) interfere, but of these only chromium is likely to occur in concentrations, that would cause interference.

PROCEDURE. Crush the sample to −80 mesh or finer and scoop 0.1 g into a 16 × 150 mm test tube. Add ~0.5 g potassium bisulfate, mix, and fuse until the flux is molten. Cool the tube, add 3 ml 4 *M* nitric acid, and place the tube in boiling water bath. Cool the tube again, dilute the sample to 10 ml with water, mix, and allow the residue to settle. Transfer a 2-ml aliquot to a 16 × 150 mm test tube calibrated at 10 ml. Add 1 ml conc. nitric acid and boil for ~5 sec. Add 0.3 ml conc. phosphoric acid and 2 ml 5% aqueous sodium tungstate solution. Dilute the solution with water up to the 10-ml mark and heat the tube again in a boiling water bath for 10 min. After cooling the tube, visually compare the color obtained with the color of the most nearly matching vanadium standard solution.

PREPARATION OF STANDARDS. To seven 16 × 150 mm test tubes calibrated at 10 ml add respectively 0, 3, 5, 10, 20, 40, and 80 μg vanadium using a standard solution of 100 ppm vanadium. Proceed with the addition of reagents as above.

PREPARATION OF STANDARD VANADIUM SOLUTION (100 ppm). Dissolve 0.2295 g ammonium metavanadate in 30 ml 4 *M* nitric acid and 100 ml of water. Dilute to 1 liter with water.

Field determination of vanadium by the second method has been tested primarily in the southern Black Hills of South Dakota. In this method, the vanadium in the rock is first reduced to trivalent vanadium with stannous chloride in the presence of citrate and EDTA. Trivalent vanadium then forms a yellow, ether-extractable complex ion with thiocyanate.

The metals potentially interfering in this reaction are molybdenum, tungsten, niobium, and uranium. They all form ether-extractable colored thiocyanate compounds, but the color of the molybdenum thiocyanate fades after half an hour. The vanadium estimation is preferentially carried out at this stage. Citrate masks tungsten quite effectively, and the solubility of uranium thiocyanate is so low that it is not considered a serious interference. Niobium, however, interferes seriously if present.

PROCEDURE (1). Crush the sample to −80 mesh or finer and scoop 0.1 g into a 16 × 150 mm test tube. Add 1 ml 9 *M* sulfuric acid and heat until the mixture starts to boil. Cool the tube, and add 4 ml 10% aqueous sodium citrate solution, 2 ml 2% disodium EDTA solution, and 3 ml of a 15% solution of stannous chloride dissolved in conc. hydrochloric acid. Shake the tube after each addition. Add 2 ml freshly prepared 20% potassium thiocyanate solution. Extract with 2 ml ethyl ether. Compare the color intensity of the organic layer with those of the standards.

PREPARATION OF STANDARDS. Add 0, 30, 60, 150, 300, 500, 1000, and 2000 μg vanadium to eight 16 × 150 mm test tubes using 1000-ppm and 100-ppm vanadium solutions. Proceed with the addition of reagents as above.

PREPARATION OF STANDARD VANADIUM SOLUTION (1000 ppm). Dissolve 1.685 g pure vanadium pentoxide in 20 ml 2.5 *M* sodium hydroxide. Neutralize with 9 *M* sulfuric acid, add 28 ml conc. sulfuric acid, and dilute to 1 liter with water.

5.3.22. Estimation of Phosphorus in Soils and Rocks

The formation of yellow phosphomolybdovanadic acid is the basis of the field evaluation of phosphate (2). The sample is fused with potassium bisulfate and the melt digested with nitric acid. The yellow color formed after the addition of the vanadate–molybdate reagent is compared visually with the colors of the standard solutions. Acid concentration should be kept below 1.6 *M*; otherwise it will slow the color development a great deal. Large concentrations of bismuth (III), thorium (IV), arsenic (V), chloride, and fluoride ions hinder the color development.

PROCEDURE. Crush the sample to −80 mesh or finer and scoop 0.1 g into a 16 × 150 mm test tube. Add about 0.5 g potassium bisulfate, mix, and fuse until 1 minute after the flux has become molten. After cooling, add 2 ml 4 *M* nitric acid and heat the tube in boiling water until the melt disintegrates. Dilute to 10 ml with water and transfer a 5-ml aliquot to a 16 × 150 mm test tube. Add 2 ml vanadate–molybdate reagent and dilute to 10 ml with water. Stopper, shake, and let stand for 30 min. Compare result with a standard series.

PREPARATION OF STANDARD SERIES. Add to fourteen 19 × 150 mm test tubes 0, 2, 4, 6, 8, 10, 15, 20, 30, 40, 50, 60, 80, and 100 μg phosphorus using a 50-ppm standard phosphorus solution to each tube. Dilute to 10 ml with water, stopper, shake, and allow to stand for 30 min.

PREPARATION OF THE STANDARD PHOSPHORUS SOLUTION (50 ppm). Dissolve 0.5782 g sodium phosphate ($Na_2HPO_4 \cdot 12H_2O$) in 800 ml water. Add 20 ml conc. nitric acid and dilute to 1 liter with water.

PREPARATION OF VANADATE–MOLYBDATE REAGENT. Dissolve 1.25 g ammonium metavanadate in 400 ml 8 *M* nitric acid (solution a). Dissolve 50 g ammonium molybdate in 500 ml water (solution b). Combine solutions a and b and dilute to 1 liter.

5.3.23. Estimation of Uranium in Geochemical Samples (1)

Most uranyl compounds fluoresce in the wavelength range of 530–600 nm. Many elements enhance the fluorescence (e.g., fluoride) or suppress it. In sampling a minute quantity (5 mg), this foreign element effect can be neglected. Increasing uranium concentrations are signaled by increasing intensities of fluorescence. Both standard and sample discs should be kept in a dessicator.

PROCEDURE. Weigh 5 mg of the sample into a platinum crucible. Add 3 g fusion mixture and heat until molten. Cool the mixture and form it into a disc. Compare it with a standard series under ultraviolet light using a yellow filter.

PREPARATION OF STANDARDS. Add 0, 0.01, 0.02, 0.03, 0.04, 0.05, 0.075, 0.1, 0.2, 0.3, 0.4, and 0.5 μg uranium to twelve platinum crucibles using a solution of 0.1 ppm uranium. Evaporate to dryness and add 3 g fusion mixture. Proceed as described above.

PREPARATION OF STANDARD URANIUM SOLUTION (1000 ppm). Dissolve 118 mg black uranium oxide in 25 ml nitric acid and dilute to 100 ml with water. Prepare additional standards by successive dilutions.

PREPARATION OF FUSION MIXTURE. Mix 455 g potassium carbonate, 455 g sodium carbonate, and 90 g sodium fluoride.

5.3.24. Field Test for the Detection of Cadmium in Prospecting Samples

Relatively high cadmium and relatively low zinc values (deviating significantly from the characteristic ratio of 1 : 500) indicate the separation of the originally coexisting metal ions in sulfide deposits by weathering in contact with oxidizing acidic ground waters. Due to the different mobilities of the two ions, zinc may be transported for considerable distances from the original sulfide source, whereas cadmium tends to remain fairly close to the original sulfide source. An estimation of <70 ppm cadmium in the ore bodies and a relatively low concentration of zinc in the sulfide deposit indicate leached massive sphalerite ore in an oxidized environment.

A simple field method (66) differentiates ores containing cadmium concentrations greater than 70 ppm from ores with lower cadmium contents. The method is used to trace dislocated secondary zinc deposits oxidized by acidic ground waters.

PROCEDURE. Crush the sample to −80 mesh or finer and scoop 90 mg of the dust into an 18 × 150 mm test tube marked at 10 ml. Add 50% hydrochloric acid to the mark, shake the tube, and let stand for 30 min. Wet a piece of Whatman no. 3 filter paper with the glass stopper from a bottle of a saturated alcoholic solution of diphenylcarbazide to form a circular spot, and dry the spot. Place a drop of the sample solution on the dried circular spot and neutralize the spot by exposing the filter paper to conc. ammonia vapor until a violet coloration appears. Wash away the red color of the reagent itself by soaking the paper for 2 min in warm water at 60°C. The remaining violet color indicates the approximate concentration of cadmium in the sample. Ore samples containing <20 ppm cadmium give no reaction, and samples containing 20–70 ppm cadmium give a very weak color, but is considered an uncertain result because of the inaccuracies and approximations inherent in this method. Ores containing >70 ppm cadmium give a distinct positive reaction.

REFERENCES

1. R. E. Stanton, *Rapid Methods of Trace Analysis for Geochemical Applications,* Edward Arnold Ltd., London (1966).
2. F. N. Ward, H. W. Lakin, and F. C. Canney, "Analytical Methods Used in Geochemical Exploration by the U. S. Geological Survey," *Geological Survey Bulletin* **1152** (1968).
3. H. E. Hawkes, *Econ. Geol.* **58,** 579 (1963).
4. F. Feigl and V. Anger, *Spot Tests in Inorganic Analysis,* 6th ed., Elsevier, Amsterdam, London, New York (1972).
5. F. Feigl and L. S. Passes de Azveda, *Chemist–Analyst* **54,** 106 (1965).
6. A. C. Vournasos, *Ber.* **2269** (1910).
7. F. Feigl and E. Jungreis, unpublished studies; see also ref. 4, p. 533.
8. F. Schwartz, *Berg u. Hütt. Jahrbuch Sprechsaal* **1953/1** (1930).
9. A. S. Komarovsky and N. S. Poluektoff, *Mikrochemie* **14,** 317 (1934).
10. F. Schwartz, *Mikrochim. Acta* **3,** 126 (1938).
11. L. Ben Dor and E. Jungreis, *Mikrochim. Acta,* 100 (1964).
12. F. Feigl and A. Caldas, *Z. Anal. Chem.* **231,** 261 (1967).
13. F. Feigl and H. A. Suter, *Ind. Eng. Chem. Anal. Ed.* **14,** 840 (1942).
14. E. Jungreis, *Project Report CS/780-3,* Anglo American Research Laboratories, Johannesburg (1980).
15. K. Agte, H. Becker-Rose, and G. Heyne, *Z. Angew. Chem.* **38,** 1124 (1925).
16. F. Feigl and A. Caldas, *Mikrochim. Acta,* 1311 (1956).
17. L. Ben Dor and G. Markovitz, *Mikrochim. Acta,* 957 (1967).
18. E. G. T. Nottes, *Erzmetall.* **29,** 337 (1976).
19. E. R. Rose, *Geol. Survey Can.* **75-16** (1976).
20. A. A. Levinson, *Introduction to Exploration Geochemistry,* Applied Publishing, Wilmette (1980).
21. D. B. Hawkins, F. C. Canney, and F. N. Ward, *Econ. Geol.* **54,** 738 (1959).
22. H. Almond and H. T. Morris, *Econ. Geol.* **46,** 608 (1951).
23. L. C. Huff, *Econ. Geol.* **46,** 524 (1955).
24. H. Bloom and H. E. Crowe, *U. S. Geol. Survey—Open file report* (1953).
25. H. Almond, *U. S. Geol. Survey Bull.* **1036-A,** 1 (1955).

26. F. C. Canney and D. B. Hawkins, *Econ. Geol.* **53,** 877 (1961).
27. R. Delwault, *J. Geochem. Explor.* **8,** 537 (1977).
28. D. Peachey, D. C. Cooper, and B. Vickers, *Rep. Inst. Geol. Sci.* (*U. K.*) **80/1,** 35 (1980).
29. H. Bloom, *Econ. Geol.* **50,** 533 (1958).
30. H. Almond, *Anal. Chem.* **35,** 166 (1953).
31. R. E. Stanton and A. J. MacDonald, *Trans. Inst. Min. Metall.* **11,** 511 (1961).
32. R. E. Stanton and J. A. Coope, *Trans. Inst. Min. Metall.* **68,** 9 (1958).
33. F. N. Ward and H. W. Lakin, *Anal. Chem.* **26,** 1168 (1954).
34. G. A. Wood, Ph.D. Thesis, London University, London (1957).
35. L. N. Lapin and V. O. Gein, *Anal. Abstr.* **4,** 3630 (1957).
36. R. E. Stanton and A. J. MacDonald, *Trans. Inst. Min. Metall.* **71,** 517 (1961).
37. H. Almond, *Anal. Chem.* **25,** 1766 (1953).
38. R. E. Stanton, *Econ. Geol.* **59,** 1599 (1964).
39. N. J. Marshall, *J. Geochem. Explor.* **10,** 307 (1978).
40. H. Bode and K. Hachmann, *Z. Anal. Chem.* **241,** 18 (1968).
41. E. Jungreis, *Project Report CS/780-4,* Anglo American Research Laboratories, Johannesburg (1980).
42. H. Schweppe, *Z. Anal. Chem.* **244,** 312 (1969).
43. F. N. Ward, *Anal. Chem.* **23,** 788 (1951).
44. W. A. Tananaev and G. A. Panchenko, *Chem. Abstr.* **24,** 566 (1930).
45. A. A. North, *Analyst* (*Lond.*) **81,** 660 (1956).
46. R. E. Stanton and A. J. Hardwick, *Analyst* (*Lond.*) **92,** 387 (1967).
47. N. J. Marshall, *Econ. Geol.* **59,** 142 (1964).
48. W. E. Baker, *Bull. Austral. Inst. Min. Metall.* **214,** 125 (1965).
49. S. J. Hoffman and M. J. Waskett-Myers, *J. Geochem. Explor.* **3,** 61 (1973).
50. B. F. Quin and R. R. Brooks, *Anal. Chem. Acta* **74,** 75 (1975).
51. G. A. Wood and R. E. Stanton, *Trans. Inst. Min. Metall.* **66,** 331 (1956).
52. E. B. Sandell, *Colorimetric Determination of Traces of Metals,* 3rd ed., Interscience, New York (1959), p. 1054.
53. H. H. Willard and L. H. Greathouse, *Am. Chem. Soc. J.* **39,** 2366 (1917).
54. H. Almond, *U. S. Geol. Survey—Open file report* (1953).
55. G. G. Karanovich, *Anal. Abstr.* **4,** 1448 (1957).
56. E. C. Hunt, R. E. Stanton, and R. A. Wells, *Trans. Inst. Min. Metall.* **69,** 361 (1959–60).
57. G. A. Wood, *Technical Communication No. 11,* Imperial College of Science and Technology Geochemical Prospecting Research Center, London (1956).
58. A. P. Marranzino and F. N. Ward, *Conference on Analytical Chemistry and Applied Spectroscopy,* Pittsburgh, Pennsylvania, March 3–7 (1958).
59. R. E. Stanton and A. J. MacDonald, *Trans. Inst. Min. Metall.* **71,** 27, (1961).
60. F. N. Ward and H. E. Crowe, *U. S. Geol. Survey Bull.* **1036 I,** 173 (1956).
61. F. N. Ward and E. H. Bailey, *Am. Inst. Mining Engineers Trans.* **217,** 343 (1960).
62. F. N. Ward and A. P. Marranzino, *Anal. Chem.* **27,** 1235 (1955).

63. L. Shapiro, *J. Res. U. S. Geol. Survey* **1,** 81 (1973).
64. N. W. Scott, *Standard Methods of Chemical Analysis,* 5th ed., vol. II, The Technical Press, London (1955), p. 2091.
65. F. N. Ward, *U. S. Geol. Survey Circ.* **119,** 4 (1951).
66. E. Jungreis, *Project Report CS/780,* Anglo American Research Laboratories, Johannesburg (1980).

CHAPTER

6

APPLICATION OF SPOT TESTS IN AIR POLLUTION CONTROL

6.1. GENERAL

The various industrial operations of the civilized world emit dusts, gases, vapors, and mists. All of these, in combination with naturally occurring airborne materials, form the basis of air pollution. Meteorological and topographical factors can aggravate the harmful effect of pollution on human health, which effect depends on the nature of the pollutant, its concentration, and the time for which the individual has been exposed to it.

To protect the workers' well-being, the concentrations of air pollutants in industrial areas must not exceed certain admissible levels. Although it is impossible to keep air in industrial areas completely pure, the concentrations of contaminants can be reduced to acceptable levels.

Analysis of the ambient air must be carried out to determine the minute quantities of harmful materials with the greatest reliability and, most importantly, in the shortest amount of time. The time consumed in ordinary laboratory methods (often requiring hours for analytical results) may endanger those working in a highly polluted industrial area. Therefore, spot test analytical methods in the form of directly read colorimetric indicators have come into use for the semiquantitative estimation of contaminant concentrations in air.

The mining industry once urgently required a method for obtaining reliable results on the presence of a harmful material that would avoid lengthy and tedious procedures, namely, a simple device for early warning of the presence of lethal carbon monoxide. The first spot test using prepared liquid reagents for the estimation of carbon monoxide in ambient air was patented in 1919 (1). The test apparatus consisted of glass tubes containing a mixture of iodine pentoxide and fuming sulfuric acid through which the air sample was sucked. The color of the reagent mixture changed to green in response to carbon monoxide in the air. The extent of the green coloration served as an indication of the carbon monoxide concentration.

Filter paper strips impregnated with chemical reagents have also been employed to detect and determine the concentrations of harmful gases. The well-known Gutzeit method was an early example for the assessment of arsine. In this method, the paper strips are simply impregnated with mercuric bromide and hung in the contaminated air. The accuracy of such procedures is naturally limited by the rather indefinite volume of the air sample. Temperature variations and air current also influence greatly the degree of the color change. Passing a measured volume of air through a defined area of the reagent paper produces more exact results. The color change is then either evaluated by comparison with color charts or recorded photoelectrically.

Glass indicating tubes containing solid reagents are convenient and compact direct-reading devices. Such tubes have recently been made available by several companies (Bacharach Industrial Instruments Co., Blawknox, Pennsylvania; Davies Engineering Equipment Inc., Baltimore, Maryland; Union Industrial Equipment Corp., Culver City, California; Mine Safety Appliances Co., Pittsburgh, Pennsylvania; Acme Protection Equipment Corp., Emerson, New Jersey; and Drägerwerk AG, Lübeck, F. R. G.). They are designed to check air quality in industrial atmospheres and to indicate toxic hazards in working areas. The operating procedure is simple, rapid, and convenient, and therefore can easily be performed by unskilled personnel. There are, however, some inherent potential errors, and the sampling procedure should be supervised constantly by trained people. The National Institute for Occupational Safety and Health (NIOSH) is currently testing the detector tubes to see if they meet necessary quality standards (2).

The simple operating procedure for the detecting tubes is as follows: The two sealed ends of the tube are broken and the tube is placed in the holder. The recommended air volume is drawn through the tube with a calibrated bellows pump, and the color change produced in the tube is compared visually against a set of standard colors. The length over which the color change has occurred in the tube is an alternative indication of the amount of airborne contaminant. Using the length of the color change reduces the errors created by variation of color perception among individuals. The reliability requirement of the NIOSH is a maximum of 35% relative error at a concentration of one-half of the threshold limit value (3).

The supporting material for the reagent system in the tube may be silver gel, alumina, ground glass, pumice, or resin. This supporting material is impregnated with the chromogen. The chemical reaction should be selective for the air contaminant being screened for, should produce a color in strong contrast with that of the unexposed reagent, and the color should not fade for at least 1 hour. The purity of the reagents in the tube should be very carefully controlled, because trace impurities sometimes serve as catalysts and change the reaction velocity. The tubes should be assembled in clean air and the dimensions of the glass tube should be carefully standardized to control the flow rate.

Some indicator tubes are composed of multiple reagent layers separated by inert materials. When a nonspecific chromogen is applied for the estimation of a given pollutant, the potentially interfering gases must first be screened out. The characterization of such interference is termed "cross-sensitivity," which is defined as the ratio of the measured value of the signal of the interfering substance to the measured value of the substance to be determined of the same concentration. Multiple-layer tubes are so designed as to reduce cross-sensitivity. The first layer consists of a precleansing chemical to filter and selectively retain interfering components. The gas to be measured is not affected by such a prelayer, and reacts as intended with the indicating layer (4). For example, in some of the carbon monoxide indicator tubes, the first reagent layer sequesters interfering hydrocarbons and nitrogen oxides. In carbon-disulfide-determining tubes, hydrogen sulfide is removed in the prelayer, and in hydrogen cyanide tubes, hydrogen chloride or sulfur dioxide are eliminated. In some other tubes the pollutant to be analyzed is first transformed in a preliminary reaction into another, more easily indicated compound. For example, the determination of trichloroethylene proceeds in two stages: In the first reaction it is oxidized to free chlorine, and in the second reaction the chlorine reacts with an appropriate reagent. The complex nature of such multiple-layer tubes means that they generally have shorter shelf-lives because of slow diffusion of the chemicals between the separate layers despite the presence of mechanisms for their isolation.

Detector tubes naturally cannot be stored for an unlimited time. A 2-year shelf-life is satisfactory for practical purposes. Small variations in impurities such as moisture content can have a very significant effect on shelf-life. Storage temperature, of course, has a great effect on shelf-life, and it is therefore recommended that the tubes be stored under refrigeration. Storage of the tubes at an ambient temperature of 30°C or higher has a definite deteriorating effect, as does storage in direct sunlight (5). The shelf-life of a tube is generally estimated from an accelerated test performed at higher temperatures. According to such estimations the shelf-life of most tubes is more than 2 years at 25°C but calibration accuracy is maintained for only 1 week at 100°C. The data plot almost on a straight line when the logarithm of the shelf-life is plotted against the reciprocal of the absolute temperature on a linear scale.

The best accuracy that can be expected from the indicator tubes is on the order of ±20%, but a total error not in excess of 25% is considered excellent, and a total error not in excess of 50% is still acceptable. [The total error is defined as the sum of the absolute error and twice the standard deviation (6)].

The calibration scales have mostly been obtained empirically, and are logarithmic with respect to concentration. Although the test gas is usually completely absorbed, equilibrium is not achieved between the reacting gas and the chromogen, due to the relatively high flow rate and the brevity of the sampling period (7). The length of the stain is determined by the kinetic rate at which the gas either reacts with the indicating chemical or is adsorbed on the silica gel. The following simplified equation shows that the stain length is proportional to the logarithm of the product of the gas concentration and the sample volume:

$$L/H = \ln (CV) + \ln (K/H)$$

where L = stain length in cm,
C = gas concentration in ppm,
V = air sample volume in cm^3,
K = a constant for a given type of indicator tube and test gas, and
H = a mass transfer proportionality factor known as the height of mass transfer unit and given in centimeters.

Temperature is also a very important factor for tube calibration. Those indicator tubes that indicate the pollutant concentration by a change in color are most sensitive to temperature changes. Accordingly, some types of carbon monoxide tubes require correction by a factor of 2 for each deviation of 10°C from the standard calibration conditions.

The detector tube method is also applicable to monitoring of the atmosphere throughout an 8-hour shift. However, no detector tube has yet been successfully developed for the estimation of particulate matter (8).

Most of the criticisms concerning the accuracy and precision of portable gas detectors have focused on their operation at high relative humidity levels. Although the portable units gave semiquantitative results similar in precision and accuracy to those obtained with high-accuracy standard detectors and were unaffected by humidity in the 5–50% range (9), their sensitivities deteriorated at higher humidity levels (10): The observed length of the stain and hence the indicated gas concentration were affected considerably at relative humidities in the 50–80% range.

A drastic color change occurs in the course of a reaction in an ideal reagent system. The concentrations of pollutant material to be ascertained by the detector tubes are sometimes extremely low. To insure a quick analysis time, a volume of a few hundred cubic centimeters should be adequate for analysis. Therefore, in most cases, only a very small amount of gas is available for the reaction within the tube.

In most detector tubes, the total length of region of color change and the concentration of the contaminant gas are evaluated according to the following relationship (8):

$$\frac{G}{C} = l - \frac{v}{A}(1 - e^{-lA/v})$$

where l = the length of the color change in mm,

v = the rate of the air sample flow in mm/sec,

A = a constant, having reciprocal time dimensions, that is, a measure of the rate at which the reagent system reacts with the gas components to be determined within the range of one layer element,

G = the total amount of gas entering into the reaction and absorbed by the tube at 20°C and 1013 mbar in μl, and

C = the absorption capacity of one layer element (length = 1 mm) at 20°C and 1013 mbar in μl.

6.2. EVALUATION OF SELECTED TOXIC AGENTS USING DETECTOR TUBES OR IMPREGNATED FILTER PAPER DISCS

6.2.1. Acetaldehyde

Acetaldehyde is a highly flammable, harmful, irritant vapor. Its maximum allowable concentration (MAC) is 200 ppm. The vapor causes irritation to the eyes, headache, and drowsiness. Repeated inhalation of the vapor can cause anemia, delirium, hallucinations, and loss of senses similar to that found in chronic alcoholism.

The indicator tube is usually constructed of a drying prelayer and an orange sulfochromic acid indicating layer. Acetaldehyde vapors reduce hexavalent chromium to trivalent ion, changing the color from orange to brownish green.

The ranges of measurement of various devices are 100–1000 ppm acetaldehyde (20°C, 1013 mbar) (Drägerwerk AG) and 25–1000 ppm (Mine Safety Appliances Co.). The volume of the gas sample taken is 2 liters.

Other reducing organic compounds interfere with the test, but the reaction is less sensitive to them.

6.2.2. Acetone

The narcotic action of acetone is rapid, but its toxicity is low. Concentrations of up to 40,000 ppm may not be inhaled for more than a few minutes, since acetone causes acute irritation to the eyes and throat. However, prolonged exposure to low concentrations does not seem to cause harmful results. The MAC of acetone is 1000 ppm.

A detector tube for acetone typically contains dinitrophenyl hydrazine, which, on contact with acetone, becomes yellow hydrazone:

$$CH_3{-}\underset{\underset{\displaystyle O}{\|}}{C}{-}CH_3 + NH_2{-}NH{-}C_6H_3(NO_2) \rightarrow$$

$$(CH_3)_2C{=}N{-}NH{-}C_6H_3(NO_2)_2$$

The ranges of measurement of some available tubes are 100–12,000 ppm (Drägerwerk AG) and 500–50,000 ppm (Union Industrial Equipment Corp.). The volume of the gas sample is typically 1 liter.

Methylethyl acetone reacts exactly the same way as acetone, and with the same sensitivity.

6.2.3. Acrylonitrile (Vinyl Cyanide)

Acrylonitrile is a highly flammable, extremely harmful poison. Its vapor may cause dizziness, nausea, and unconsciousness. Inhalation of low concentrations of vapor may cause flushing of the face, nausea, giddiness, and jaundice. Its MAC is 20 ppm.

In a preliminary layer of the detection tube, the acrylonitrile is first cleaved to hydrocyanic acid with chromate. In the indicating layer, the weak hydrocyanic acid is transformed with mercuric chloride to strong hydrochloric acid. The latter reacts with methyl red indicator, producing a red reaction product.

Range of Measurement	Manufacturer
0.5–1 ppm	Drägerwerk AG
1–20 ppm	
5–30 ppm	
5–150 ppm	Mine Safety Appliances Co.
10–500 ppm	Union Industrial Equipment Corp.
1000–3500 ppm	

The volume of the gas sample applied varies from 1 to 2 liters.

Acid vapors in the atmosphere such as HCl, HCN, H_2S, and SO_2 naturally interfere with the tests. Aliphatically bound organic nitriles provide a similar reaction, and thus disturb the results.

6.2.4. Aniline

Aniline produces harmful vapors, the inhalation or absorption through the skin of which causes headache, drowsiness, cyanosis, and mental confusion. Prolonged exposure to the vapor affects the nervous system and the blood, causing fatigue and dizziness. The MAC is 5 ppm.

The chromogen in the detecting tube is furfurol, which forms a red product with aniline. The range of measurement is 1–20 ppm aniline. The volume of the gas sample varies between 0.5 and 2.5 liters.

$$C_6H_5{-}NH_2 + \underset{\text{furfurol}}{(CH{=}CH{-}CH{=}C(CHO){-}O)} \longrightarrow (CH{=}CH{-}CH{=}C{-}O){-}CH{=}N^{+}{-}C_6H_5 + C_6H_5{-}NH_2 \longrightarrow C_6H_5{-}HN{-}CH{=}CH{-}CH{=}C(OH){-}CH{=}N{-}C_6H_5$$

The cross-sensitivity of ammonia in this reaction is 0.05. Dimethylaniline does not interfere at all.

6.2.5. Arsine

Arsine is a very harmful gas even at a dilution of 1:200,000. Only a few inhalations may be fatal, with death caused by anoxia or pulmonary edema. The MAC for prolonged exposure is 0.05 ppm.

A typical detector tube for the estimation of arsine includes a pale-blue precleansing layer that retains all the potentially interfering gases such as hydrogen sulfide, hydrogen selenide, and the mercaptans. This prelayer contains cupric ions and precipitates the reducing gases. The indicating layer is loaded with auric gold compound, which oxidizes the arsine. This reaction is indicated by the appearance of violet-gray colloidal elemental gold (11):

$$AsH_3 + 2AuCl_3 + 3H_2O \rightarrow 2Au^0 + 6HCl + H_3AsO_3$$

Range of Measurement	Manufacturer
0.05–3 ppm	Drägerwerk AG
0.025–1 ppm	Mine Safety Appliances Co.
5–160 ppm	Union Industrial Equipment Corp.

The volume of the gas sample varies between 0.1 and 20 liters. Using a 20-liter sample, arsine concentrations as low as 0.005 ppm can be determined.

Phosphine and antimony hydride also reduce gold compounds to metallic gold, but these reactions are not as sensitive.

6.2.6. Benzene

Benzene is a highly flammable, harmful liquid. The inhalation of its vapors causes irritation of mucous membranes, convulsions, excitement, and depression. The resulting respiratory failure can cause death, and repeated inhalation of low concentrations over a considerable period can cause fatal blood disease. The MAC is 25 ppm.

A typical indicator tube (12) for the estimation of benzene includes an acid- and aldehyde-containing prelayer that absorbs potentially interfering toluene, xylene, and naphthalene. The white indicating layer is composed of concentrated sulfuric acid containing formaldehyde. Benzene first condenses with formaldehyde (13), and the resulting diarylmethylene is oxidized to the reddish brown *p*-quinoidal compound.

$$2\,C_6H_6 + HCHO \longrightarrow C_6H_5{-}CH_2{-}C_6H_5 \quad + 2H_2SO_4 \longrightarrow C_6H_5{-}CH{=}C_6H_4{=}O + 3H_2O + 2SO_2$$

Range of Measurement	Manufacturer
0.5–10 ppm	Drägerwerk AG
5–40 ppm	
15–420 ppm	
5–200 ppm	Mine Safety Appliances Co.
10–310 ppm	Union Industrial Equipment Corp.

The volume of the gas sample varies between 0.2 and 4 liters.

Because the potentially interfering aromatic compounds are removed by the precleansing layer, the determination can be considered specific.

6.2.7. Carbon Disulfide

Carbon disulfide vapor is extremely harmful. High concentrations inhaled produce narcotic effects and may cause unconsciousness. Repeated inhalation of the vapor may cause severe damage to the nervous system, including failure of vision, mental disturbance, and paralysis.

The determination of carbon disulfide in indicator tubes is based on the formation of dimethyldithiocarbamate as a product in a reaction between carbon disulfide and dimethylamine. Dimethyldithiocarbamate precipitates cupric ions as a brown, water-insoluble copper salt.

$$\begin{array}{lclcl} & \mathrm{CH_3} & & & \mathrm{S} \\ & \quad\diagdown & & & /\quad\diagdown \\ \mathrm{CS_2}\ + & \qquad\mathrm{NH} & +\ \tfrac{1}{2}\mathrm{Cu^{2+}} \rightarrow & \mathrm{CS} & \qquad\mathrm{Cu}/_2 \\ & \quad / & & \quad\diagdown & \\ & \mathrm{CH_2} & & & \mathrm{N(CH)_3} \end{array}$$

Yellowish-green copper dimethyldithiocarbamate

The tube also contains a cupric-salt-containing precleansing layer to retain potentially interfering hydrogen sulfide.

Range of Measurement	Manufacturer
13–288 ppm	Drägerwerk AG
32–3200 ppm	
5–500 ppm	Mine Safety Appliances Co.
10–200 ppm	Union Industrial Equipment Corp.

An alternative rapid method for rapid screening for low concentrations of carbon disulfide in the air is based on reaction between carbon disulfide and piperazine, and precipitation of the resulting brown cupric salt (14).

PROCEDURE. Prepare the indicator tube reagent by treating 100 g purified silica gel with a mixture of 150 ml 15% ethanolic piperazine solution and 25 ml 0.05% ethanolic cupric acetate solution. To test for carbon disulfide, pass the air sample at 10 ml/sec through the indicator tube containing the dried reagent until the resulting color matches that of the tube containing the reference standard mixture. The volume of air drawn is then used to calculate the carbon disulfide content.

Preparation of the Reference Standard Mixture. Treat 10 g silica gel with 16 ml of a mixture of 8.8 ml 0.005% auramine solution and 3.2 ml 0.005% methylene blue solution diluted with water to 25 ml. Dry the gel.

This method will detect 3.5 mg carbon disulfide per cubic meter of air.

Carbon disulfide content in the air can also be tested under field conditions by sucking the air through an absorbent solution containing a secondary amine and cupric ions (15).

Procedure. Aspirate 500 ml of the atmosphere to be tested through 10 ml absorbent solution. After 10 min, transfer the solution in an uncontaminated atmosphere to a color-comparison tube and compare with standards prepared from solutions of carbon disulfide in ethanol.

Preparation of the Absorbent Solution. Add 5 ml triethylamine and 2 ml diethylamine to 100 ml 0.01% cupric acetate monohydrate in ethanol, and dilute to 500 ml with ethanol.

This method is suitable for the determination of 10–40 ppm carbon disulfide in air.

6.2.8. Carbon Monoxide

Carbon monoxide is an extremely harmful gas that causes unconsciousness due to anoxia resulting from the combination of carbon monoxide with hemoglobin. In lower concentrations it causes headache, throbbing of the temples, and nausea, followed sometimes by collapse. The MAC is 100 ppm.

The vast variety of carbon-monoxide-estimating tubes mostly utilize the selenium-dioxide-catalyzed reduction of iodine pentoxide by carbon monoxide to elemental iodine in the presence of fuming sulfuric acid.

$$5CO + I_2O_5 \xrightarrow{SeO_2} I_2 + 5CO_2$$

The color change produced by this reaction is from white to brownish green.

This reaction is far from specific: Acetylene, petroleum distillates, benzene, and hydrogen sulfide react with iodine pentoxide as well. These potentially interfering materials must be selectively oxidized by chromate in a precleansing layer and thus hindered from entering into the indicating layer. In the presence of high concentrations of hydrocarbons, a special active-charcoal container is attached to absorb these interfering gases.

Range of Measurement	Manufacturer
8–150 ppm	Drägerwerk AG
10–300 ppm	
100–3000 ppm	
0.1–1.2 vol%	
0.3–4 vol%	
0.5–7 vol%	
5–10,000 ppm	Bacharach Industrial Instruments Co.
10–3,000 ppm	Mine Safety Appliances Co.
25–6,000 ppm	Union Industrial Equipment Corp.

The volume of the gas sample varies between 100 ml and 1 liter.

6.2.9. Chlorine

Chlorine is an extremely harmful gas that causes severe lung irritation and damage. The MAC in the air for prolonged exposure is 1 ppm, and that for 1-hour exposure is 4 ppm. Brief exposure to 1000 ppm causes rapid death.

The chlorine indicator tube contains a precleansing and drying agent. The chromogen in the indicating layer is *o*-tolidine, which is oxidized by chlorine into a full-quinoidal yellow compound.

H_3C, CH_3, H_2N–(ring)–(ring)–$NH_3 + Cl_2$ ⟶ HN=(ring)=(ring)=NH, H_3C, CH_3, $\cdot HCl$, $\cdot HCl$

o-tolidine

Quinoidal yellow reaction product

Range of Measurement	Manufacturer
0.2–3 ppm	Drägerwerk AG
0.5–20 ppm	Mine Safety Appliances Co.
1.0–4.0 ppm	Union Industrial Equipment Corp.

The volume of the sample size can be arbitrarily varied between 100 cm and 1 liter, and at extremely low concentrations a sample of 10 liters can be taken by successive strokes of the calibrated pump.

The cross-sensitivity for bromine is approximately the same as that for chlorine, but is considerably higher than (approximately twice) that for chlorine dioxide.

Another rapid method of chlorine determination at industrial locations is based on the familiar liberation of free iodine from iodide by chlorine and subsequent formation of the blue starch–iodine complex. (16).

PROCEDURE. Pass 400 ml of the air sample through a detector tube containing a silica gel column impregnated with cadmium iodide, mercuric iodide as stabilizer, and starch. The length of the resulting stain is then measured and calibrated.

PREPARATION OF THE CHROMOGEN. Treat 1 g silica gel (0.25–0.315 mm particle size) with 0.5 ml 0.2% starch solution containing 5–10 mg mercuric iodide/100 ml. Dry for 35–40 min at 35–45°C. Then add 1 ml 0.5% cadmium iodide solution, mix it in uniformly, and dry again, for 1 hour at 35–45°C. Store the powder in evacuated and sealed ampoules wrapped in black paper.

The sensitivity of the detector tube is 0.5 mg chlorine/m^3 air.

6.2.10. Cyanogen Chloride

Cyanogen chloride vapors are extremely poisonous, causing paralysis, unconsciousness, convulsions, and respiratory arrest. The MAC is less than 0.5 ppm.

The determination of cyanogen chloride in indicator tubes is based on the formation of glutanonic aldehyde by hydrolytic opening of the pyridine ring. This takes place instantly after the intermediate addition of cyanogen chloride to pyridine. Glutaconic aldehyde then condenses with barbituric acid to give pink polymethyl dyestuffs (17).

CN-Cl + (pyridine) ⟶ (N-chloro-N-cyano pyridinium: Cl, CN) ⟶ HO−CH=CH−CH=CH−CHO +
Glutaconaldehyde

NH − CO, OC, CH, NH − CO (+ Barbituric acid) ⟶ [HO−(pyrimidine: N, OH, OH)−CH=CH−CH=CH−CH(pyrimidine: HO, N, N, OH)]$^+$ Cl^-

Barbituric acid

The range of measurement is 0.25–5 ppm cyanogen chloride (Drägerwerk AG).

Cyanogen bromide also reacts, but only in higher concentrations.

6.2.11. Diethyl Ether

Diethyl ether is a highly flammable harmful vapor, the inhalation of which may cause drowsiness, dizziness, mental confusion, faintness, and at high concentrations, unconsciousness. Its repeated inhalation may lead to an "ether habit," the symptoms of which resemble those of chronic alcoholism.

The indicating reaction in the detector tube is based on the oxidation of diethyl ether to ether peroxide with sulfochromic acid. The color change is from the orange-colored sulfochromic acid to greenish-brown Cr (III) compounds.

Range of Measurement	Manufacturer
100–4000 ppm	Drägerwerk AG
40–14,000 ppm	Union Industrial Equipment Corp.

The volume of the gas sample varies between 1 and 5 liters.

Interference is recorded in the presence of readily oxidizable organic compounds, although the sensitivity of the indication is different.

6.2.12. Dimethylformamide

The prolonged inhalation of dimethylformamide in a concentration of 100 ppm in a contaminated working area has been reported to produce damage to the liver.

The indicator tube for the estimation of dimethylformamide is composed of a white conversion layer containing sodium hydroxide and an isolated indicating layer containing bromophenol blue indicator. The dimethylformamide is converted by sodium hydroxide to an amine, which is indicated by a change in the color of the bromophenol blue indicator to blue.

Range of Measurement	Manufacturer
2.5–40 ppm	Drägerwerk AG
5–150 ppm	Mine Safety Appliances Co.

Other basic compounds, such as ammonia, amines, and hydrazine interfere with this test. The cross-sensitivity of ammonia gas is 10 relatively to dimethylformanide.

6.2.13. Dimethylsulfate

Dimethylsulfate forms extremely harmful vapors that cause severe irritation of the respiratory system, with possible severe lung injuries after a latent period. The MAC is 1 ppm.

The indicating layer of the detector tubes is located with 4-(*p*-nitrobenzyl)-pyridine, which forms a blue semiquinoidal compound with dimethylsulfate via a colorless protonated intermediate (18).

$$(CH_3)_2SO_4 + N\langle C_5H_4\rangle{-}CH_2{-}\langle C_6H_4\rangle{-}NO_2 \longrightarrow R{-}\overset{\oplus}{N}\langle C_5H_4\rangle{-}CH_2{-}\langle C_6H_4\rangle \xrightarrow{-H^+}$$

$$\longrightarrow R{-}N\langle C_5H_4\rangle{=}CH{-}\langle C_6H_4\rangle{-}NO_2$$

Range of Measurement	Manufacturer
0.2–5 ppm	Drägerwerk AG
1–50 ppm	Mine Safety Appliances Co.

The volume of the gas sample varies between 0.2 and 5 liters.

Other alkylating agents interfere with the test to various extents.

6.2.14. Ethyl Benzene

Vapors of ethyl benzene may be irritating to the eyes, skin, and mucous membranes, and at high concentrations have a narcotic effect.

The detector tube for ethyl benzene is equipped with a drying-agent-containing precleansing layer and an isolated indicating layer composed of iodic acid and fuming sulfuric acid. The appearance of brown elemental iodine indicates the ethyl benzene concentration.

The range of measurement is 30–400 ppm ethyl benzene (Drägerwerk AG). The volume of the gas sample applied is 0.6 liters.

The cross-sensitivity of some aromatic compounds is rather significant: That of toluene is ~1.5; that of xylene, ~1.

6.2.15. Ethylene Oxide

Brief exposures to high concentrations of ethylene oxide can be tolerated; however, long exposures to fairly low concentrations may cause vomiting and difficulty in breathing. The MAC is 100 ppm.

The indicating layer of the detector tube for ethylene oxide is composed of an orange chromate compound that oxidizes ethylene oxide. The color change is from orange [Cr(VI)] to pale green [Cr(III)].

Range of Measurement	Manufacturer
25–500 ppm	Drägerwerk AG
100–35,000 ppm	Union Industrial Equipment Corp.

The volume of the gas sample applied varies between 0.5 and 3.0 liters.

Propylene oxide reacts similarly to ethylene oxide, but the cross-sensitivity to it is lower, ~0.5.

6.2.16. Formaldehyde

Formaldehyde vapors irritate all parts of the respiratory system. High concentrations of the vapors inhaled for long periods can cause laryngitis, bronchitis, and bronchial pneumonia. The MAC is 5 ppm.

The determination of formaldehyde in the detector tube is based on the condensation reaction between formaldehyde and xylene, which leads to the formation of *bis*-(dimethyl)-diphenylmethane. The latter is transformed in the indicating layer to a pink semiquinoidal compound by sulfuric acid.

$$HCHO + C_6H_4(CH_3)_2 \longrightarrow (CH_3)_2{-}C_6H_3{-}CH_2{-}C_6H_3{-}(CH_3)_2 +$$

$$+ H_2SO_4 \longrightarrow (H_3C)_2C_6H_3{-}CH{=}C_6H_4{=}\overset{\oplus}{C}H_2$$

Range of Measurement	Manufacturer
0.5–10 ppm	Drägerwerk AG
1–100 ppm	Mine Safety Appliances Co.

The volume of the gas sample varies between 0.1 and 2.0 liters.

Other aldehydes such as acetaldehyde and acrolein turn the indicating layer brown; furfuryl alcohol gives a reddish-brown indication.

6.2.17. Hydrazine

Hydrazine is a violent poison having a strong caustic effect on the skin and mucous membranes. It can disturb certain enzyme systems and thus lower metabolism. The MAC is 5 ppm.

The measurement of hydrazine in indicator tubes is based on titration of a premeasured acid in the indicating layer in the presence of bromophenol blue indicator. The color change is from yellow to blue.

Range of Measurement	Manufacturer
0.25–3 ppm	Drägerwerk AG
0.5–20 ppm	Mine Safety Appliances Co.

The volume of the gas sample applied varies between 0.5 and 2 liters; however, the range of measurement can be extended to 0.05–0.6 ppm hydrazine by pumping 5 liters of gas sample into the tube.

All bases react in the same way as hydrazine in this test. The cross-sensitivity of 1,1-dimethylhydrazine, ammonia, ethylene imine, and propyleneimine is 1.0; that of ethylene diamine, 0.5; and that of diethylene triamine, only 0.1.

6.2.18. Hydrocarbons

Hydrocarbons irritate the respiratory tract and in high concentrations are narcotics.

The basis for the estimation of various hydrocarbons lies in their reducing power. The indicating layer is composed of iodine pentoxide, selenium dioxide, and fuming sulfuric acid. The appearance of brownish-green elemental iodine indicates the presence of hydrocarbons. All aliphatic hydrocarbons except methane and ethane react in the same way. The range of measurement is 500–2500 ppm hydrocarbons (Drägerwerk AG).

The gas volume taken for measurement varies between 0.1 and 1.0 liters.

6.2.19. Hydrocyanic Acid

The vapors of hydrocyanic acid are extremely toxic and can cause headache, dizziness, a feeling of suffocation, nausea followed by unconsciousness, and even death. The MAC is 10 ppm.

The principle behind the determination of hydrocyanic acid in indicator tubes is the transformation of the weak hydrocyanic acid to the strong hydrochloric acid. The hydrocyanic acid is complexed with a mercury ion and an equivalent amount of hydrochloric acid is liberated. The hydrochloric acid content is then indicated by methyl red by a color change from yellow to red.

Because an acid–alkali indicator is used as a reagent, all the potentially interfering gases such as hydrogen sulfide, hydrogen chloride, sulfur dioxide, nitrogen dioxide, and ammonia must be removed before the sample gas reaches the indicator layer. This is done by a precleansing layer containing a lead compound.

Range of Measurement	Manufacturer
2–30 ppm	Drägerwerk AG
100–30,000 ppm	Union Industrial Equipment Corp.

Hydrogen cyanide vapors in the atmosphere can also be screened for by drawing air across properly impregnated filter paper and visually comparing the resulting Prussian blue with standards.

PROCEDURE (19). Pass 360 ml of the air sample through cotton wool to remove dust. Draw the sample through a reagent paper rectangle for 1 min. Immerse the paper in a 30% (v/v) sulfuric acid bath, and compare any blue color developed within 60 sec with a standard chart or with glass standards (Tintometer). Although this comparison may be made while the paper is still in the acid, the result is more accurate if the paper is rinsed well with water and air-dried before evaluation of the stain.

PREPARATION OF THE REAGENT PAPER. Soak 3.5×2.5 cm rectangles of filter paper in 10% $FeSO_4 \cdot 7H_2O$, dry in warm air, soak in 20% aqueous sodium hydroxide solution, and dry in a vacuum desiccator.

Chlorine inhibits the reaction, but there is no interference from up to 50 ppm sulfur dioxide or hydrogen sulfide, nor from up to 900 ppm hydrochloric acid or ammonia.

6.2.20. Hydrochloric Acid

The inhalation of hydrochloric acid may cause coughing, choking, and inflammation and ulceration of the respiratory tract. The MAC is 5–10 ppm.

The color change of bromophenol blue from blue to yellow is the basis of the indicator reaction. At high relative humidities, at which hydrochloric acid mist might form, a drying agent is placed in a precleansing layer before the indicating layer.

Range of Measurement	Manufacturer
1–10 ppm	Drägerwerk AG
50–500 ppm	
2–500 ppm	Mine Safety Appliances Co.

The volume of the gas sample applied varies between 0.1 and 10 liters.

Elemental chlorine interference is indicated by a pale blue discoloration. The cross-sensitivity of the chlorine is ~0.5.

6.2.21. Hydrogen Fluoride

The vapor of hydrogen fluoride is extremely harmful. It irritates all parts of the respiratory system and leads to pulmonary inflammation and congestion. Repeated inhalation may cause fluorosis, indicated by weight loss, malaise, anemia, discoloration of teeth, and osteosclerosis.

The estimation of hydrogen fluoride is carried out in the detector tube on the basis of the demasking reaction of red zirconium quinalizarine lake by hydrofluoric acid (4). The extremely stable zirconium fluoride anion complex decomposes the pale blue adsorption compound quite selectively, causing the color change to the pink quinalizarine.

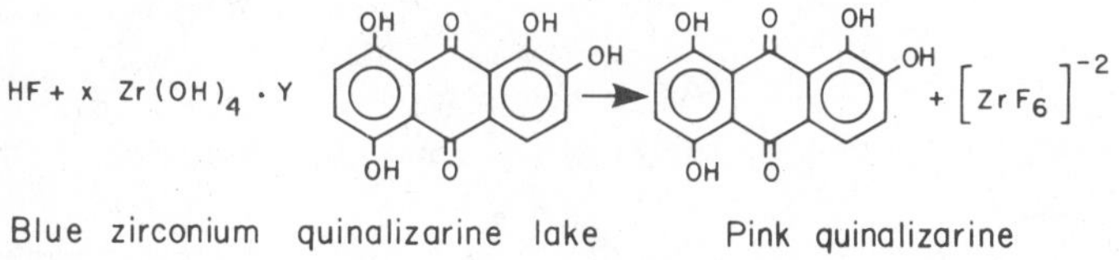

$$HF + x\ Zr(OH)_4 \cdot Y \ \text{[quinalizarine: OH, O, OH, OH, OH, O]} \longrightarrow \text{[quinalizarine: OH, O, OH, OH, OH, O]} + [ZrF_6]^{-2}$$

Blue zirconium quinalizarine lake — Pink quinalizarine

Range of Measurement	Manufacturer
1.5–15 ppm	Drägerwerk AG
0.5–5 ppm	Mine Safety Appliances Co.

The volume of the gas sample applied varies between 2 and 5 liters.

The reaction is specific for hydrofluoric acid, although at high relative humidities a mist is produced that is not entirely indicated in the tube.

6.2.22. Hydrogen Sulfide

Hydrogen sulfide is an extremely hazardous gas. Collapse and coma from respiratory failure may come within a few seconds of inhalation of a high concentration of this gas. Although its offensive odor is a good signal of its presence, one's sense of smell might not be functioning and therefore fail to warn of high concentrations of this gas. Low concentrations produce irritation of conjunctival and mucous membranes.

The indication of hydrogen sulfide in the indicator tubes is based on precipitation of insoluble dark-colored lead or copper sulfide or on the reduction of elemental iodine. The latter mechanism is used to measure mixtures of hydrogen sulfide and sulfur dioxide.

$$H_2S + I_2 \rightarrow 2HI + S^0$$

$$SO_2 + I_2 + 2H_2O \rightarrow 2HI + H_2SO_4$$

Range of Measurement	Manufacturer
1–20 ppm hydrogen sulfide	Drägerwerk AG
10–200 ppm hydrogen sulfide	
100–2000 ppm hydrogen sulfide	
0.2–7 vol% ppm hydrogen sulfide + sulfur dioxide	
1–650 ppm hydrogen sulfide	Bacharach Industrial Instruments Co.
1–800 ppm hydrogen sulfide	Mine Safety Appliances Co.
5–160 ppm hydrogen sulfide	Union Industrial Equipment Corp.
100–1700 ppm hydrogen sulfide	

The volume of the gas sample varies between 0.1 and 10 liters. The gas is drawn with a hand-operated bellows pump.

The cross-sensitivity of sulfur dioxide in the lead-salt-containing tube is ~0.5, whereas in the copper-salt-containing tubes, no interference by sulfur dioxide was recorded.

Another detection tube (20) makes use of the demasking effect of hydrogen sulfide on the $K[Ag(CN)_2]$ complex and precipitation of black silver sulfide. The silica gel in the detection tube is packed with the complex. To determine >1 μg hydrogen sulfide/liter of air, the length of the colored portion of the packing is measured after intake of a given volume of air. Lower concentrations are determined by measuring the volume of air passed through the system up to the moment, when the first coloration appears.

6.2.23. Mercury

Harmful mercury vapors in high concentrations may cause nausea, abdominal pain, vomiting, diarrhea, and headache. The effects of chronic inhalation of mercury vapor include inflammation of the mouth and gums, swelling of the salivary glands, loosening of the teeth, kidney damage, a jerking gait, spasms of the extremities, and personality changes. The MAC is 0.1 mg mercury/m^3.

The principle behind the estimation of mercury vapor in detector tubes is the formation of the red cupro tetraiodo mercuriate complex by reaction of mercury vapor with colorless cuprous iodide (21):

$$2Cu_2I_2 + Hg^0 \rightarrow \underset{\text{red}}{Cu_2[HgI_4]} + 2Cu^0$$

Range of Measurement	Manufacturer
0.05–2 mg/m^3	Drägerwerk AG
0.5–2 mg/m^3	Mine Safety Appliances Co.
0.1–2 mg/m^3	Union Industrial Equipment Corp.

The volume of the gas sample varies between 0.1 and 4 liters.

Chlorine interferes with this test by reducing the signal given by mercury.

The same reagent system was used in an impregnated-test-paper form (22) for the detection of mercury vapor in the workplace atmosphere at concentrations up to 100 μg/m^3. In this test, a measured volume of air is drawn through a piece of silica-gel-loaded filter paper coated with a mixture of sodium carboxymethylcellulose and copper (I) iodide.

PROCEDURE. Draw the atmosphere being tested at a flow rate of 500 ml/min through a piece of the test paper. After 2 min, compare the resultant pink coloration visually with the color standards.

PREPARATION OF TEST PAPER. Spread a uniform 0.25-mm-thick layer of impregnating solution on Whatman SG 81 silica-gel-loaded filter paper. The impregnating solution is a freshly prepared thin paste of 5 g copper (I) iodide with 10 ml 3% sodium carboxymethylcellulose.

PREPARATION OF COLOR STANDARDS. Suspend strips of Whatman 3MM paper in an aqueous mixture of dyes for 10 sec and dry the strips in air. The dying solutions are prepared from aliquots of 0.044% (m/v) Biebrich scarlet (solution A), 0.084% (m/v) Pontamine Sky Blue (solution B), and 0.083% (m/v) Metanil Yellow (solution C), diluted to 100 ml with water to produce the different paper standards. For the preparation of a 100 $\mu g/m^3$ mercury standard dip the paper in a mixture of 1.1 ml A, 0.2 ml B, and 2.0 ml C diluted to 100 ml with water. The 50-$\mu g/m^3$ standard is prepared with an A/B/C ratio of 0.4 : 0.1 : 1.5 (ml), whereas the 25-$\mu g/m^3$ standard corresponds to the dye ratio 0.3 : 0.1 : 1.1. The blank standard is prepared by dipping the paper into a mixture of 0.3 ml A, 0.2 ml B, and 1.0 ml C diluted to 100 ml.

6.2.24. Methyl Bromide (Bromomethane)

Brief exposure to high concentrations of methyl bromide vapor causes headache, dizziness, nausea, vomiting, and weakness, which may be followed by mental excitement, convulsions, and even acute mania. Chronic exposure to lower concentrations of methylbromide may lead to bronchitis and pneumonia. The MAC is 20 ppm.

In the detector tube, methyl bromide is first oxidized in the conversion layer to elemental bromide. This is done by fusing sulfuric acid and potassium permanganate. The free bromine oxidizes the greenish-gray *o*-dianisidine in the indicating layer to a brown quinoidal compound.

$$Br_2 + H_2N\text{-}C_6H_3(OCH_3)\text{-}C_6H_3(OCH_3)\text{-}NH_2 \longrightarrow HN{=}C_6H_3(OCH_3){=}C_6H_3(OCH_3){=}NH + 2HBr$$

greenish grey — brown

Range of Measurement	Manufacturer
5–50 ppm	Drägerwerk AG
50–100 ppm	Mine Safety Appliances Co.
10–500 ppm	Union Industrial Equipment Corp.

Free halogens, hydrogen halides, and halogenated hydrocarbons interfere seriously with this determination.

6.2.25. Monostyrene (Phenylethylene)

Styrene in the air may be irritating to the eyes and mucous membranes, and in high concentrations it can have a narcotic effect.

The determination of monostyrene in detector tubes is based on the polymerization process. When it is treated with fuming sulfuric acid, a mixture of pale yellow polystyrene and yellow sulfonated product results.

Range of Measurement	Manufacturer
10–200 ppm	Drägerwerk AG
50–400 ppm	
10–900 ppm	Mine Safety Appliances Co.
1–300 ppm	Union Industrial Equipment Corp.

6.2.26. Nickel Tetracarbonyl

The vapors of nickel tetracarbonyl may cause irritation, congestion, and edema of the lungs. The MAC is 1 ppm for prolonged periods of exposure.

The reaction principle behind the quick determination of nickel tetracarbonyl in detector tubes is its conversion with elementary iodine to nickel iodide and the estimation of the amount of nickel (II) ion by assessment of the formation of the red dimethylglyoxime complex. The quantitation is done by simple color comparison with a graduated color scale.

The range of measurement is 0.1–1 ppm nickel tetracarbonyl (Drägerwerk AG).

6.2.27. Nitrogen Dioxide

Nitrogen dioxide is an extremely harmful gas with a MAC of 25 ppm. It causes headache, nausea, dizziness, abdominal pain, and cyanosis; in severe cases, convulsions and asphyxia may follow.

The detector tube for the quick estimation of nitrogen dioxide contains a drying agent and pale gray diphenylbenzidine as the indicating layer. Nitrogen dioxide oxidizes the diphenylbenzidine to the bluish-gray quinoidal reaction product (23).

$$NO_2 + C_6H_5\text{-}NH\text{-}C_6H_4\text{-}C_6H_4\text{-}NH\text{-}C_6H_5 \longrightarrow$$

Diphenylbenzidine

$$\longrightarrow C_6H_5\text{-}N{=}C_6H_4{=}C_6H_4{=}\overset{+}{N}H\text{-}C_6H_5 + NO + H_2O$$

Diphenylbenzidine blue

Range of Measurement	Manufacturer
0.5–10 ppm	Drägerwerk AG
5–25 ppm	
5–100 ppm	
0.1–50 ppm	Mine Safety Appliances Co.
1–1000 ppm	Union Industrial Equipment Corp.

Nitrogen monoxide does not react with the diphenylbenzidine, but ozone and chlorine do, both with cross-sensitivities of ~0.5 relative to nitrogen dioxide.

6.2.28. Nitrous Fumes (Nitrous Oxide and Nitrogen Dioxide)

The estimation of the combined amount of nitrous and nitric oxide in a working-area atmosphere is done using a detector tube in which any nitrous oxide present is first oxidized to nitric oxide by an oxidation layer containing chromate. The nitrogen dioxide formed then oxidizes *o*-dianisidine to a reddish-brown product (5).

NO_2 + H_2N–(C$_6$H$_3$(OCH$_3$))–(C$_6$H$_3$(OCH$_3$))–NH_2 → [HN=(C$_6$H$_3$(OCH$_3$))=(C$_6$H$_3$(OCH$_3$))=NH · H_2N–(C$_6$H$_3$(OCH$_3$))–(C$_6$H$_3$(OCH$_3$))–NH_2]

Range of Measurement	Manufacturer
0.5–10 ppm	Drägerwerk AG
2–50 ppm	
20–500 ppm	
100–1000 ppm	

Nitrous fumes are alternatively screened for by passing the air to be tested through a bubbler containing sodium hydroxide, sodium metaarsenite, and sulfanilic acid (24). Addition of *N*-1-naphthyl ethylenediamine causes diazotization, and the resulting color is compared with standards prepared from aqueous methylene violet (C. I. 50210).

Sulfur dioxide interferes with this test, but can be removed by passing the sample through cotton wool impregnated with lead acetate.

6.2.29. Olefins

Olefins may be narcotic in high concentrations. Their estimation in detector tubes is based on their oxidation with potassium permanganate. A color change to pale brown indicates the presence of olefins. Although the olefin compounds discharge the color of acidified permanganate readily, other organic compounds with a carbon–carbon double bond are also indicated by this test.

The range of measurement is 1–55 mg propylene or butylene/liter (Drägerwerk AG).

6.2.30. Ozone

High concentrations of ozone may cause severe irritation of the respiratory tract and eyes. In the detector tube constructed for the determination of ozone the indicating layer containing indigo changes color from blue to colorless due to oxidation by ozone.

$$2O_3 + \text{Indigo (blue)} \longrightarrow 2\ \text{Isatine (colorless)}$$

Range of Measurement	Manufacturer
0.05–1.4 ppm	Drägerwerk AG
0.05–5.0 ppm	Mine Safety Appliances Co.

Chlorine and nitrogen dioxide also discharge the color of indigo when present in concentrations higher than 5 ppm.

6.2.31. Perchloroethylene (Tetrachloroethylene)

Perchloroethylene forms very harmful vapors, the inhalation of which may cause dizziness, nausea, vomiting, and, in high concentrations, stupor. Its MAC is 200 ppm.

The determination of perchloroethylene in detector tubes is based on the oxidation of the gas to chlorine by permanganate in the preliminary layer and the estimation of free chlorine in the indicating layer with *N,N′*-diphenylbenzidine.

$$CCl_2=CCl_2 \xrightarrow{[O]} 2Cl_2 + 2CO_2$$

$Cl_2 + H_2O +$ diphenylbenzidine →

diphenylbenzidine

→ Blue quinoidium salt of diphenylbenzidine $+ 2HCl$

Blue quinoidium salt of diphenylbenzidine

Range of Measurement	Manufacturer
10–500 ppm	Drägerwerk AG
25–400 ppm	Mine Safety Appliances Co.
10–400 ppm	Union Industrial Equipment Corp.

Free halogens, hydrogen halides, and easily cleaved halogenated hydrocarbons react similarly to perchloroethylene in this determination.

6.2.32. Phenol

Phenol vapor irritates the respiratory system and eyes. The inhalation of phenol over a long period may cause digestive disturbances, nervous disorders, skin eruptions, and damage to the liver and kidneys. Its MAC is 5 ppm.

The condensation of phenol with 2,6-dibromoquinone-4-chlorimide produces colored indophenols (22, 25).

HO–C₆H₄ + Cl–N=(2,6-dibromoquinone) =O → HO–C₆H₄–N=(dibromoquinone)=O + 2HCl

2,6-dibromo-quinone chloride   Blue indophenol dye

The range of measurement is 0.5–10 ppm phenol (Drägerwerk AG).

Other aromatic compounds containing a phenol group and a free *para*-position and not containing a CHO, NO_2, NO, or COOH group in the *ortho*-position to the phenolic group give a positive reaction as well.

6.2.33. Phosgene

Phosgene, an extremely harmful gas, produces secretion of fluid into the lung when inhaled and there may be a delay of several hours before symptoms develop. These include breathlessness, cyanosis, and the coughing up of a frothy fluid. The MAC is 1 ppm.

The detector tube for the selective estimation of phosgene is composed of a drying-agent prelayer and an indicating layer containing dimethylaminobenzaldehyde and diethyl aniline. Phosgene reacts with dimethylaminobenzaldehyde to form a product that then reacts with diethyl aniline to form a bluish-green meriquinoidal compound containing one molecule of amine and one molecule of imine (26).

$$COCl_2 + (CH_3)_2N\text{-}C_6H_4\text{-}CHO \rightarrow (CH_3)_2N\text{-}C_6H_4\text{-}CHCl_2 +$$

$$+ (C_2H_5)_2N\text{-}C_6H_5 \rightarrow \left[\begin{array}{c} (CH_3)_2N\text{-}C_6H_4\text{-}CH\text{-}C_6H_4\text{-}N(C_2H_5)_2 \\ (CH_3)_2N\text{-}C_6H_4\text{-}CH{=}C_6H_4{=}\overset{\oplus}{N}(C_2H_5)_2 \end{array}\right] Cl^{\ominus}$$

Blue meriquinoidal end product

Carbonyl bromide and acetylchloride react similarly to phosgene in this test.

Range of Measurement	Manufacturer
0.1–10 ppm	Mine Safety Appliances Co.
0.04–1.5 ppm	Drägerwerk AG
0.25–15 ppm	
0.05–50 ppm	Union Industrial Equipment Corp.

6.2.34. Phosphine

The MAC of phosphine is 0.05 ppm. It causes weakness, vertigo, bronchitis, edema, lung damage, convulsions, coma, and ultimately death.

Its rapid determination in detector tubes is based on the high reduction potential of phosphine. It reduces gold (III) compounds to finely divided violet metallic gold:

$$PH_3 + 2AuCl_3 + 3H_2O \rightarrow 2Au^0 + 6HCl + H_3PO_3$$

Range of Measurement	Manufacturer
0.1–4 ppm	Drägerwerk AG
50–1000 ppm	
0.025–1.0 ppm	Mine Safety Appliances Co.
5–50 ppm	Union Industrial Equipment Corp.
20–800 ppm	

Arsine and antimony hydride react similarly to phosphine, though with a lower sensitivity. Other common reducing gases, such as hydrogen sulfide, hydrogen selenide, and mercaptans, are retained in the precleansing layer of the detector tube, which contains a copper (II) compound:

$$H_2S + Cu^{2+} \rightarrow CuS\downarrow + 2H^+$$

$$2RSH + Cu^{2+} \rightarrow (RS)_2Cu\downarrow + 2H^+$$

The selective indicator paper used for the detection of phosphine (27) utilizes the liberation of hydrochloric acid in the reaction between phosphine and mercuric chloride.

Preparation of the Reagent Strips. Dip filter paper into a 1% solution of mercuric chloride dissolved in 0.05% ethanolic dimethyl yellow (C.I. Solvent Yellow 2) and air-dry.

A red color forms immediately when the paper is exposed to 0.3 ppm phosphine in the air, and a pink color within 7 min. Ammonia interferes with the test.

6.2.35. Sulfur Dioxide

The vapors of sulfur dioxide irritate the respiratory system and may cause bronchitis and asphyxia. High concentrations of sulfur dioxide vapor irritate the eyes and may cause conjunctivitis. The MAC is 10 ppm.

The indicating reaction for the estimation of sulfur dioxide in the various indicator tubes may involve formation of the extremely stable anion complex $[Hg(SO_3)_2]^{2-}$. The indicating reagent in the tube is composed of disodium tetrachloromercuriate and methyl red. Hydrochloric acid liberated during the reaction changes the indicator color.

$$2SO_2 + Na_2[HgCl_2] + 2H_2O \rightarrow Na_2[Hg(SO_3)_2] + 4HCl$$

Acidic compounds in the air may interfere with this reaction.

The reaction of free iodine with sulfur dioxide serves as an indicating reaction by decolorizing the blue iodine–starch complex.

$$SO_2 + I_2 + 2H_2O \rightarrow SO_4^{2-} + 2I^- + 4H^-$$

Range of Measurement	Manufacturer
0.1–3 ppm	Drägerwerk AG
0.5–5 ppm	
1–25 ppm	
20–200 ppm	
1–40 ppm	Mine Safety Appliances Co.
1–300 ppm	
5–300 ppm	Union Industrial Equipment Corp.
0.02–0.3%	
0.1–4.0%	

Hydrogen sulfide is indicated with the same sensitivity as is sulfur dioxide when a detector tube without a cupric-compound-containing precleansing layer is used.

Another type of sulfur dioxide indicator tube uses the spot test with zinc nitroprusside. Sulfur dioxide forms a red reaction product with the reagent, the composition of which has not been determined. Possibly a loose addition compound is formed between the two.

Preparation of the Detector Tube (28). Add to 200 g of the carrier (silica gel deposited on powdered glass) 1 ml each of 40% hexamine solution, 30% sodium nitroprusside dissolved in 2 : 1 water–glycerol, and 40% zinc chloride in 0.01 *N* hydrochloric acid. Agitate the mixture for 10 min, dry the material at room temperature in a 1-cm-thick layer, and pack into glass tubes.

When exposed to sulfur dioxide, the indicator produces a pinkish-red zone, the length of which is proportional to the sulfur dioxide concentration.

The same reaction is also used in a test-paper method for the detection of sulfur dioxide in air (29).

PREPARATION OF THE TEST PAPER. Dip filter paper strips successively into vessels containing 1 *M* zinc acetate dissolved in 19 : 1 water–glycerol and 1% sodium nitroprusside in 50% ethanol; then dry at 50°C. Such reagent papers are stable for a few days in stoppered dark-glass bottles.

Sulfites are rapidly determined by using detector tubes containing triphenylmethane dyestuffs.

Neutral sulfite solutions instantly decolorize aqueous solutions of malachite green. The quinoidal structure is destroyed; consequently, the color is lost (30).

$(CH_3)_2N$ $(CH_3)_2N$ $C=$ $=\overset{+}{N}(CH_3)_2\,Cl^-$ $\xrightarrow{[SO_3]^{2-}}$ $(CH_3)_2N$ $(CH_3)_2N$ C $N(CH_3)_2$ SO_3H

malachite green leuco malachite green

PROCEDURE (31). Pack 0.1 g dry detection reagent into an 11-cm section of glass tube and plug both ends of the tube with cotton wool. Open one end and soak the detector tube in a solution containing sulfite. The bluish-pink packing turns pink, and the length of the pink zone is proportional to the concentration of sulfite ions.

The range of measurement is 10–100 ppm sulfite.

PREPARATION OF THE DETECTION REAGENT. Impregnate cellulose powder (containing 0.2% sodium carboxymethylcellulose) with a 1 : 1 mixture of 0.05% aqueous solutions of malachite green (C.I. Basic Green 4) and Magdala Red (C.I. Basic Red 6). After 10 min, filter off the treated powder and dry it at 100°C.

Sulfide, pyrosulfate, and nitrite ions seriously interfere with this method.

6.2.36. Liquid and Solid Aerosols (32)

Detector tubes for aerosol determination employ a two-stage procedure. The pollutant is first deposited on a suitable medium in the tube (e.g., paper, a silica gel filter layer, fabric) and then brought into contact with the reagent solution, contained in an ampoule, by breaking the ampoule. The color reaction then takes place, and the intensity of discoloration is compared against color standards supplied with the tube.

Oil mist and chromium (III) oxide, cyanide, and sulfuric acid aerosols have been successfully screened with such tubes, but with such high standard deviations that concentrations could be only roughly estimated.

REFERENCES

1. A. B. Lamb and C. R. Hoover, U. S. Patent 1321062, 1919.
2. *Federal Register* **38,** no. 88 (1973).
3. K. Leichnitz, *Can. Occup. Safety News* **Jan.–Feb.** (1975).
4. *Detector Tube Handbook: Air Investigations and Technical Gas Analysis with Drager Tubes,* 4th ed., Drägerwerk AG, Lübeck, F. R. G. (1979).
5. K. Leichnitz, *National Safety News* **Oct.,** 67 (1974).
6. E. F. McFarren, R. J. Lishka, and J. H. Parker, *Anal. Chem.* **42,** 358 (1970).
7. B. E. Saltzman, *Am. Ind. Hyg. Assoc. J.* **23,** 112 (1962).
8. K. Leichnitz, *Chem. Ztg.* **97,** 638 (1973).
9. H. V. Carrol jun and F. E. Armstrong, *Rep. Invest. U. S. Bureau Mines* **R1 7811** (1973).
10. H. B. Carrol jun and F. E. Armstrong, *Rep. Invest. U. S. Bureau Mines* **BF8010** (1975).
11. V de Sachy, *Ann. Fals. Fraudes* **50,** 53 (1957).
12. C. S. Pop, *Rev. Chim.* **24,** 44 (1973).
13. A. L. Le Rosen, R. T. Moravek, and J. K. Carlton, *Anal. Chem.* **24,** 1335 (1952).
14. I. A. Osin, V. Zhukor, E. V. Utenkova, V. I. Vostrikov, and M. I. Bukovskii, *Zav. Lab.* **40,** 1119 (1974).
15. E. C. Hunt, W. A. McNally, and A. F. Smith, *Analyst (Lond.)* **98,** 585 (1973).
16. M. I. Kolesnik, M. M. Sidei, N. N. Bukorskii, and T. F. Lakeeva, *Zav. Lab.* **40,** 1119 (1974).
17. V. Anger and S. Ofri, *Mikrochim. Acta,* 987 (1964).
18. J. Epstein, R. W. Rosenthal, and R. J. Ess, *Anal. Chem.* **27,** 1435 (1955).
19. *Health and Safety Executive, Booklet No. 2,* H. M. Factory Inspectorate, H. M. Stationary Office, London (1973).
20. K. Stepankova and J. Juranek, *Chemicky Prum.* **23,** 36 (1973).
21. P. Artman, *Z. Anal. Chem.* **60,** 81 (1921).

22. H. D. Gibbs, *J. Biol. Chem.* **72,** 649 (1927).
23. B. A. Denenberg and R. Kriesel, *Am. Ind. Hyg. Assoc. J.* **37,** 246 (1976).
24. *Health and Safety Executive, Booklet No. 5,* H. M. Factory Inspectorate, H. M. Stationary Office, London (1975).
25. F. Feigl and E. Jungreis, *Anal. Chem.* **31,** 2099, 2101 (1959).
26. J. Matoušek and J. Tomaček, *Analyse Synthetischer Gifte,* Deutscher Militärverlag, Berlin (1965), p. 65.
27. M. Mysore, S. Majumdar, and H. A. Parpia, *J. Agric. Food Chem.* **21,** 184 (1973).
28. V. S. Velichkova, E. Lambreva, and D. Gradeva, *Khim. Ind.* **45,** 160 (1973).
29. J. Nair and V. Gupta, *J. Indian Chem. Soc.* **54,** 841 (1977).
30. E. Votoček, *Ber.* **40,** 414 (1907).
31. F. Goshima and I. Ichiara, *Benseki Kagaku* **30,** 796 (1981).
32. K. Leichnitz and J. Walton, *Ann. Occup. Hyg.* **24,** 43 (1981).

CHAPTER

7

WATER QUALITY SCREENING

7.1. GENERAL

In 1974 the Safe Drinking Water Act was passed by the U. S. Congress and the Environmental Protection Agency (E.P.A.) took responsibility for ensuring the safety of water supplies. The E.P.A. proposed a list of materials that carry potential hazards when present in drinking water. Maximum allowable levels were set for 10 inorganic chemicals and 6 organic pesticides, and turbidity and the amount of coliform bacteria were also limited. Additionally, secondary regulations were issued in 1977 that deal with the aesthetic qualities of water. These are connected with the sulfate, iron, and manganese contents of drinking water, and cannot be legally enforced.

Water pollution measurement encompasses a multitude of physical, chemical, and biochemical procedures. Recent water pollution controls rely heavily on advanced instrumental methods and automated analytical techniques. Spot tests based on simple kits for quick and inexpensive water testing still find wide application, however, and a number of commercial companies (Hach Chemical Manufacturing, Loveland, Colorado; E. Merck, Darmstadt, F. R. G.; Gallard–Schlesinger Chemical Manufacturing, Carle Place, New York; La Motte Chemical Co. Ltd., Towson, Maryland; Taylor Chemical Co. Ltd., Baltimore, Maryland; Hellige Inc., Garden City, New York) market variations of test kit methods that are approved by the E.P.A. for Safe Drinking Water Act reporting and recommended for onsite testing.

The reliabilities of spot-test-based water analysis kits for ecological purposes have been evaluated (1). Statistical analysis of the results for a large series of samples showed that for routine purposes, there was good agreement ($r > 0.85$) with the results of standard methods. When high accuracy and precision were needed, however, the kits were unsuitable.

A multitude of physical, chemical, and biological interactions determine water quality, and a dynamic state of change prevails in all water sources. The aquatic ecosystem is frequently disturbed by humans. Such ecological disturbance is described in terms of pollution.

Modern civilization classifies natural bodies of fresh water according to intended use, for example, public water supply, fish propagation, recreation, agricultural use, industrial water supply, hydroelectric power, and so forth. The aims of analysis are to determine the suitability of water for its intended use and to evaluate possible water treatment methods in view of water reuse.

The parameters of significance for water quality characterization are listed in Tables 1 and 2.

Table 1. Quality Criteria for Domestic Water Supplies[a]

Parameter	Level	
	Permissible	Desirable
Color ((cobalt-platinum) scale)	75 units	<10 units
Odor	Virtually absent	Virtually absent
Taste	Virtually absent	Virtually absent
Turbidity	Virtually absent	Virtually absent
Inorganic Chemicals		
pH	6.0–8.5	6.0–8.5
Alkalinity ($CaCO_3$)	30–500 mg/liter	30–500 mg/liter
Ammonia	0.5 mg/liter	<0.01 mg/liter
Arsenic	0.05 mg/liter	Absent
Barium	1.0 mg/liter	Absent
Boron	1.0 mg/liter	Absent
Cadmium	0.01 mg/liter	Absent
Chlorides	250 mg/liter	<25 mg/liter
Hexavalent chromium	0.05 mg/liter	Absent
Copper	1.0 mg/liter	Virtually absent
Cyanide	0.01 mg/liter	Absent
Dissolved oxygen	>4.0 mg/liter	Air-saturated
Fluorides	0.8–1.7 mg/liter	1.0 mg/liter
Iron (filterable)	<0.3 mg/liter	Virtually absent
Lead	<0.05 mg/liter	Absent
Manganese (filterable)	<0.05 mg/liter	Absent
Nitrates + nitrites	<10 mg/liter	Virtually absent
Phosphorus	10–50 μg/liter	10 μg/liter
Selenium	0.01 mg/liter	Absent
Silver	0.05 mg/liter	Absent
Sulfate	250 mg/liter	<50 mg/liter
Sulfides	0.01 mg/liter	Absent
Total dissolved solids	500 mg/liter	<200 mg/liter
Zinc	1.5 mg/liter	Virtually absent

Table 1 (*Continued*)

Parameter	Level	
	Permissible	Desirable
Organic Chemicals		
Carbon–Chloroform extract	0.15 mg/liter	<0.04 mg/liter
Methylene-blue-active substances	0.5 mg/liter	Virtually absent
Pesticides		
Aldrin	0.017 mg/liter	Virtually absent
Chlordane	0.003 mg/liter	Virtually absent
DDT	0.042 mg/liter	Virtually absent
Dieldrin	0.017 mg/liter	Virtually absent
Endrin	0.001 mg/liter	Virtually absent
Heptachlor	0.018 mg/liter	Virtually absent
Heptachlor epoxide	0.018 mg/liter	Virtually absent
Lindane	0.056 mg/liter	Virtually absent
Methoxychlor	0.035 mg/liter	Virtually absent
Organic phosphate	0.1 mg/liter	Virtually absent
Taxophane	0.005 mg/liter	Virtually absent
Herbicides	0.1 mg/liter	Virtually absent

[a] Reproduced from ref. 2.

Table 2. Quality Criteria for Agricultural Water Use[a]

Element	Maximum Allowable Level for:	
	Continuous Water Use (mg/liter)	Short-term Water Use (mg/liter)
Aluminum	1.0	20.0
Arsenic	1.0	10.0
Beryllium	0.5	1.0
Boron	0.75	2.0
Cadmium	0.005	0.05
Chromium	5.0	20.0
Cobalt	0.2	10.0
Copper	0.2	5.0
Lead	5.0	20.0
Lithium	5.0	5.0
Manganese	2.0	20.0
Molybdenum	0.005	0.05

Table 2 (*Continued*)

Element	Maximum Allowable Level for: Continuous Water Use (mg/liter)	Short-term Water Use (mg/liter)
Nickel	0.5	2.0
Selenium	0.05	0.05
Vanadium	10.0	10.0
Zinc	5.0	10.0

[a] Reproduced from ref. 2.

7.2. TESTING FOR INORGANIC MATERIALS IN WATER

7.2.1. Acidity–Alkalinity

Acidity in surface or ground water is caused by humic acid extracted from swamps or peat beds and by industrial pollution. Excessive acidity causes corrosion of zinc from galvanized pipes, and can be detrimental to fish populations.

Acidity (base comsumption up to pH 8.2) is tested by drop count titration with sodium hydroxide in the presence of phenolphthalein and bromocresol green–methyl red indicator. Titration against the mixed indicator gives an estimate of free mineral acid only, whereas the drop count against phenolphthalein measures the free mineral acids and the free carbon dioxide.

The presence of weak acids having buffering effects, such as humic acid, phosphoric acid, citric acid, and carbonic acid, may disturb the results of these titrations.

Alkalinity in water is caused mainly by the presence of carbonates, bicarbonates, and hydroxides. A high alkalinity may sometimes indicate pollution by industrial wastes.

Although it is not considered a health hazard, high alkalinity is unwanted in many food manufacturing processes. On the other hand, alkaline water is beneficial as boiler feed water, because it inhibits corrosion in industrial boilers.

The spot tests used for alkalinity estimation (Hach Co.; E. Merck) also use drop count titrations with standardized mineral acid. Hydroxide and carbonate alkalinities are neutralized simultaneously, and phenolphthalein is used as the indicator.

$$Ca(OH)_2 + H_2SO_4 \rightarrow CaSO_4 + 2H_2O$$

$$2CaCO_3 + H_2SO_4 \rightarrow Ca(HCO_3)_2 + CaSO_4$$

Neutralization of the bicarbonate occurs after the hydroxide and carbonate species have already been neutralized, and is determined using the combined methyl red–bromocresol green indicator. The color change is from blue-green to yellow-red (3).

7.2.2. Aluminum

Aluminum occurs in natural waters as a result of contact with rocks, soil, and clay. Water treatment plants employing aluminum sulfate as a precipitating agent may also contribute to the aluminum content of water.

Aluminum content in water is determined semiquantitatively by its conversion with alkali into aluminate, followed by treatment with the ammonium salt of aurin-tricarboxylic acid (4).

Potentially disturbing ions are first precipitated with sodium hydroxide, and then test strips (Merck Co.) incorporating the reagent are dipped into the supernatant fluid. After removal of the test strip, the red color is compared with a calibrated color scale.

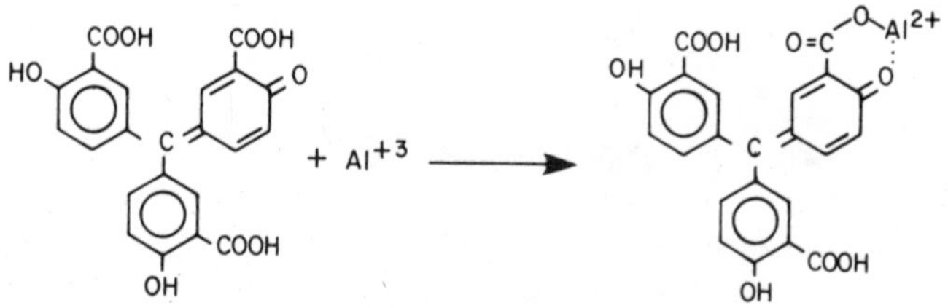

The test is interfered with by beryllium, copper (II), iron (II) and (III), and vanadyl (II) ions. Large quantities of cyanide, chromate, vanadate, citrate, succinate, fluoride, thiosulfate, permanganate, tungstate, phosphate, sulfide, and sulfite cause the estimation to show an aluminum content less than that actually present.

7.2.3. Ammonium

Ammonium ions detected in surface and ground water almost invariably indicate the presence of impurities in water that represent a health hazard. Drinking water must not contain any detectable amounts of ammonium compounds. The situation is different with mineral water, in which the presence of ammonium compounds need not signal contamination problems.

Generally, by monitoring ammonium levels even semiquantitatively it is possible to determine how well nitrogen-containing wastes are being stabilized.

Nesslerization is an established colorimetric procedure for ammonia determination in several commercial test kits (Merckoquant®; Merck Co.; Aquaquant®; Merck Co.; Hach Co.; Hellige Co.). The amount of yellow-brown reaction product formed between potassium iodomercurate and ammonia is visually compared with comparator blocks or five-step color-graduated cubes molded of acrylic plastic. The range of this determination is 0.05–10 ppm ammonia.

Indophenol, in the presence of ammonia, forms indophenol blue (5).

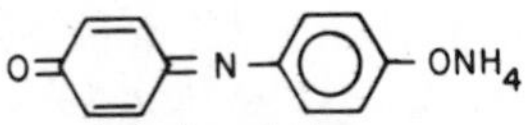

$$O{=}C_6H_4{=}N{-}C_6H_4{-}ONH_4$$

Indophenol blue

This reaction is utilized in the Nanocolor testing system (Macherey-Nagel, Düren, F. R. G.). Reagent sets are available for the estimation of ammonium in the 0.02–0.7 ppm range.

7.2.4. Arsenic

Severe toxicity has been reported after ingestion of only 100 mg arsenic, and chronic toxicity develops with lower intake (6). Arsenic occurs geologically in natural water as arsenate and arsenite, but industrial discharges may increase the arsenic content of water. The maximum allowable concentration (MAC) of arsenic in ground water is 0.1 ppm. Mineral waters, however, may contain somewhat higher concentrations.

Mercuric bromide test paper is used for the semiquantitative evaluation of arsenic (E. Merck; Macherey-Nagel). Test paper impregnated with mercuric bromide is inserted for 30 min into the reaction tube, in which arsine is liberated. The resulting brown reaction product is compared with a 0/0.1/0.5/1.0/1.7/3.0-ppm color scale.

7.2.5. Barium

In trace concentrations, barium acts as a cardiac stimulator, but it is considered lethal to humans at dosages of 550–600 mg. Barium concentrations average about 0.05 ppm in drinking waters, but may range as high as

0.9 ppm in some natural waters. The MAC of barium in drinking water is 1 ppm.

Barium content is determined turbidimetrically after barium sulfate precipitation. The finely divided barium sulfate particles are held in suspension by a protective colloid.

7.2.6. Boron

Boron is an essential element for plant growth, but is toxic to many plants at levels as low as 1 ppm. The Public Health Service has established an upper limit of 1 ppm, which provides a good factor of safety for humans. The ingestion of larger amounts of boron can affect the central nervous system, and protracted ingestion may result in a clinical syndrome known as borism.

Boron can be tested for by combining acidified water samples with curcumin. The boric acid changes the yellow curcumin into its red-brown isomer rosocyanine. The resulting red color is compared visually with standards. Ions commonly found in water do not interfere with this method, with the exception of nitrate at concentrations above 20 ppm, as nitrate–nitrogen. The sensitivity of this test is 0.2 μg boron.

7.2.7. Cadmium

The cadmium content of drinking water varies between 0.4 and 50 μg/liter. Environmental Protection Agency drinking water standards limit the concentration of cadmium in potable water to 10 μg/liter. At present no spot-test-like screening tests are available to detect such low concentrations.

7.2.8. Chloride

A high chloride content in drinking water has no toxic effects on man; on the contrary, it is essential in the human diet. The maximum allowable chloride concentration of 250 ppm in drinking water was established for reasons of taste rather than as a safeguard against physical hazard.

Several quick, semiquantitative chloride tests are available. The Quantab® (Ames Co.) chloride titrators are thin, chemically inert plastic strips. Laminated within the strip is a capillary column impregnated with silver dichromate. When the strips are placed in the test solution, fluid rises in the column by capillary action and continues to progress for as long as chloride solution enters the column. The reaction of silver dichromate with chloride produces a white-colored area within the capillary column:

$$Ag_2Cr_2O_7\ \text{(brown)} + 2NaCl \rightarrow Na_2Cr_2O_7 + 2AgCl\ \text{(white)}$$

The length of the white area in the capillary column is proportional to the chloride concentration.

The titration ranges of the available Quantab strips are 60–480, 300–3600, and 3600–38,400 ppm chloride. The test is complete within 4–6 min.

In the drop count titration test available from the Hach Co., the sample is titrated with mercuric nitrate solution, with diphenylcarbazone present as an indicator. Mercuric nitrate first reacts selectively with all the chloride present. An excess of mercuric ions forms at the end point of the "titration" (7) and reacts with the indicator.

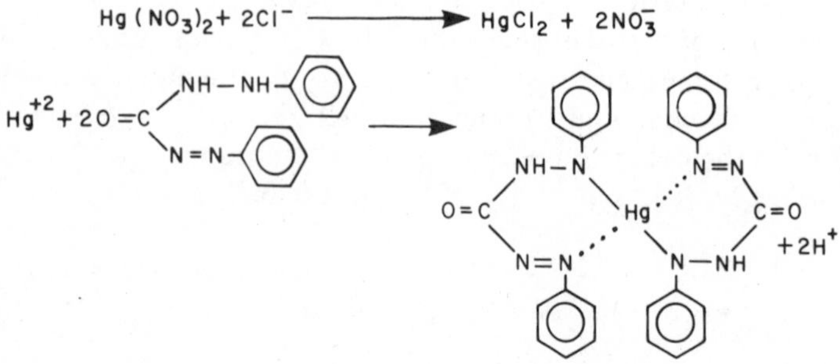

$$Hg(NO_3)_2 + 2Cl^- \longrightarrow HgCl_2 + 2NO_3^-$$

$$Hg^{+2} + 2\,O{=}C\begin{matrix}NH-NH-C_6H_5\\ N{=}N-C_6H_5\end{matrix} \longrightarrow \text{(Hg complex)} + 2H^+$$

A rapid, on-the-spot form of the familiar Mohr argentometric titration consists of drop-counting of the standardized silver nitrate solution using a powdered chromate indicator containing the proper pH buffer. This test is marketed in powdered pillow form (Hach Co.).

Another simple screening test for chloride (Aquaquant®, E. Merck) uses the demasking reaction of mercuric thiocyanate. The free thiocyanate ions liberated are indicated by ferric nitrate. The amount of orange-red iron (III) thiocyanate formed is visually compared with a standardized color scale.

$$2Hg(SCN)_2 + 6Cl^- \rightarrow HgCl_2 + HgCl_4^{2-} + 2SCN^-$$

$$SCN^- + Fe^{3+} \rightarrow Fe(SCN)^{2+}\ \text{(red)}$$

A simple, semiquantitative chloride determination method, called by its authors a "micro-titration on paper" (8), uses a demasking reaction carried out by chloride ions with a colored silver salt, thus measuring decolorization (Saltesmo paper, Machery-Nagel). The test permits the semiquantitative determination of chloride ions in the following concentrations: 0, 0.25, 0.5, 1, 2, 3, 4, and 5 g/liter.

7.2.9. Chlorine

Chlorine is usually added to the public water supply and to public and private swimming pools. It is the most widely used bactericide in water. It is normally present in two active forms, the "free active chlorine" present in the water as free chlorine, hypochlorous acid, or hypochlorite ions; and the "bound active chlorine," meaning inorganic and organic chloramines still having oxidizing power, but to a lesser degree than free chlorine. Chlorine added to public drinking-water supplies to destroy bacteria and viruses must yield an excess of 1 ppm free chlorine to control bacteria without causing a noxious color or taste. In swimming pools, 0.3 ppm free active chlorine must be present at every point in the swimming pool to maintain the bactericidal action as new bacteria are introduced by bathers.

A number of applied spot tests are used to determine rapidly, but reliably, both the total chlorine and the free chlorine content. There are three different reaction mechanisms applied in chlorine screening.

1. The reaction of *N,N*-diethyl-*p*-phenylenene-diamine (DPD) with chlorine is the most popular one for control analysis. The reaction is stable, the rate of color development is high, and this reaction makes it possible to distinguish between free and bound chlorine.

The DPD test indicates free chlorine by producing a magenta-colored quinoidal oxidation compound (9).

$$H_5C_2{-}N{-}C_2H_5\text{-}C_6H_4\text{-}NH_2 \xrightarrow{Cl_2} H_5C_2{-}\overset{+}{N}{-}C_2H_5{=}C_6H_4{=}NH + 2Cl^- + 2H^+$$

In the estimation test for both free and bound chlorine, codyes are added to the reagent system. With the bound chlorine compounds, chloramine oxidizes iodide to iodine, which then reacts with DPD together with the free chlorine. Thus, the color formed is directly proportional to the chlorine concentration.

Residual chlorine content in drinking water is estimated by a stable tablet test (10) that does not involve the use of hazardous materials. The tablet is simply added to 10 ml of the sample. The amount of the resulting blue starch–iodine complex is visually compared with a calibrated color chart.

The tablet contains equal parts by weight of potassium iodide, potassium bitartrate, and soluble starch, and is sealed in foil.

Bromine, iodine, ozone, and other oxidants react similarly to chlorine in this test, but their presence is rather unlikely in water. The E.P.A. accepts the use of the visual comparison method using color-step cubes, (Hach Co.) or continuous color discs (Hach Co.; E. Merck; Hellige Co.).

2. Total chlorine (both free and bound) content can be screened for with the *o*-tolidine reagent (Aquamerck®, E. Merck; Hellige Co.). This reagent is oxidized by chlorine to the yellow quinoidal oxidation product:

$$H_2N{-}C_6H_3(CH_3){-}C_6H_3(CH_3){-}NH_2 \xrightarrow{[O]} HN{=}C_6H_3(CH_3){=}C_6H_3(CH_3){=}NH$$

o-tolidine yellow oxidation product

3. For high chlorine contents, the quick drop count titration with phenyl arsine oxide may be used (11).

$$Cl_2 + 2H_2O + C_6H_5{-}As{=}O \longrightarrow 2HCl + C_6H_5{-}As({=}O)(OH)_2$$

A simple phenyl arsine–starch indicator titration kit that uses a digital titrator instead of conventional burets, and registers a color change from dark blue to colorless in the presence of chlorine, is commercially available (Hach Co.).

7.2.10. Chromium

The presence of chromium in public drinking supplies is objectionable because it is a suspected carcinogen. The MAC of chromium is 0.05 ppm. It may be present in water in the hexavalent or the trivalent form, the latter very rarely. Hexavalent chromium enters the water supply from cooling towers, waste water plants, plating operations, and the tanning industry. Chromium concentrations in drinking water above 0.003 ppm clearly indicate the presence of industrial wastes.

Most of the quick spot tests for chromium are based on a redox reaction between chromate ions and diphenylcarbazide. Chromate ions oxidize diphenylcarbizide in acid solution to diphenylcarbazone, which is further oxidized to diphenylcarbadiazone. The divalent chromium ions formed react with the enol form of diphenylcarbazone to yield a red-violet complex (see section 4.3.8).

For total chromium analysis the trivalent chromium is oxidized in alkaline solution with hypobromite to chromate, and then the excess of hypobromite is removed (12). The on-the-spot estimation kits use comparator blocks (Aquaquant®, E. Merck), calibrated continuous color discs, and color cubes (Hach Co.; Hellige Co.).

Test strips for the semiquantitative determination of chromate ions using the same reaction mechanism include the Quantofix-Chromium (Macherey-Nagel) and Merckoquant® (E. Merck). strips. The concentrations represented by the color scale of the latter are 0, 3, 10, and 30 ppm chromate.

7.2.11. Copper

A rather narrow range of copper content is admissible in water. Copper concentrations of 0.1 ppm are needed for plant growth and for normal body metabolism. In the total absence of copper in drinking water, nutritional anemia may develop in children. Concentrations greater than 0.1 ppm in ponds and rivers control aquatic plant life, diminishing algae and plankton growth, but at the same time are toxic to some sorts of fish.

Although copper is not considered a health hazard, a MAC of 1 ppm copper is set for potable water, to minimize the bitter taste of the water.

Various spot tests are applied for the semiquantitative estimation of copper content in water. A simple reagent kit (Aquaquant®, E. Merck) for the range 0.05–0.4 ppm copper uses the spot test based on the specific reaction between copper (II) ions and *bis*-cyclohexanone oxalylhydrazone.

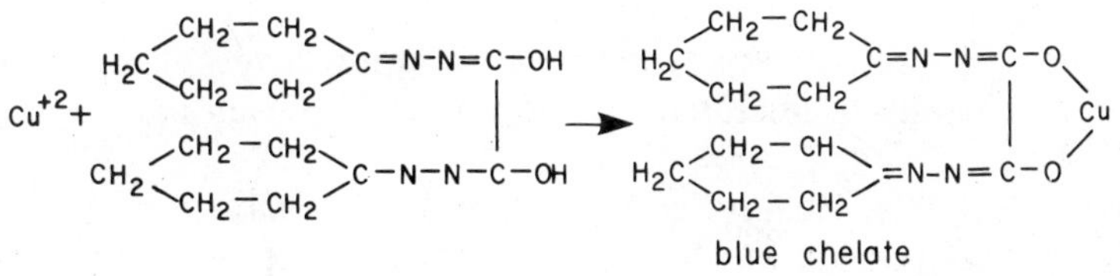

```
        CH2—CH2                                   CH2—CH2
   H2C          C=N-N=C—OH                   H2C          C=N—N=C—O
        CH2—CH2         |                         CH2—CH2         |     \
Cu+2 +                  |          →                              |      Cu
        CH2—CH2         |                         CH2—CH          |     /
   CH2          C—N—N—C—OH                   H2C          =N-N=C—O
        CH2—CH2                                   CH2—CH2
                                                  blue chelate
```

Although the test is interfered with by large concentrations of iron, nickel, cobalt, and chromium salts, only the iron content usually reaches significant concentrations. In the Aquaquant kit the potential interference of iron salts is reduced by a masking agent, and the blue color is evaluated by comparison with a color block.

The formation of the purple intercalation complex between 2,2′-biquinoline-4,4′-dicarboxylic acid and the monovalent copper ion is the basis of another screening test for copper content in water (13, 14).

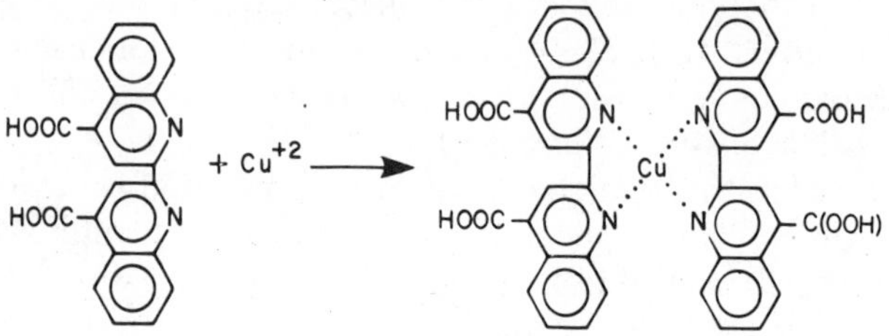

The commercially available variation of this test contains a prepacketed reagent, reducing material, and buffer system, and the color produced is readily compared with calibrated continuous color discs or a five-step test cube. The estimation range of this system is 0–5 ppm copper.

A "dip and read" strip test for the semiquantitative estimation of traces of monovalent, divalent, and metallic copper (Merckoquant®, E. Merck) is used to control the quality of potable water, swimming pool water, waste water, galvanizing solutions, copper etching baths, and so forth.

The chemical reaction behind the method is very similar to that of the previously described test. Monovalent copper ions form a red intercalation compound with 2,2′-biquinoline (cupron) at acidic pH levels. The reducing agent and the buffer system are incorporated into the reagent zone of the strip along with the reagent. The strip is simply immersed in the test solution, and after about 30 sec the test zone is compared against a graduated color scale representing 0, 10, 30, 100, and 300 ppm copper.

7.2.12. Cyanides

Cyanide is toxic to aquatic organisms even at low levels. The toxicity is due predominantly to liberated hydrocyanic acid, which inhibits oxygen metabolism.

The U. S. Public Health Service recommends a maximum limit of 0.2 ppm cyanide in potable water. Such cyanide originates from metal cleaning and electroplating baths, gas scrubbers, and the like. These chemical processes constitute potential hazards when waste from them is carelessly disposed of. Natural waters do not contain cyanide; thus its presence always indicates contamination by industrial effluent.

A very sensitive test for cyanides is based on polymethine dye formation (15). For quick screening tests, the chloramine T–pyridine pyrazolone reagent system is used. The cyanide is oxidized to cyanogen chloride by chlorine liberated from the chloramine T; the cyanogen chloride then forms an intermediate nitrile. The nitrile hydrolyzes promptly to glutaconaldehyde. The latter reacts with pyrazolone to form a blue polymethine dye. Cyanide–metal complexes are not detected by this reaction.

$$2CN^- + Cl_2 \longrightarrow 2\,CNCl$$

$$CN{-}Cl + \text{pyridine} \longrightarrow \text{N-cyanopyridinium} + 4H_2O \longrightarrow O{=}CH{-}CH{=}CH{-}CH_2{-}CHO + 2NH_3 + CO_2 + HCl$$

$$O{=}CH{-}CH{=}CH{-}CH_2{-}CHO + 2\ \text{(3-methyl-1-phenyl-pyrazol-5-one)} \longrightarrow$$

$$\longrightarrow \text{(pyrazolone)}{=}CH{-}CH{=}CH{-}CH_2{-}CH{=}\text{(pyrazolone)}$$

The condensation of pyrazolone with glutaconaldehyde is used by Hach Co. in their cyanide test kit, whereas a similar condensation between *N,N'*-dimethyl barbituric acid and glutaconic acid is the essential part of the Aquaquant (E. Merck) cyanide estimation system. Both methods are simple color-comparison methods. Thiocyanate interferes in this reaction, but because cyanide can be driven off and the remaining thiocyanate determined, consecutive determinations can be used to obtain an accurate cyanide estimate.

7.2.13. Dissolved Oxygen

The amount of Dissolved oxygen in lakes or streams is probably the most important water quality parameter. Several water samples have to be taken at different times, locations, and depths to obtain representative data.

Drinking and industrial water should contain at least 2 ppm oxygen, but the oxygen content for water that must sustain aquatic life should be much higher, ~4–5 ppm. In fish ponds, the dissolved oxygen content should reach the 8–15 ppm range.

The classical Winkler titration is the chemical basis of the on-the-spot evaluation of oxygen content in water. Manganese (II) ions react with the dissolved oxygen in an alkaline medium to form tetravalent manganese, which liberates elementary iodine from iodides. Drop count titration with sodium thiosulfate using a graduated titrating pipet measures the iodine quantity, which is equivalent to the amount of dissolved oxygen in the water (Aquamerck, Hellige Co.).

$$2Mn^{2+} + O_2 + 2H_2O \rightarrow 2MnO(OH)_2 \downarrow$$

$$MnO(OH)_2 + 6I^- + 4H^+ \rightarrow Mn^{2+} + 2I_3^- + 3H_2O$$

$$I_3^- + 2S_2O_3^{2-} \rightarrow 3I^- + S_4O_6^{2-}$$

Such tests should be carried out immediately after sampling, since oxygen reactions continue in the sample bottle.

The very same principle of titration of equivalent liberated iodine liberation is somewhat modified in another quick test for dissolved oxygen (16). Nitrite ion can interfere with the iodine-liberation test because it oxidizes iodide to free iodine as well as tetravalent manganese does. When hydrazoic acid is incorporated into the reagent system, it destroys the nitrite beforehand:

$$HN_3 + HNO_2 \rightarrow N_2 + H_2O + N_2O$$

Because of the relative instability of thiosulfate, phenyl arsine may be substituted for thiosulfate in the titration:

$$2H_2O + I_3^- + C_6H_5{-}As{=}O \longrightarrow 2HI + I^- + C_6H_5{-}As(=O)(OH)_2$$

The use of prepackaged powder pillows and digital titrators facilitates this semiquantitative determination.

7.2.14. Fluoride

The Safe Drinking Water Act permits fluoride ion concentrations in potable water in the 1.4- to 2.4-ppm range. In many municipal areas a 1-ppm level is artificially maintained in public water supplies to prevent dental caries. Excessive amounts of fluoride cause fluorosis and discoloration of the tooth enamel.

A simple spot test uses the phenomenon of prevention of fluorescence of metal oxinates for fluoride estimation in water. Qualitative filter paper contains small amounts of calcium, magnesium, and aluminum oxide. These highly dispersed metal oxides form highly fluorescent inner complex compounds with oxine (8-hydroxy quinoline). If, however, vapors of hydrofluoric acid act on qualitative filter paper, the exposed areas of the paper cannot be transformed to fluorescent oxinates because of the formation of calcium and magnesium fluoride, which masks the fluoride. Thus the fluoride-exposed areas of the paper will be quenched, whereas the rest of the paper will exhibit strong fluorescence.

PROCEDURE. Mix one drop of the water sample with three drops of sulfuric acid in a microcrucible. Cover the mouth of the crucible with a piece of qualitative filter paper larger in diameter than the crucible. Heat the crucible for 5 min at 50–60°C. Treat the paper with a 0.05% solution of 8-hydroxy quinoline in chloroform and examine under ultraviolet light. If fluoride is present in the sample, the exposed area will show no fluorescence or subdued fluorescence as compared with the surrounding portions of the paper.

The sensitivity of this test is 1 ppm fluoride. The water sample should be diluted twofold, threefold, and fourfold to enable semiquantitative evaluations.

7.2.15. Iron

Iron is normally present in natural water in minor amounts in the form of hydrogen carbonate. Subterranean water contains 1–3 ppm iron and flowing water ~0.3 ppm, but mineral water iron content may reach 10–50 ppm.

Iron exists in water in the soluble ferrous and the insoluble ferric forms. It can enter the water system by leaching from natural iron deposits or iron-containing industrial wastes.

Although not a potential health hazard, iron's presence in water may cause many aesthetic problems. The taste of drinking water is highly objectionable when the threshold concentrations of 0.1 ppm bivalent and 0.2 ppm trivalent iron are exceeded. In domestic water supplies the oxidation of ferrous iron to ferric iron shows up as rust spots on clothing and rust stains on sinks. Water containing high iron concentrations can cause pipe encrustation in industry.

Formation of the red intercalation complex of ferrous ions with 2,2′-bipyridine or 1,1′-phenanthroline is used in a number of commercially available tests for iron (Aquaquant®, E. Merck; Ferro-Ver®, Hach Co.; Macherey-Nagel).

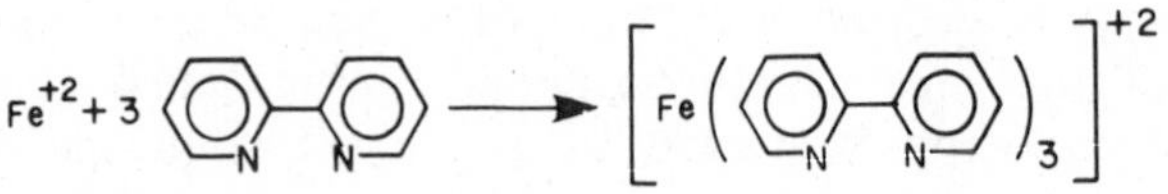

$$Fe^{+2} + 3\ \text{(2,2'-bipyridine)} \longrightarrow \left[Fe(\text{2,2'-bipyridine})_3\right]^{+2}$$

After addition of iron-solvating and -reducing thioglycolic acid buffer, the sample is treated with the reagent and the iron content is assessed by visual comparison with calibrated color cubes, continuous color discs, or comparator blocks. The complex formation is not disturbed by the presence of other iron-complexing ions, such as chloride, phosphate, and oxalate. These tests may cover a wide range of iron contents (0.1–50 ppm).

Formation of the familiar red ferric thiocyanate complex is the basis used of another test for iron in water (Hellige Co.). Ten calibrated glass color standards represent concentration values of 0, 0.5, 1.0, 1.5, 2.0, 2.5, 3.0, 3.5, 4.5, and 5.5 ppm iron.

Using the ultrasensitive and specific iron reagent 3-(2-pyridyl)-5,6-*bis*(4-phenylsulfonic acid)-1,2,4-triazine monosodium salt (FerroZine Iron Reagent®, Hach Co.) (17), very low concentrations of iron can be determined. The amount of purple complex formed is directly proportional to the ferrous iron concentration.

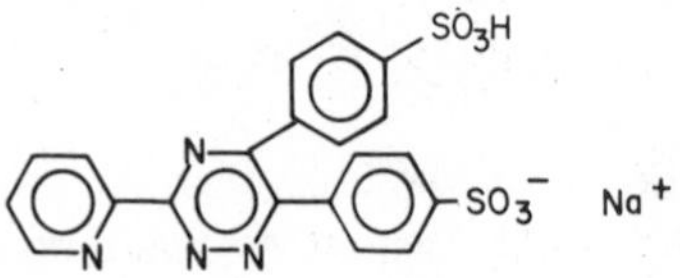

SO_3H SO_3^- Na^+

This colorimetric estimation method is used for trace iron analysis in such ultrapure waters as high-pressure steam boilers water, turbine feed water, and nuclear reactor cooling water.

7.2.16. Lead

Lead in drinking water may arise from a number of sources, in particular the use of lead pipes or of plastic pipes stabilized with lead compounds. It is toxic to aquatic organisms, but the degree of toxicity varies greatly depending on water quality characteristics as well as the species being considered. The MAC of lead in drinking water is 50 ppm. The metal accumulates in the bone structure when more than 300 μg/day is ingested. It is seldom found in natural waters in more than trace quantities; accumulation of significant lead salt quantities in water is hampered because chlorides and carbonates tend to precipitate insoluble lead salts whenever the lead concentration reaches solubility product levels.

Decisive identification of traces of lead in water is very important from the medical and public health points of view. Direct tests for the detection of lead in water in the range 10–100 ppm are not reported in the literature. A "trace catcher" pretreatment is required to obtain a sample for which the classical lead tests will succeed. Minute quantities of lead that alone would not be precipitated as sulfides readily coprecipitate with mercuric sulfide. The mercuric ion can then be driven out and the lead determined by the sodium rhodizonate reaction.

The following procedure will detect as little as 0.001 ppm lead in a 100-ml sample.

PROCEDURE. Add 0.25 ml sodium mercury sulfide solution to a 10-ml water sample. Add ~1 g ammonium chloride.

$$Na_2[HgS_2] + 2NH_4Cl \rightarrow 2NaCl + (NH_4)_2S + HgS$$

Add 5 ml 3% hydrogen peroxide solution. Heat the precipitate in a crucible on a hot plate until the black mercuric sulfide disappears. Heat the crucible over a burner for 1 additional minute. After cooling, add five drops of 0.2% freshly prepared aqueous sodium rhodizonate solution. Add several drops of pH 2.79 buffer solution. In the presence of lead, red particles of lead rhodizonate appear.

PREPARATION OF SODIUM MERCURY SULFIDE SOLUTION. Dissolve 13.6 g mercuric chloride in 60 ml hot water. Add solid sodium sulfide until the black mercuric sulfide precipitate dissolves, and dilute to 100 ml.

PREPARATION OF pH 2.79 BUFFER SOLUTION. Dissolve 1.5 g tartaric acid and 1.9 g sodium bitartrate in 100 ml water.

The nonspecific reaction of lead ion with dithizone is specific in the presence of potassium cyanide, which masks all other heavy metal ions. The amount of the chloroform-extractable red salt formed can be visually compared with a series of permanent color standards (Hellige Co.).

7.2.17. Manganese

The presence of manganese in potable water does not represent a health hazard. An increased manganese content causes only taste problems. Concentrations exceeding 0.1 ppm produce an objectionable bitter taste in drinking water and leave gray or black stains on laundry as a result of manganese dioxide precipitation.

The paper industry requires absolutely manganese-free water supplies, and manganese content has to be controlled also in the beverage, textile, dying, and food-processing industries.

The maximum recommended manganese content in public water supplies is 0.05 ppm, with total iron and manganese content not to exceed 0.3 ppm.

The classical spot test for the determination of manganese by acid oxidation of the divalent ion to violet permanganate with periodate in acidic media (18) has been used successfully in designing test kits for semiquantitative estimation.

$$3H_2O + 2Mn^{2+} + 5IO_4^- \rightarrow 2MnO_4^- + 5IO_3^- + 6H^+$$

The intensity of the purple permanganate color is directly proportional to the concentration of manganese present in the water sample. Visual color comparison is done by means of continuous color discs. The range of the determination is 0–1.0 ppm manganese (Hach Co., Hellige Co.).

For detection of traces of manganese in water supplies, tetraphenylarsonium chloride is used to precipitate and concentrate permanganate (19):

$$MnO_4^- + (C_6H_5)_4As^+ \rightarrow (C_6H_5)_4AsMnO_4$$

The formation of an ion association complex incorporating a large organic molecule causes solubility in organic solvents. The extraction of this complex in dichloromethane greatly enhances the sensitivity of the test.

Another very old spot test (20) uses autooxidation of manganese (II) hydroxide to tetravalent manganese:

$$2Mn(OH)_2 + O_2 \rightarrow 2MnO_2 \cdot H_2O$$

The resulting highly dispersed manganese dioxide can be made visible by its action on reducing chromogens (benzidine, *o*-tolidine).

In the test strip form of the same spot test (Merckoquant, E. Merck), a paper test zone impregnated with *o*-tolidine is immersed briefly into the water sample, bathed for 15 sec in 1 *N* sodium hydroxide, and then bathed for 45 sec in 10% acetic acid solution. In the presence of manganese a blue quinoidal oxidation product is formed; the intensity of the color is then visually compared with a 0/5/25/100/500-ppm-manganese graduated color scale.

The *o*-tolidine also responds to other oxidizing cations and anions present in water. Their presence is recognized, however, by a change in the color of the test zone before the alkaline treatment.

7.2.18. Nitrate Nitrogen

The appearance of nitrate nitrogen in water usually indicates the final stage of waste decomposition, although nitrates can be formed directly from atmospheric nitrogen and oxygen by the action of lightning. Nitrites are converted to nitrate by nitrate-forming bacteria under aerobic conditions.

High levels of nitrate may indicate biological waste material in the final stages of stabilization or contamination of the water by run-off from fertilized agricultural areas. One of the negative effects of nitrate is that it encourages the excessive growth of algae. The MAC of nitrate nitrogen in drinking water is 10 ppm by the U. S. Public Health Service Drinking Water Standards. Higher nitrate concentrations appear to cause methemoglobemia in infants ("blue babies").

Most of the quick tests for nitrate content are based on the rapid reduction of nitrate to nitrite and subsequent azo dye formation. Cadmium metal may be used for the reduction of nitrate (21, 22). According to the classical Griess test, sulfanilic acid is reacted with nitrous acid to form the appropriate diazonium compound. The latter is coupled with gentisic acid to form an amber-colored compound whose concentration is directly proportional to the nitrate concentration of the water sample. For quantitation, this color can be compared visually against a continuous color disc (Hach Co.).

$$NO_3^- + Cd + 2H^+ \longrightarrow NO_2^- + Cd^{+2} + H_2O$$

$$NO_2^- + H_2N\text{-}C_6H_4\text{-}SO_3H + 2H^+ \longrightarrow HO_3S\text{-}C_6H_4\text{-}N^+{=}N + H_2O \longrightarrow$$

$$HO_3S\text{-}C_6H_4\text{-}N{=}N + C_6H_3(OH)_2\text{-}COOH \longrightarrow HO_3S\text{-}C_6H_4\text{-}N{=}N\text{-}C_6H_2(OH)_2\text{-}COOH + H^+$$

The estimation range of this test is 0–50 ppm nitrate nitrogen.

For more sensitive nitrate testing (concentrations down to 0.02 ppm), the same reduction technique and diazotization procedure are used, but the diazonium salt is coupled with chromotropic acid, yielding a red-orange complex.

$$HO_3S\text{-}C_6H_4\text{-}\overset{+}{N}{=}N + C_{10}H_4(OH)_2(SO_3H)(SO_3Na) \longrightarrow NaO_3S\text{-}C_6H_4\text{-}N{=}N\text{-}C_{10}H_3(OH)_2(SO_3Na)_2 + 2H^+$$

Test strips for the semiquantitative determination of nitrate also use the diazotization with an aromatic amine of nitrous acid produced *in situ* at the reaction zone by a reducing agent and coupled with *N*-(1-naphthyl)-ethylenediamine. The intensity of the red-violet azo dye formed is visually compared with a 0/10/30/100/250/500-ppm-nitrate graduated color scale (Merckoquant).

Interfering nitrite ions may be destroyed with aminosulfonic acid before the test (23):

$$MeNO_2 + NH_2SO_3H \rightarrow MeHSO_4 + H_2O + N_2$$

Oxidizing anions discolor the test zone, but this interference is recorded only at such concentrations of chromate, ferricyanate, iodate, periodate, permanganate, and vanadate as are not usually present in surface waters.

The rapid reduction of nitrate to nitrite and subsequent azo-dye formation is the basis of a screening test in which all the chemicals needed are incorporated into a single, stable reagent tablet (24). Formation of a very stable complex with sulfosalicylic acid greatly enhances the reducing power of zinc dust. These materials are compatible in the tablet with the chromogenic reagents, sulfanilamide and *N*-(1-naphthyl) ethylenediamine 2HCl.

PROCEDURE. Put 5 ml of the water sample in a 16 × 150 mm test tube. Drop the reagent tablet into the solution and shake until the tablet dissolves (~60–90 sec). At 2 min after beginning shaking, visually compare the color produced with a 1/2/5/10/20-ppm-nitrate graduated color chart, making sure to judge the color when looking through the test tube from the top to the bottom.

7.2.19. Nitrite Nitrogen

Nitrite must not be present in potable water. Its presence indicates pollution and in most cases demonstrates the presence of fecal impurities. It occurs in water as an intermediate stage in the biological decomposition of nitrogen-containing organic compounds. Nitrite-forming bacteria may convert ammonia to nitrite under aerobic or anaerobic conditions.

Although nitrites should not be present at all in drinking water, nitrite-containing additives are used in cooling circuits to prevent corrosion. Since nitrites are oxidized by oxygen, their presence in water reduces its oxygen content. Drinking-water nitrite concentrations seldom exceed the 0.1-ppm level.

All of the portable test kits for the estimation of nitrite are based on diazo fixation reactions (Hach Co.; Aquaquant, Aquamerck, E. Merck; Nitrotesmo, Machery-Nagel).

7.2.20. Phosphates

The determination of phosphates is very important in analyzing surface and waste water, since excessive phosphate content leads to progressive atrophy in lakes. Testing is necessary in boiler-water-treatment and cooling-tower plants. Phosphates and polyphosphates form a thin protective film of calcium iron phosphate on the walls of industrial pipelines.

Phosphates enter the water supply from domestic biological wastes and residues and from agricultural fertilizer run-off. Additional sources of phosphates in water are corrosion-control materials, scale-control additives, and detergents and surfactants. The permissible upper limit of phosphorus content in drinking water is 50 ppm. The principle of most analytical procedures for phosphates is the heteropolyacid formation that occurs when ammonium molybdate reacts with phosphate solutions. The phosphomolybdic acid formed, $H_7P(Mo_2O_7)_6$, is reduced to molybdenum blue. Germanic, arsenic, and silicic acids behave analogously, forming the corresponding heteropolyacids.

Metaphosphates are often used in boiler water treatment systems, and numerous organophosphorous compounds are applied as pesticides. All of these phosphorous compounds must be transformed to orthophosphates in various pretreatment steps (25). Metaphosphates are hydrolyzed by acids:

$$Na_4P_2O_7 + 2H_2SO_4 + H_2O \rightarrow 2H_3PO_4 + 4Na^+ + 2SO_4^{2-}$$

and organophosphorous compounds can undergo oxidative splitting:

$$R{-}O{-}\overset{\displaystyle O}{\overset{\|}{\underset{\underset{\displaystyle OH}{|}}{P}}}{-}OR' + K_2S_2O_8 + H_2SO_4 \rightarrow$$

$$H_3PO_4 + 2K^+ + 2SO_4^{2-} + \text{organic fragments}$$

The orthophosphates form a yellow complex, phosphomolybdic acid, with sodium molybdate. The phosphomolybdic acid may be reduced with a variety of reducing agents such as benzidine, stannous compounds, sulfites, ascorbic acid, and metol:

$$12Na_2MoO_4 + H_2PO_4^- + 24H^+ \rightarrow [H_2PMo_{12}O_{40}]^- + 24Na^+ + 12H_2O$$

$$[H_2PMo_{12}O_{40}]^- + 2e^- \rightarrow [H_2PMo_{12}O_{40}]^{3-}$$

The phosphorus concentration is estimated by visual comparison of the resulting blue color with graduated color discs, a five-step color cube, or color comparator blocks (Hach Co., E. Merck, and Hellige Co., respectively).

7.2.21. Selenium

Selenium is extremely toxic to humans and animals, causing inflammation of the lungs and disturbances of the digestive and nervous systems. It is also suspected of causing dental caries and of being carcinogenic. The MAC of selenium in potable water is 0.01 ppm. Concentrations of selenium in water higher than 0.01 ppm are caused by contamination by industrial waste or dissolution of selenium-containing soils.

One of the most sensitive tests for selenous acid (26) involves formation of a yellow piazselenol complex by condensation of diaminobenzidine with selenous acid.

$$(H_2N)_2C_6H_3{-}C_6H_3(NH_2)_2 + H_2SeO_3 \longrightarrow (H_2N)_2C_6H_3{-}C_6H_3N_2Se + 3H_2O$$

For the estimation of organic selenium compounds in water, pretreatment with acid permanganate is necessary to convert all selenium compounds to selenates. These are reduced to selenites in warm 4 *N* hydrochloric acid.

The optimum pH for the formation of piazselenol is approximately 1.5. Addition of formic acid adjusts the sample to this pH. The sample is then buffered to pH 10–11 and the complex extracted into toluene (27). Visual comparison of the color with standards allows estimation of selenium contents in drinking water in the 0- to 1.5-ppm range.

7.2.22. Silver

Silver content may be selectively estimated in potable water on the basis of formation of a deep-blue triple complex. The silver ion forms a water-soluble, positively charged intercalation complex with 1,10-phenanthroline that reacts with bromopyrogallol red to form the blue compound silver(phenanthroline),2(dibromopyrogallol red) (28).

Potential interference by ferrous ions is eliminated by the addition of 1,10-phenanthroline as a masking agent. EDTA is used as a complex-binder for the masking of other potentially interfering ions.

7.2.23. Sulfate

The MAC of sulfate in potable water is 250 ppm, according to the Public Health Service water standards. The main adverse effect of sulfate is the possible increased dissolution of lead from lead pipes by water containing more than 250 ppm sulfate. Public water supplies may be used even when containing higher concentrations, but may have a laxative action, especially in new users. The taste threshold of magnesium sulfate is 400–600 ppm, and that of calcium sulfate 250–800 ppm.

An orientative rapid test for the approximate estimation of sulfate content is based on the demasking of the red thorin [1-(2-arsenophenylazo)-2-hydroxy-3,6-naphthalene disulfonic acid sodium salt] barium complex. In the reaction, white barium sulfate is formed and the red color is discharged.

Test strips for the rapid, semiquantitative estimation of sulfate (Merckoquant®, E. Merck) contain three test zones impregnated with varying amounts of the red thorin–barium complex. The amount of color change from red to yellow indicates the sulfate concentration in the test solution:

First zone decolorized	200–300 ppm sulfate
Second zone decolorized	400–500 ppm sulfate
Third zone decolorized	800–900 ppm sulfate

The pH of the test solution in which the test strips are dipped should be in the range 4–8.

Another fast test for the evaluation of sulfate content is based on the precipitation of finely divided barium sulfate particles while keeping the suspension stable with colloidal additives (29). The sulfate content may then be assessed by turbidimetric measurement (Hach Co., Hellige Co.).

7.2.24. Sulfide

Hydrogen sulfide is an extremely hazardous gas. Collapse, coma, and death from respiratory failure may ensue within a few seconds of its intake. The toxicity of hydrogen sulfide is about the same as that of hydrogen cyanide. Its offensive odor is detectable long before toxic levels are reached, but in many cases an individual's sense of smell may be fatigued and fail to give warning of high concentrations of hydrogen sulfide.

Hydrogen sulfide is a by-product of the anaerobic decomposition of organic matter, so it indicates heavy contamination of the water. The MAC is 20 ppm.

An extremely sensitive spot test for the detection of soluble and insoluble sulfides in water is possible after preconcentration of the sulfides on the surface of metallic mercury (30). A 10-ml water sample is vigorously shaken with a drop of mercury, the water is discarded, and the localized formation of mercuric sulfide is detected by the sodium azide–iodine catalytic reaction:

$$2NaN_3 + I_2 \rightarrow 2NaI + 3N_2$$

The reagent for this test is 3 g sodium azide dissolved in 100 ml 0.1 *N* iodine. The detection of nitrogen bubbles on the surface of the mercury indicates the presence of at least 0.005 ppm hydrogen sulfide.

Metallic sulfides in water may be detected simply by the appearance of the characteristic odor of hydrogen sulfide on acidification of the water sample with hydrochloric acid or by suspending a moistened piece of lead-acetate-impregnated paper in the atmosphere above the water sample in a closed vessel. The appearance of brown lead sulfide signals the presence of sulfide in the water.

Hydrogen sulfide may also be detected through the formation of methylene blue (31–33). In the presence of mild oxidizing agents such as potassium dichromate, hydrogen sulfide and acid-soluble metallic sulfides react with *N,N*-dimethyl-*p*-phenylenediamine and form methylene blue. The color formed may be compared with color blocks to obtain estimations in the 0.025–0.3 ppm range (Aquaquant, Merck Co.).

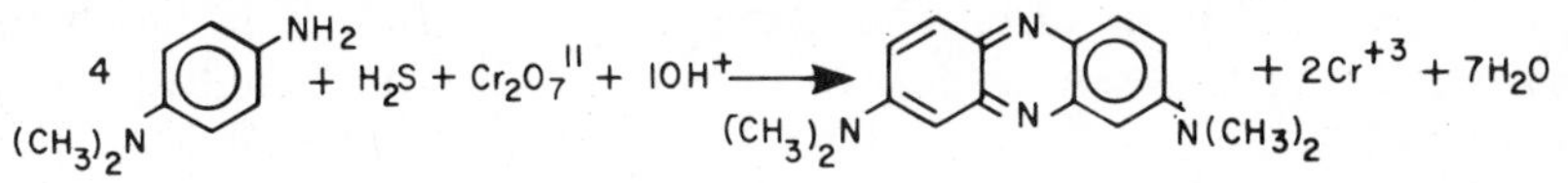

$$4\,(CH_3)_2N\text{-}C_6H_4\text{-}NH_2 + H_2S + Cr_2O_7^{\,II} + 10H^+ \longrightarrow \text{methylene blue [}(CH_3)_2N \ldots N(CH_3)_2\text{]} + 2Cr^{+3} + 7H_2O$$

7.2.25. Zinc

High concentrations of zinc in potable water have no toxic effects on humans, but a zinc content above 5 ppm gives water a bitter taste. The maximum recommended zinc concentration in public water supplies was set by the E.P.A. at 5 ppm. Higher concentrations may act as temporary stomach irritants. The presence of unusually high concentrations of zinc may indirectly suggest the presence of lead and cadmium originating from deteriorating galvanized iron.

A quick qualitative spot test selectively identifies the zinc with dithizone after masking of interfering heavy metals with sodium thiosulfate. The green solution of dithizone turns a purplish red on being shaken with a zinc solution.

Procedure. Acidify ~50 ml of the water sample with a few drops of hydrochloric acid and treat with 30% aqueous sodium acetate until a pH of 5 is obtained. Add 3 ml sodium thiosulfate and shake vigorously with 3 ml dithizone solution. A red color is produced at zinc concentrations at or above 5 ppm.

Preparation of Dithizone Solution. Dissolve 10 mg dithizone in 400 ml carbon tetrachloride.

Preparation of Sodium Thiosulfate Solution. Dissolve 50 g sodium thiosulfate pentahydrate in 30 ml deionized water (34).

Dithizone-based test strips for the semiquantitative estimation of zinc are based on the reaction between hydroxyzincate and the reagent paper (Merckoquant®). Since dithizone reacts with many ions, sodium hydroxide is added to the sample prior to immersion of the strip to precipitate out interfering ions and to convert zinc ions into the zincate form.

PROCEDURE. Add two sodium hydroxide pellets to 5 ml of the solution to be tested. Filter off the hydroxide and immerse the test strip in the solution for ~2 sec. Compare the color of test zone with the color scale, which is calibrated at 0, 10, 40, 100, and 200 ppm zinc.

This test kit has proven suitable for use in quick screening (35).

The chemical basis of an alternative test is the formation of a blue-colored complex when zincon (2-carboxy-2′-hydroxy-5′-sulfoformazyl benzene) is added to the pH-9-buffered zinc-containing sample. As the complex formation is nonselective for zinc, all the metal ions present must first be complexed with cyanide. The next step is selective liberation of zinc by cyclohexanone, followed by formation of the blue complex (36–38).

$$Zn^{+2} + 4CN^{-} \longrightarrow Zn(CN)_4^{-2}$$

The minimum detectable concentration of zinc in this test is 0.02 ppm.

7.2.26. Hardness

The hardness of water is defined by its content of alkaline earth ions, mainly calcium and magnesium ions. The total hardness is estimated by the sum of all calcium and magnesium ions, the concentration of both is expressed in terms of calcium carbonate concentration per liter. Hardness may be caused by natural enrichment of calcium and magnesium salts by leaching of the soil and geological formations, but manmade operations may also cause increases in hardness.

Originally, hardness was defined by the phenomenon of precipitation of soap by hard water. In routine water analysis, water hardness is among the most frequently determined parameters.

Boiler feed water must be freed of magnesium and calcium salts to prevent scaling. In the brewing industry, pasteurization takes place in soft water. High hardness levels inhibit the action of soap or detergents in washing machines and dishwashers, and cause spotting of glassware in dishwashers. Hard water may also irritate the skin of bathers. Hardness is an undesirable property in photography, food processing, the pulp and paper industry, dying, metal finishing, and machine tooling. Hardness levels above 500 mg/liter are undesirable for domestic water; most drinking water supplies average ~250 mg/liter.

The terms total hardness, carbonate hardness, and noncarbonate hardness are used in practical water analysis. The SI unit of hardness is millimoles per liter (mmol/liter) of alkaline earth ions, but other units are also widely used (Table 3).

There are several simple spot tests for the qualitative and semiquantitative estimation of hardness in water. Qualitative differentiation between hard and soft water is possible by the redox reaction between manganese (II) ions and silver (I) ions in the presence of alkali (39):

$$Mn^{2+} + 2Ag^{-} + 4OH \rightarrow MnO_2 + 2Ag + 2H_2O$$

Since calcium carbonate and calcium oxide react similarly to free alkali in this procedure, it is possible to use this test for the visual estimation of hardness.

Table 3. Conversion Table for Hardness Units

	Equivalent				
Unit	Parts per Million	Grains per U. S. Gallon	Clark Degrees	French Degrees	German Degrees
1 ppm	1.0	0.058	0.07	0.1	0.056
1 grain per U. S. gallon	17.1	1.0	1.2	1.71	0.958
1 Clark deg.	14.3	0.829	1.0	1.43	0.8
1 French deg.	10.0	0.583	0.70	1.0	0.56
1 German deg.	17.9	1.044	1.24	1.78	1.0

PROCEDURE. Treat the residue from the evaporation of a drop of water with a drop of reagent solution and compare the resulting gray or black spots with a standard series prepared from water samples of known hardness. Such a series may be preserved for permanent reference. Equal-sized drops are obtained by carrying out the tests on slides frosted with hydrofluoric acid.

PREPARATION OF THE REAGENT SOLUTION. Dissolve 2.87 g manganese nitrate in 40 ml water and add to it a solution of 3.35 g silver nitrate in 40 ml water. Dilute to 100 ml and add dilute alkali dropwise until a black precipitate is formed. Filter off the precipitate. The reagent solution keeps well in a dark bottle.

Test strips for the evaluation of total hardness are based on the complexing reaction between the calcium and magnesium ions and ethylenediaminetetracetic acid disodium salt. The test strip has four test zones of gradually increasing reagent content. The test strip is dipped into the water sample, and the water hardness is given by the number of test strips changing color from green to red-violet (Merckoquant®). Similar dip-and-read test systems have been designed by Macherey-Nagel (Aquadur®) and Ames Co. (Softix™).

Drop count complexometric titration constitutes a feasible rapid, semiquantitative screening method for both total hardness and calcium hardness.

For estimation of total hardness the red Calmagite–calcium complex is demasked by the gradual addition of EDTA in a pH-10-buffered system (40). The end point is signaled by a color change from red to blue.

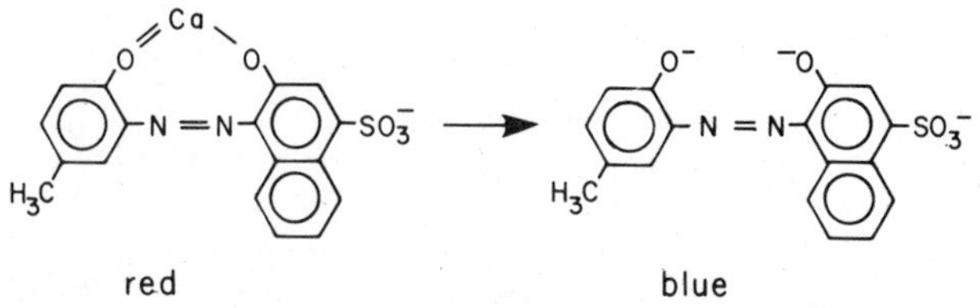

red blue

The drop count titration used in the calcium hardness test (Cal Ver® II, Hach Co.) precipitates magnesium in an alkaline medium of pH 12–13 and binds calcium in a stable complex with hydroxynaphthol blue. Gradual addition of EDTA removes calcium from this complex, making the blue color of the indicator visible.

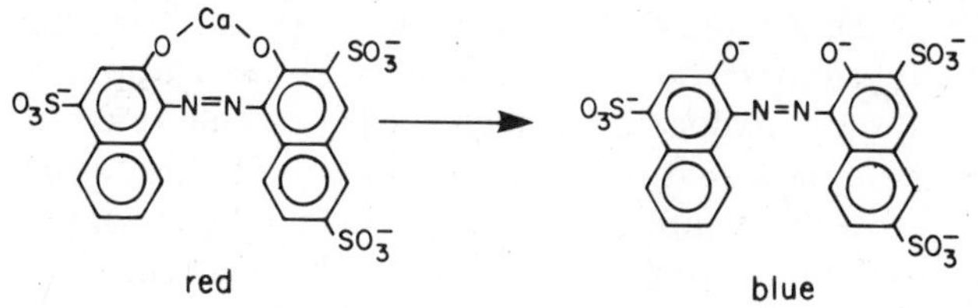

7.3. TESTING FOR ORGANIC MATERIALS IN WATER

7.3.1. Biochemical Oxygen Demand

The biochemical oxygen demand (BOD) is a very important parameter showing the amount of oxygen required by a water sample for biochemical degradation of organic pollutants in association with microorganisms. The classical dilution method for determining BOD is generally carried out over a period of 5 days (BOD_5) at 20°C in the dark. This is a standard method of the American Public Health Association (APHA) and is approved by the E.P.A. Another method is a direct, simplified manometric technique not yet approved by the E.P.A.

The classical dilution methods determine BOD by sealing a sample in a completely filled container for 5 days at 20°C and then measuring the decrease in the amount of dissolved oxygen. The decrease in the oxygen content is caused by the bacterial oxidation of the organic pollutants present.

Although BOD does not reliably determine the amount of organic matter present in the water, it is a highly valuable parameter for indicating the degree of water pollution. Sewage treatment plants routinely compare the BOD of incoming sewage with that of the effluent.

The simplified manometric method does not require chemical analysis of the oxygen, and is suited to routine monitoring. A manometer attached to the sample bottle continuously measures the drop in the air pressure in the bottle. Because no dilution of the sample water is required, the sample is maintained under its natural conditions.

PROCEDURE. Place a measured sample of water in an amber-colored bottle connected to a closed-end manometer. The sample should contain ammonium chloride as a nutrient salt. Above the sample, introduce a volume of 21%-oxygen-containing air. The oxygen content in the water sample is reduced by the bacterial oxidation of the organic matter, and the oxygen consumed in the solution is replenished from the air in the closed system. Keep the sample at 20°C, and stir constantly to maintain rapid transfer of oxygen to the liquid. Measure the pressure drop caused by this procedure with the manometer, which should be calibrated in mg BOD/liter. The carbon dioxide formed by the oxidation of the organic material should be absorbed with lithium hydroxide so that it will not interfere with the pressure difference–oxygen consumption correlation. Periodically record and plot the manometer readings on a BOD versus time graph.

The Hach Manometric BOD apparatus employs five bottles independently equipped with mercury manometers.

A statistical comparison of the manometric method with the classical dilution method using the *t*-test (41) showed no significant difference between experimental values and theoretical values for a 95% confidence interval.

7.3.2. Chemical Oxygen Demand

The chemical oxygen demand (COD) is an approximate measure of the amount of dissolved oxygen required by a water sample to oxidize all water-soluble organic substances to carbon dioxide and water in a reaction with a strong chemical oxidizing agent such as potassium dichromate in a 50% sulfuric acid solution (42). The oxidation is carried out in the presence of a silver compound as a catalyst, and to avoid potential interference caused by the oxidation of chloride ions, mercuric sulfate is added. After the oxidation step is completed, the amount of dichromate can be estimated by a relatively quick colorimetric procedure.

The organic materials are oxidized by dichromate to carbon dioxide, while any hydrogen present is converted to water. Taking potassium acid phthalate as an example, in the oxidative digestion the following reaction occurs:

$$2KC_8H_5O_4 + 10K_2Cr_2O_7 + 41H_2SO_4 \rightarrow 16CO_2 + 46H_2O + 10Cr_2(SO_4)_3 + 11K_2SO_4$$

It is best to run the COD determination immediately after sampling. If this is not feasible, the pH should be lowered to 2 or less by adding 2 ml sulfuric acid per liter sample. Samples should be analyzed within 7 days after collection.

PROCEDURE USING MICRODIGESTION. Add a 2-ml water sample, 2.5 ml sulfuric acid, 0.0245 g potassium bichromate, 0.03 g silver sulfate, and 0.03 g mercuric sulfate to the reaction tubes (43, 44). Close tubes with caps and place in the cavities of a preheated (150 ± 1°C) block heater (Hach Co). After 2 hours, remove the tubes and monitor the digestion colorimetrically, either by following the disappearance of the yellow-colored dichromate at 420 nm or by following the appearance of the green-colored Cr ions at 620 nm.

7.3.3. Bacteria in Water

The presence of bacteria has been used to determine the sanitary quality of water for the last 100 years. Coliform bacteria are normal inhabitants of feces, soil, and vegetation. Differentiation between fecal and nonfecal coliforms is crucial, since the disease-producing potential of water is linked primarily with the fecal coliform content.

The presence of fecal coliform organisms indicates recent and possibly dangerous pollution. The E.P.A. drinking water standards set a limit of 10,000 coliforms/100 ml water. The APHA multiple fermentation technique is used for the determination of total and fecal coliform bacteria.

Coliform testing tube assembles that have been sterilized and packaged in a sealed outer plastic bag and are ready for use are available commercially (Hach Co.).

For the presumptive test, lauryl tryptose broth is used, whereas brilliant green–lactose broth is the medium of the confirmatory coliform test. Fecal coliforms are tested in EC medium, and incubated at 44.5 ± 0.2°C.

7.3.4. Anionic Detergents

Estimation of the concentration of anionic detergents (alkyl benzene sulfonates) in water is feasible through measurement of methylene blue activity. The test sample (50 ml) is measured into a simplified extraction bottle, methylene blue reagent and chloroform are added, and the intensity of the colored chloroform layer is compared with colored glass standards representing the 0- to 2-ppm alkyl benzene sulfonate range (Hellige Co.).

7.3.5. Tannin and Lignin

Lignin is often discharged as a waste product during the manufacture of paper pulp. Another plant constituent, tannin, may enter the water supply from tanning industry discharges. It is also widely applied as a scale reducer in boiler feed water treatment.

Both compounds are reducing materials containing aromatic hydroxyl groups, and form blue products with tungstophosphate and molybdophosphoric acid. Because other reducing materials react similarly, this test cannot be considered specific.

PROCEDURE (45). Add 2 ml reagent to 50 ml sample solution, and after 5 min add 10 ml sodium carbonate solution and mix. After 10 min, compare visually with simultaneously prepared standards (45).

PREPARATION OF REAGENT. Dissolve 100 g sodium tungstate dihydrate, 20 g molybdophosphoric acid, and 50 ml 85% phosphoric acid in 750 ml distilled water. Boil the liquid under refluxing for 2 hours; cool and make up to 1 liter with distilled water.

PREPARATION OF SODIUM CARBONATE SOLUTION. Dissolve 200 g anhydrous sodium carbonate in 500 ml warm distilled water and dilute to 1 liter.

PREPARATION OF STANDARD SOLUTIONS. Dissolve 1 g tannic acid, tannin, or lignin in 1000 ml distilled water to make a stock solution. Prepare a series of standard solutions ranging between 10 and 50 ppm tannin or lignin.

7.3.6. Organophosphorous Compounds

A very sensitive test for organophosphorous compounds can be carried out with two detector strips. The test is based on the inhibition of cholinesterase by organophosphorous compounds (46–48). One detector strip (A) is impregnated with cholinesterase present in horse-liver acetone powder. It metabolizes the 1-naphthyl acetate present on the second strip (B) to naphthol. The latter couples with fast blue B in a diazotization reaction to form a magenta-colored compound. Methyl parathion and other organophosphorous compounds inhibit this reaction, causing a white spot to appear on the detector strip.

PROCEDURE (49). Using a microcapillary pipet, drop onto detector strip A one drop of the test solution and dry the spot in a stream of air. Place this strip horizontally on a clean glass microslide and wet with distilled water. Warm the slide for 2 min at 40°C (a cigarette lighter flame will serve for this purpose under field conditions). Place a 2.5-cm^2 portio of detector strip B on another glass microslide, moisten it, and place strip A below strip B between the two slides. The presence of organophosphorous compounds is indicated by the appearance of a white spot.

Compound	Detection Limit
Methyl paraoxon	0.1 μg
Omethoate	0.1 μg
Fenitrooxon	0.5 μg

PREPARATION OF DETECTOR STRIP A. Dip a 7 × 2.5 cm strip of Whatman no. 3 filter paper into 2% horse-liver acetone powder suspended in acetone. Dry it in the horizontal position at room temperature.

PREPARATION OF DETECTOR STRIP B. Dip a 7 × 2.5 cm strip of Whatman no. 3 filter paper in a 1% acetone solution of 1-naphthyl acetate. Dry the paper in air and dip it into 1% aqueous fast blue B solution. Dry at room temperature or in a 50°C oven.

7.3.7. Hallucinogens in Water

Many basic esters of benzilic and glycolic acids are known hallucinogens. A simple, specific, and nondestructive test has been devised for their detection in water (50). The hallucinogens are pretreated with sodium tetraphenylboron, collected on glass-fiber filter paper, and sprayed with 2-diphenylacetyl-1,3-indandione-1-imine derivatives. The latter form weakly fluorescent molecular complexes with these hallucinogens in the solid state (51). The visual observation of fluorescence indicates the presence of the basic esters of benzilic and glycolic acids.

PROCEDURE. Pass 100 ml of the water to be tested through a CM-Sephadex C-25 column under gravity flow. Elute the hallucinogenic compound from the column with 100 ml 0.3 *M* sodium chloride. Add 1 ml 0.01 *M* potassium hydrogen phthalate and 4 ml 4% sodium tetraphenylboron. The insoluble potassium tetraphenylboron (TPB) serves as a carrier for the hallucinogen TPB precipitate. After 20 min, collect the precipitate on a glass-fiber filter, and after removal dry the filter with hot air. Spray it with detector spray II, redry, and examine for fluorescence under long-wave ultraviolet light. A pink-orange fluorescence indicates a positive test. Spraying with detector spray I leads to yellow-green fluorescence in the event of positive reaction.

PREPARATION OF DETECTOR SPRAY I. Dissolve 10 mg 2-diphenylacetyl-1,3-indandione-1-dibutylamidinohydrazone in 10 ml tetrahydrofuran acidified with four drops of concentrated hydrochloric acid.

PREPARATION OF DETECTOR SPRAY II. Dissolve 12 mg 2-diphenylacetyl-1,3-indandione-1-(*p*-dimethyl amino) benzaldizine in 100 ml benzene. Use this reagent under a fume hood.

REFERENCES

1. C. E. Boyd, *Trans. Am. Fish. Soc.* **109,** (2), 239 (1980).
2. *Analytical Chemistry: Key to Progress in National Problems, National Bureau of Standards Special Publication 351* (1972).
3. *APHA Standard Methods for the Examination of Water and Wastewater,* 14th ed., American Public Health Association, Washington, D.C. (1976), p. 278.
4. *APHA Standard Methods for the Examination of Water and Wastewater,* 12th ed., American Public Health Association, Washington, D.C. (1965), p. 53.
5. L. N. Lapin, *Anal. Abstr.* **3,** 2043 (1956).
6. M. J. Suess, *Examination of Water for Pollution Control,* vol. 2, Pergamon Press, New York (1982), p. 179.
7. See ref. 3, pp. 303, 304.
8. F. J. Förg and M. Staub, *Chem. Rundschau* **17,** 19 (1964).
9. See ref. 3, pp. 33, 327.
10. N. Thorsell, E. Malm, and M. Mikiver, *J. Am. Water Works Assoc.* **69,** 209 (1977).
11. See ref. 3, p. 318.
12. See ref. 3, p. 192.
13. S. Nakano, *Chem. Abstr.* **58,** 3390e (1963).
14. V. M. Ostroskaya, M. S. Aksenova, V. V. Suvorov, O. M. Shtern, and E. Y. Kostrikina, *Energetik* **7,** 10 (1978).
15. *APHA Standard Methods for the Examination of Water and Wastewater,* 13th ed., American Public Health Association, Washington, D.C. (1971), p. 404.
16. *EPA Methods for Chemical Analysis of Water and Wastes,* (1974), p. 51. EPA Washington, D.C.
17. L. L. Stokey, *Anal. Chem.* **42,** 779 (1970).
18. H. H. Willard and L. H. Greathouse, *J. Am. Chem. Soc.* **39,** 2366 (1917).
19. M. L. Richardson, *The Analyst* (*Lond.*) **87,** 435 (1962).
20. F. Feigl, *Z. Anal. Chem.* **60,** 24 (1921).
21. See ref. 3, p. 418.
22. T. Ichikawa, K. Kiyoshi, and H. Kakihana, *Nippon Kagaku Kaishi* **7,** 1186 (1981).
23. L. Baumgarten and I. Marggraff, *Ber.* **63,** 1019 (1930).
24. E. Jungreis, *Report PDS-73-17,* Ames Research Laboratory, Elkhart, Indiana (1973).

25. See ref. 3, pp. 473, 474, 481.
26. L. Barcza and L. Sommer, *Z. Anal. Chem.* **192,** 3404 (1962).
27. See ref. 3, p. 328.
28. R. M. Dagnall and T. S. West, *Talanta* **11,** 1533 (1964).
29. See ref. 3, p. 277.
30. F. Feigl and L. Weidenfeld, *Mikrochemie* (*Emich Festschrift*) (1930), p. 132.
31. E. Fischer, *Ber.* **16,** 2234 (1883).
32. M. Fujimoto, *Anal. Abstr.* **4,** 1918 (1957).
33. See ref. 3, p. 503.
34. *The Testing of Water,* Merck Co., Darmstadt (1976).
35. A. Tschiri, O. Marks, F. A. Saroff, and P. Fehr, *Oberfläche-Surf.* **21,** 81 (1980).
36. J. A. Platte and V. M. Marcy, *Anal. Chem.* **31,** 1274 (1959).
37. See ref. 3, p. 268.
38. D. G. Miller, *J. Water Pollution Contr.* **51,** 2402 (1979).
39. Y. Kondo, *Mikrochim. Acta* **1,** 154 (1937).
40. See ref. 3, p. 202.
41. P. L. Alessandrano and W. G. Charaklis, *Water and Sewage Work Magazine Sept.* (1972).
42. A. M. Jirka and M. J. Carter, *Anal. Chem.* **47,** 1397 (1975).
43. H. Z. Kelkenberg, *Wasser Abwasser Forschung* **8,** 146 (1975).
44. Environmental Protection Agency, *Federal Register,* 2681 (1980).
45. See ref. 4, p. 303.
46. C. E. Mendoza, *Residue Rev.* **43,** 105 (1972).
47. C. E. Mendoza, P. J. Wales, H. A. McLeod, and W. P. McKinley, *Analyst* (*Lond.*) **93,** 34 (1968).
48. N. V. Nanda Kumar, K. Visweswaraiah, and S. K. Majumdar, *J. Agric. Biol. Chem.* **40,** 431 (1976).
49. N. V. Nanda Kumar and Y. Prameli Dev, *JAOAC* **64,** 841 (1981).
50. E. J. Poziomek and E. V. Crabtree, *Anal. Lett.* **14,** 1185 (1981).
51. E. J. Poziomek, E. V. Crabtree, and R. A. Mackay, *Anal. Lett.* **13,** 1429 (1980).

CHAPTER

8

RAPID SCREENING TESTS OF SOILS AND PLANT TISSUES

8.1. GENERAL

Soils are composed of organic and mineral materials on the Earth's surface in which physical, chemical, and biological differentiation into incipient or perceptible horizontal layers have taken place. Plant life and the various phases of crop production are intimately correlated with the quality of soils.

The upper layer of the Earth's solid crust, to a depth of about 10 miles, is about 98% composed of eight chemical elements. These are, in order of decreasing abundance, oxygen, silicon, aluminum, iron, calcium, magnesium, potassium, and sodium. The first four elements contribute to the acidic or weakly basic character of the soil; the last four form the strong bases.

These elements are bound into simple oxides and carbonates or, to a much greater extent, into complex aluminosilicates or iron silicates of the four strong bases mentioned.

These primary silicate rocks may decompose during the weathering of the soil, and the strong bases then appear as water-soluble compounds, such as hydroxides, carbonates, and bicarbonates. These are carried downward by water and leave the soil unless they are changed again into insoluble matter by further chemical reaction. The remaining secondary clay minerals contain much less of the strongly basic elements calcium, magnesium, sodium, and potassium. Simpler compounds such as hydrated oxides of silicon, aluminum, and iron appear in the residual clay, and form from the primary minerals during weathering.

Plant and animal materials decompose on the Earth's surface, forming the organic substances of the soil.

The world's human population is mostly supported by crops grown in surface soils of humid and semihumid regions; therefore the chemical characteristics of these soils are the most important from the agricultural point of view.

Primary minerals that do not undergo weathering changes constitute some 70% of the soil mass. These are the complex silicates, silicon dioxide, calcium carbonate, and dolomite.

Secondary clay minerals are composed of aluminum and iron silicates. These minerals form extensive colloids with enormously high surface areas. The soluble chemical compounds accumulating at the surface of these colloids greatly intensify the chemical processes occurring in the soil.

The ion-exchange property of the clay minerals is one of the most important characteristics of the soil. Because the weathering process causes the split-off of soluble bases from the primary silicates, the secondary colloidal clays are negatively charged. Metallic ions such as calcium or, when there is a deficiency of bivalent metals, hydrogen, are held by the cation-exchange capacity of the soil, and thus are not leached out with water. Consequently, plant nutrient ions are protected in the soil against excessive loss in drainage water. The retention of the bases by the clay minerals decreases in the order calcium > magnesium > potassium > sodium, and their relative abundances in exchangeable form in the soil are usually in the same order.

Soil pH is a very important characteristic, because crop plants vary widely in their requirements and tolerances in this respect. For example, alfalfa and many clovers require essentially neutral soils (a minimum pH of 6.2–6.5). At the other extreme, cranberries require very high acidities around pH 4. The degree of acidity of a soil depends mainly on the ratio of exchangeable hydrogen ions to basic ions held on the clay surfaces. Soils in humid regions gradually become more acidic because calcium and other cations are slowly replaced in the colloids and removed from the soil by leaching. The replacing ions are chiefly hydrogen ions from carbonic acid and other acids formed from decomposing organic matter in the soil. The application of limestone or of slaked lime to acidic soils reverses this process. Clay minerals of the carbonate type possess anion-exchange properties, and phosphate ions may displace the mobile hydroxyl group layer of the day.

The soil's organic content derives from the decomposition of plant and animal material. The rate of this decomposition is fairly rapid at the beginning, then slows as residue is reached that is resistant to further decomposition. This residue forms the fairly permanent stock of soil organic matter. Some of the essential nutrient elements, such as nitrogen, phosphorus, and sulfur are found organically bound in the soil. The major elements required for growth and reproduction of plants are carbon, oxygen, hydrogen, nitrogen, phosphorus, potassium, calcium, and magnesium. Trace elements such as iron, manganese, boron, zinc, and copper are needed for plant growth as well.

Most of these elements (the exceptions are carbon, hydrogen, oxygen, and part of the nitrogen) are absorbed as free ions through the root system. The quality of soil varies according to the concentration of these ions in the soil. The exchangeable forms of the basic elements constitute the principal immediate source from which the nutrients in the soil solution are renewed.

The clay minerals contain basic elements such as potassium and magnesium in their complex structure, not only in the exchangeable form. A clay mineral rich in potassium is illite. Although not immediately available, potassium in the complex mineral serves as a reservoir for the ions, and their abundance is a measure of durability of the soil's chemical fertility. The concentration of complex phosphate is a similar indication of the future fertility of the soil.

The nonmetallic nutrients nitrogen, phosphorus, and sulfur are often bound to organic molecules, and the decomposition of the organic material furnishes these elements in a form available to plants.

The soil chemical tests measure mainly the water-soluble nutrients and extracts of the exchangeable cations and anions. The amounts of exchangeable potassium, calcium, and magnesium, of the acid-soluble phosphates, and of ammonia and nitrate nitrogen are very important indicators of soil quality.

8.2. SCREENING TESTS OF SOILS

Rapid screening tests of soils and plant tissue are similar in their orientative importance to clinical testing of urine and blood in clinical diagnosis. The simplification of complex soil chemical analyses became necessary to make them performable by less trained personnel and agricultural agents. Modern farming requires the screening of soil quality in conjunction with testing of the plant sap (1).

Parallel testing of plant sap and soil may be carried out in the field. A number of commercial companies market portable kits (Hellige Inc., Long Island City, N.Y.; La Motte Chemical Products, Towson, Maryland; Edwards Laboratory, Cleveland, Ohio; Hach Chemical Co., Loveland, Colorado) for rapid spot testing of plant tissues and soils. Although such field tests are only semiquantitative, systematic screening in the field that they make possible gives an immediate diagnostic picture of soil nutrient problems to the farmer. The application of commercial soil test kits for field use has been evaluated (2).

The most important rapid soil tests are carried out in order to measure aeration and porosity, to determine the soil pH and lime requirements, and to estimate the amounts of extractable soil phosphorus, calcium, magnesium, potassium, and nitrate ions. Saline soils require rapid testing for chloride, sulfate, and sodium.

8.2.1. Spot Test for Soil Aeration

A high concentration of oxygen in the soil is required for the satisfactory growth of most agricultural crops, including wheat, corn, alfalfa, and oats. On the other hand, a number of plants are able to function with a very low activity of soil oxygen. These plants, such as cypresses, weeping willows, cranberries, and rice, possess the ability to absorb oxygen from the air above the ground and transfer it to the roots. In most other plants, varying degrees of soil oxygen concentration are needed in the life processes of the plant. The presence of soil oxygen enables the uptake of the nitrogen-, phosphorus-, and potassium-based fertilizers, and the ammonium ion is oxidized in the root area if oxygen is present. The decomposition of organic material in the soil is possible only if excess oxygen is present.

A simple test for aeration is based on the ferrous–ferric redox reaction (3). Good oxygen supply is indicated by the presence of ferric ion in the soil, whereas an oxygen deficiency is shown by the occurrence of ferrous ions.

PROCEDURE. Sample the soil successively at 10, 20, 30, and 40 cm depths. Carry out the test with a freshly taken sample before sampling at the next depth. Place two fresh soil samples of several centigrams each on opposite ends of a piece of filter paper. Add two drops of 20% aqueous hydrochloric acid to each of the samples. Fold the paper and treat one of the wet areas with a drop of 10% aqueous potassium thiocyanate solution, and the second with 0.5% potassium ferricyanate solution. The appearance of a red spot indicates good oxygen activity, whereas a blue spot indicates poor oxygen supply. If both colors appear, the oxygen deficiency is not severe.

The necessity to carry out this test very quickly can be demonstrated by leaving soil samples exposed to the air for a few minutes before repeating the test. A negative ferric ion test rapidly becomes a positive one.

A simple test for the porosity of the soil may give an alternative indication of its oxygen activity. The soil is sampled with an open-faced sampler and calcium carbonate suspension (10 ml dry, powdered precipitated calcium carbonate dispersed in 50 ml water) is added dropwise to its surface. The extent of penetration of the chalk particles and of their disappearance from the surface indicates the degree of porosity and accordingly the degree of aeration. If the white chalk particles remain on the surface, fine porosity and a poor oxygen supply are indicated.

8.2.2. Tests of Soil pH

The most important of all soil screening tests is that for pH, since the availabilities of calcium, magnesium, boron, phosphorus, iron, and manganese are greatly influenced by the pH. The measurement of pH may determine whether pretreatment of the soil by addition of ground limestone is necessary. Battery-operated pH-meters, pH combination electrodes, and any of several systems of colorimetric pH indicators may be used for this determination.

The soil chosen for screening should be subjected to an extraction procedure using mineral acid, neutral salt, or a buffer mixture. The acetic acid extractant should naturally not contain material that could interfere with the particular spot test used.

A number of field testing kits still contain colorimetric indicators, which give reliable if not accurate results. A pH range of 4.0–7.5 is covered by the use of a mixed indicator of bromocresol green, bromocresol purple and bromocresol red (0.05%, 0.1%, and 0.02%, respectively) (4).

Procedure. Place three drops of the indicator on a white spot plate and disperse some grains of the dried soil sample in the fluid. Stir for 1 min by blowing the suspension around with a dropper pipet, and then examine the color of the soil against the white background.

Soil Acidity	Soil pH	Indicator Color
Very strongly acidic	4.0	Yellow
Strongly acidic	4.5	Greenish yellow
Acidic	5.0	Yellowish green
Moderately acidic	5.5	Light green
Slightly acidic	6.0	Bluish green
Very slightly acidic	6.5	Greenish blue
Neutral	7.0	Dark blue
Alkaline	7.5	Purple

Ground and purified barite mineral is added to mask the color of the soil and bring out the indicator color. Such three-component indicators blended with pure precipitated barium sulfate powder for colorimetric pH testing are marketed by Hellige Inc.

Rapid differentiation between acid and neutral soils is possible by the classical thiocyanate test. A 4% potassium thiocyanate solution in 95% ethanol is shaken with an equal volume of soil in a glass tube. Acid soils develop a pink to red color after 10 min, the color depending on the degree of acidity. Soils of pH 6.5 or higher give a colorless solution.

The other rapid test for acidic-soil screening is based on the liberation of hydrogen sulfide from neutral zinc sulfide by the action of acidic soils (5).

PROCEDURE. Place 9 ml of the soil sample in a 300-ml conical flask. Add 0.8 g of a mixture of barium chloride and zinc sulfide (1 : 10 by weight) and 100 ml deionized water. Boil the suspension for 1 min; then place a moistened piece of lead-acetate paper over the flask. Acidic soils are indicated by the appearance of black lead sulfide deposits on the paper; the extent of the blackening is proportional to the soil acidity.

Different crops have different pH preferences. The desirable soil pH range for cranberry, for example, is 4.2–5.4, whereas spinach prefers 6.4–7.7 and eggplant 5.4–6 (6).

For excessively acidic soils, ground lime is applied to elevate the pH. The quantity of limestone needed per acre depends on the soil type and on the crop to be planted in the particular soil.

8.2.3. Test for Phosphorus in Soil

Phosphates react with molybdate to produce salts of the complex phosphomolybdic acid. In these polyheteroacids the molybdic acid has an enhanced oxidizing power toward many inorganic and organic compounds. When stannous chloride is oxidized the reduction product of molybdic acid, called molybdenum blue, is formed.

PROCEDURE (7). Fill a glass tube graduated at 10 ml with molybdate reagent. Add half a level teaspoon of soil, and shake the tube for ~1 min. Centrifuge, and to a 5-ml portion of the supernatant add some grains of dry, powdered stannous chloride. Mix vigorously, and compare the resultant color with the phosphate test chart. High concentrations of phosphorus give a dark blue color, whereas a low phosphorus content does not lead to color formation. This test does not work on alkaline soils.

PREPARATION OF MOLYBDATE REAGENT. Dissolve 8 g ammonium molybdate in 200 ml deionized water. Add to this solution a mixture of

126 ml conc. hydrochloric acid and 74 ml deionized water, with constant stirring. Prepare from this stock solution a working solution by diluting it just before running the test.

The sodium bicarbonate test indicates the presence of exchangeable or surface-adsorbed soil phosphorus.

PROCEDURE (8). Place 5 g dry soil and 1 teaspoon of carbon black 250-ml Erlenmayer flask with 100 ml 0.5 *M* sodium bicarbonate adjusted to pH 8 with sodium hydroxide and shake for 30 min. Filter through a piece of Whatman no. 40 filter paper. Place an aliquot of the filtrate in a 25-ml volumetric flask. Add 5 ml molybdate reagent (see above) and mix. Add 1 ml dilute stannous chloride solution, mix immediately, and bring volume to 25 ml. Read the color intensity by comparing the color with the standard color chart. (The standard chart is prepared using 5 ml of the sodium bicarbonate solution and the standard phosphate solution.)

PREPARATION OF DILUTE STANNOUS CHLORIDE SOLUTION. Dissolve 10 g stannous chloride dihydrate in 25 ml conc. hydrochloric acid to make a stock solution. Add 0.5 ml of the stock solution to 66 ml deionized water to make the dilute solution.

PREPARATION OF STANDARD PHOSPHATE SOLUTION. Dissolve 0.2195 g dried KH_2PO_4 in water and dilute to 1000 ml. This solution contains 50 ppm phosphorus.

For quick soil testing for phosphorus, a reagent system in a stable powdered formulation is marketed by Hach Chemical Co. The use of these compact powder pillows eliminates the necessity of handling strongly acidic solutions.

8.2.4. Rapid Test for Exchangeable Potassium

A screening test for water-soluble and exchangeable potassium is necessary to decide whether additional potassium fertilizer is needed.

The precipitation of potassium cobaltinitrite and assessment of the ensuing turbidity using a standard chart is still a very common field technique.

PROCEDURE (9). Fill a graduated test tube to 10 ml with sodium cobaltinitrite reagent. Add 1 level teaspoon of air-dried soil and shake the tube for ~1 min. Filter the solution, and to a 5-ml aliquot add 2.5 ml anhydrous isopropyl alcohol. Compare with the standard potassium chart after 3 min.

PREPARATION OF SODIUM COBALTINITRITE REAGENT. *Stock solution:* Dissolve 5 g sodium cobaltinitrite and 30 g sodium nitrite in 80 ml deionized water. Add 5 ml glacial acetic acid and make up the volume to 100 ml. This stock solution is stable only for a few days. *Working solution:* Dilute 5 ml of the stock solution in 100 ml 15% sodium nitrite solution, and adjust the pH to 5.0 with acetic acid.

Not only the exchangeable potassium, but also the potassium measured by acid extraction (with 0.5 *N* HCl) is of importance, because this additional potassium can be released by weathering.

The Hach Portable Soil Test Laboratory estimates potassium content using powder pillows containing sodium tetraphenylborate. The water-soluble reagent is a sensitive precipitant for potassium in a neutral or acetic acid solution. Because heavy metal ions interfere with this test a second powder pillow is included that contains the appropriate complex-binding agents.

8.2.5. Rapid Tests for Nitrate

The level of nitrogen-containing compounds (nitrates, nitrites, and ammonia) in the soil is very much dependent on the climatic conditions prevailing immediately prior to the test.

A classical spot test for the estimation of nitrate uses diphenylamine, which is oxidized in concentrated sulfuric acid medium to a blue semiquinoidal compound.

PROCEDURE. (10). Place a drop of the soil extract on a white spot plate and add four drops of a 0.2% solution of diphenylamine in conc. sulfuric acid. Stir the mixture. After 2 min compare the intensity of the resulting blue color with a calibrated color chart.

A noncorrosive, stable reagent tablet may also be used for the semiquantitative estimation of nitrate in soil. The reagent tablet contains sulfanilamide, *N*-1-naphthylethylenediamine dihydrochloride, zinc dust, sulfosalicylic acid, boric acid, and magnesium stearate. Rapid reduction of nitrate to nitrite is accomplished by formation of the stable complex between zinc dust and sulfosalicylic acid, with subsequent azo dye formation as the chromogenic reaction.

PROCEDURE (11). Place 5 ml of the soil extract in a standard test tube. Drop a reagent tablet into the test solution and shake until it dissolves. Compare the color 2 min after beginning shaking with a standardized color chart, being sure to judge color while looking through the test tube from the top to the bottom.

Nitrate nitrogen concentrations can be rapidly estimated in field soils by using commercial paper test strips (E. Merck, Darmstadt, F. R. G.). The strips can be calibrated with standard nitrate solutions. (12)

Test for Nitrate Nitrogen Released in Soils

Apart from soluble nitrates, there are bound nitrates in the soil, and their estimation is crucial for evaluation of the fertility of the soil. The soil subjected to such a test should be incubated for a given period of time under specified conditions.

In a test developed in the Iowa State College Soil Testing Laboratory, a sample of soil mixed with vermiculite is placed in a filter tube and carefully washed free of soluble nitrate with deionized water. The leaching with water serves a dual purpose. It removes soluble nitrates, and it ensures the optimal moisture conditions for microbial activity. The sample is incubated for 2 weeks in a humid chamber at 35°C. The vermiculite ensures moisture control due to its high water-holding capacity. After the 2-week incubation, the nitrate produced is estimated by the phenyldisulfonic acid method.

PROCEDURE (13). Place 10 g of soil mixed with an approximately equal volume of exfoliated vermiculite in a test tube through which water can be percolated. Wash the sample free of nitrates with three successive 20-ml portions of deionized water, draining each portion before adding the next. Apply suction to remove the excess water. Place the sample in an incubator set at 35°C. A high level of humidity is maintained in the incubator by placing a perforated plastic cap on the tube. This simple procedure minimizes moisture loss during the incubation.

After 2 weeks, leach the sample again as above. Evaporate 2 ml of the leachate to dryness and add 1 ml saturated calcium hydroxide solution and 1 ml phenoldisulfonic acid reagent. Allow to stand for 10–15 min. Add 14 ml water and 5 ml ammonium hydroxide and read the concentration by visual comparison against a series of prepared standards.

PREPARATION OF PHENOLDISULFONIC ACID REAGENT. Dissolve 25 g pure white phenol in 150 ml conc. sulfuric acid. Add 75 ml fuming sulfuric acid (15% sulfur trioxide), stir well, and heat in a boiling water bath for 2 hours.

The estimation of nitrate is based on the nitration of position 6 of 2,4-phenoldisulfonic acid in fuming sulfuric acid.

8.2.6. Test for Calcium in Soil

The following quick turbidimetric test for calcium in the soil using oxalate as the precipitant is disturbed by many factors, such as pH level and presence of organic material. The soil pH test generally renders testing for soil calcium unnecessary.

PROCEDURE (4). In a micro test tube add two drops of 5% ammonium oxalate solution to 1 ml soil extract and mix by shaking. Compare the extent of white precipitate formation with a turbidity chart.

8.2.7. Test for Magnesium in Soil

Magnesium is an essential plant nutrient. There are extensive agricultural areas lacking sufficient magnesium concentrations. Quick field tests usually use the titan yellow procedure. This involves the precipitation of magnesium hydroxide, which forms a magnesium lake with the reagent.

PROCEDURE. (15). Add one drop of titan yellow solution (0.15 g titan yellow dye dissolved in a mixture of 90 ml ethyl ether or isopropyl alcohol and 10 ml deionized water) to 1 ml soil extract in a micro test tube, and mix by shaking. Add to this mixture one drop 5% sodium hydroxide solution. Shake thoroughly. Compare with an appropriate test color chart. The color ranges from light orange to peach red, with a yellow color indicating no magnesium.

When dolomite limestone is used for adjustment of the soil pH, magnesium testing is unnecessary. If the liming material is calcic limestone, however, soil magnesium testing is essential.

8.2.8. Rapid Test for Soil Salinity

The most important determination for soil salinity appraisal is the total soluble salt content. This is estimated roughly by either simple evaporation of a soil extract and weighing of the residue or by measurement of the extract's electrical conductivity. The gypsum requirement (Section 8.2.9.) gives another important indication of the exchangeable sodium status of the soil.

Total soluble solids are estimated by evaporating a sample of saturation extract or a 1 : 5 soil–water extract. A saturated soil paste is prepared by adding deionized water to a sample of soil under constant stirring. At saturation the soil paste glistens as it reflects light, flows slightly when the container is tipped, and slides freely and cleanly off a spatula. The sample is allowed to stand for an hour, after which the criteria for saturation should be rechecked. No free water should collect on the soil surface and the paste should not stiffen. A 250-g sample is generally sufficient to prepare a representative sample.

PROCEDURE. Weigh 250 g of air-dried soil and add sufficient water to it to form a saturated soil paste. Transfer the paste to a Buchner filter funnel with a piece of filter paper in place and apply a vacuum. If boron is also to be determined, use acid-weathered Pyrex glass as a collecting bottle. Refilter turbid filtrates through the soil. Take an aliquot of the clear extract and evaporate it in a tared porcelain dish. After evaporation to dryness, dry in an oven at 105°C, cool in a desiccator, and weigh the residue.

There is no rigid limit for salinity of the soil; it is a function of the type of crop, type of soil, moisture regime, climate, and so forth. Above certain salinity levels, however, crops might be injured. The usual range of salinity encountered in soils is 100–3000 ppm, but a satisfactory range for most plants is less than the 100–1000 ppm soluble salt in a saturation extract.

The salt range can also be indicated by the conductivity of the extract. The standard solution used is 0.1 *N* aqueous potassium chloride solution, which has a conductivity of 0.01288 mho at 25°C. The satisfactory range of conductivity of the soil extract for plants is $0.1–1.0 \times 10^3$ mho.

8.2.9. Test for Gypsum Requirement

A very indicative screening test for exchangeable sodium and magnesium in soils is the rough measurement of the amount of gypsum required to replace the sodium. To a weighed amount of soil is added a given amount of a saturated solution of gypsum and the combined amount of calcium and magnesium left in solution is estimated by drop count titration. The amount of calcium absorbed by the soil (initial calcium in solution minus calcium and magnesium in solution after equilibration with soil) is the gypsum requirement.

PROCEDURE. Place 5 g of air-dried soil in a flask and add 100 ml of saturated gypsum solution. Stopper and shake for 10 min. Filter, and determine the amount of calcium plus magnesium in a 5-ml aliquot as follows: Add a few crystals of sodium diethyldithiocarbamate. Swirl to dissolve. With a calibrated sampling scoop add ~50 mg buffer indicator reagent, and carry out a drop count titration with 0.01 *N* EDTA reagent until the color changes from wine red to blue. Use a similar procedure to titrate 5 ml saturated gypsum solution.

When *x* ml 0.01 *N* EDTA is used for titration of 5 ml gypsum solution, and *y* ml was consumed by 5 ml soil extract, (*x*-*y*) ml 0.01 *N* EDTA calcium equivalent was absorbed by 0.25 g of soil.

Each milliliter of EDTA used indicates a requirement of 8 tons of 85% gypsum for 1 acre-foot of soil. Because a small amount of exchangeable sodium in the soil is permissible, a reduction of 4–5 tons of gypsum applied per acre-foot can be made.

8.2.10. Test for Carbonates in the Soil

A simple test for free lime in soil makes use of the effervescence produced when lime is treated with acid. An additional indication is the pH of a soil extract. If the pH of the soil in a 1 : 25 soil–water extract is not higher than 8.5, the effervescence can be attributed to the presence of calcium carbonate or a mixture of calcium and magnesium carbonates.

PROCEDURE. To a spoonful of soil in a watch glass add 33% hydrochloric acid dropwise unitl fizzing is detected.

8.2.11. Test for Lime Requirement

The amount of lime needed to adjust the pH value of the soil to a desired level is called the lime requirement. A simple but not very accurate method for measuring the lime requirement utilizes the depression of the pH of a buffer solution when soil is added (16). A fairly large number of samples can be screened relatively quickly.

PROCEDURE. Add 20 ml buffer solution to 10 g soil in an Erlenmeyer flask and shake for 10 min. Transfer to a 50-ml beaker and measure the pH with a pH meter. The lime requirement is indicated by the depression in pH of the buffer according to Table 1. The indicated amounts of pure calcium carbonate have to be added per acre if mixed with 6.5 inches of soil having a bulk density of 1.35.

With some sandy, highly acidic soils, a pH higher than 6.5 is obtained. In these cases the procedure has to be repeated with 50 g soil and 20 ml buffer. The lime requirement then has to be divided by 5.

PREPARATION OF THE BUFFER SOLUTION. Dissolve 1.8 g *p*-nitrophenol,

Table 1.

Soil Buffer pH	Lime Requirement (Tons $CaCO_3$/acre)
6.7	1.6
6.6	2.2
6.5	2.8
6.4	3.4
6.3	4.0
6.2	4.5
6.1	5.2
6.0	5.8
5.9	6.4
5.8	7.0
5.7	7.6
5.6	8.2
5.5	8.9
5.4	9.5
5.3	10.1
5.2	11.0
5.1	11.7
5.0	12.4
4.9	13.2
4.8	14.0

2.5 ml triethanolamine, 3.0 g potassium chromate, 2.0 g calcium acetate dihydrate, and 40.0 g calcium chloride dihydrate in ~800 ml deionized water. Adjust the pH to 7.5 using hydrochloric acid or sodium hydroxide solution, and dilute to one liter.

8.3. RAPID CHEMICAL TESTING IN PLANT TISSUE

The sap of fresh plant tissue may be screened in the field for nitrate, phosphorus, and potassium content. These spot tests are very useful as guides to the interpretation of the relative amounts of nutrients taken up by the plants.

8.3.1. Nitrate Nitrogen Spot Test in Plant Tissue Sap (17)

The nitrate concentration in plant tissue depends on the stage of growth, the weather conditions, and the concentrations of other nutrients. In case of deficiency of other nutrients, the nitrate test can easily give high results even when the soil nitrate content is moderate. Because the nitrate concentration shows a fairly wide variation in different parts of the plant, the test should be standardized on an appropriate part of the plant.

In corn, the nitrate content in the stalk varies greatly: The highest concentration is at the base; the lowest, at the top. The nitrate concentration during the ear development is very high. The localization of nitrate to below the ear shank is so strong that a strongly positive test at this point could be accompanied by a negative test above this point. Thus the location of the test becomes crucial.

PROCEDURE. Cut a vertical slice of nodal tissue with a clean sharp knife. Place two drops of 1% diphenylamine solution in conc. sulfuric acid on the freshly cut surface. An abundant supply of nitrogen is indicated by the appearance of a dark blue color. Nitrogen deficiency, on the other hand, is indicated by the lack of blue color (the plant tissue may be charred by sulfuric acid to a brown color). The intensity of the blue coloration indicates the nitrogen concentration.

The test can be standardized against known amounts of nitrogen. A 100 ppm solution of nitrogen in the nitrate form is prepared by dissolving 0.072 g potassium nitrate in 100 ml deionized water.

These extremely simple tests can be carried out all along the stalk in order to determine the area in which peak level of nitrogen is present and to estimate total nitrogen concentration. The base of the leaf midrib is also a favorite site for the test.

Before a decision is made for a particular soil treatment, six to eight representative stalks should be chosen and the nitrate content estimated in each of them from base to tassel.

It has been found that the nitrate level is highest from about 11 a.m. to 3 p.m. (18).

In oats, the stem test for nitrates is carried out at a node, since more sap is available there. A drop of the diphenylamine reagent is placed on the freshly cut surface of the node. Evaluation of the result is similar to that for the corn.

The stem test for nitrates in soybeans shows a situation completely inverse to that in corn. The concentration of nitrate is the highest at the top of the stem and gradually decreases toward the base. The stem should be diagonally at an appropriate height and the freshly cut surface treated with two drops of diphenylamine reagent.

Nitrate content in grass blades is estimated by chopping a representative sample. The mashed blades are placed in a depression of a white spot plate and treated in the usual way.

Noncorrosive reagents for the estimation of nitrate content have been developed by many commercial companies. The reactive materials of a typical noncorrosive test for nitrate (Nitraver® VI Reagent Pillow, Hach Co.) include a reducing metal (generally finely powdered zinc or cadmium) in a mild acid medium. The powder also contains chromogens for the estimation of nitrites, which are formed as reduction products from nitrates (11).

PROCEDURE. Make a diagonal cut into the conductive tissues of the stalk. Insert a filter paper strip into the cut, and wet the paper by pressing it against the fresh cut. Treat the paper with a small amount of the reagent powder or reagent pillow. Observe the eventual color development after 1 min. The oldest formulation of such reagent powder (20) is a mixture of 4 g sulfanilic acid, 2 g α-naphthylamine, 10 g magnesium sulfate, 2 g finely powdered zinc, 100 g barium sulfate, and 25 g citric acid.

8.3.2. Screening Test for Phosphorus in Plant Tissue Sap

Although plants absorb phosphorus mainly in the inorganic form, the plant metabolism incorporates it into organically bound form. Organically bound phosphorus is not detected by the usual molybdenum blue reaction, but the detection of inorganic phosphorus in the plant still has important diagnostic value. It indicates the abundance of the reserve of inorganic phosphorus in the plant sap.

In cases of limited supplies of nitrogen and potassium, a high phosphorus test result may be found. On the other hand, plants showing phosphorus deficiency usually have high nitrogen contents.

The spot test for phosphate is based upon production of the complex phosphomolybdic acid with sodium molybdate and the reduction of this heteropolyacid to molybdenum blue with a tin rod.

Because phosphate accumulates to differing extents in different parts of plants, the tested position on the plant should be standardized: near the tassel in corn, at the upper stem or a node in oats, and at the upper part of the stem in soybeans.

PROCEDURE. Slice the plant tissue with a sharp knife, and wet a filter paper strip with the plant juice, squeezing the tissue if necessary (4). Treat the wetted area with a drop of the molybdate reagent. Press a tin rod to the spot for ~10 sec, and read the resulting color as follows:

No color: plant very deficient in phosphorus
Slight blue: plant deficient in phosphorus
Medium blue: plant slightly deficient in phosphorus
Dark blue: plant adequately supplied with phosphorus

PREPARATION OF THE MOLYBDATE REAGENT. Dissolve 8 g ammonium molybdate in 200 ml deionized water. Add to this solution a mixture of 126 ml conc. hydrochloric acid and 74 ml deionized water, with constant stirring. Dilute the concentrated molybdate reagent with four volumes of deionized water just before use (20).

In the classical Purdue test the available phosphates are extracted from the plant tissues into the ammonium molybdate–acid solution, and the reduction of the resultant phosphomolybdic acid to molybdenum blue is achieved by the addition of some relatively stable stannous oxalate powder.

This test in corn is usually applied to the leaf blades, with the leaf just below the lower ear node generally being taken as the sample. A representative sample should be composed of material from at least six to eight stalks. In oats, similarly located leafs are collected from the upper part of the plant. The sample for phosphorus screening in soybeans is prepared from the terminal leaflet from the third node below the growing point. Leaflets from 15 to 20 different plants constitute the representative sample.

PROCEDURE. Grind the sample tissue with a blender and place a quarter-teaspoon of the ruptured cells in a test tube filled to the 10-ml mark with the diluted molybdate reagent (see above). Stopper and shake with a Vortex mixer for exactly 1 min. Add about a pinhead of solid stannous oxalate powder, shake again, and observe the color. Colorlessness or a yellow tint signals a very poor phosphorus supply; green or bluish green, a deficient phosphorus concentration; light blue, a medium phosphorus supply; medium blue, an adequate phosphorus concentration; and dark blue, abundant phosphorus.

Arsenic is a potentially interfering chemical in this test, because in the presence of arsenic compounds the similar arsenomolybdic acid is formed and further reduced to the blue compound. This test is thus not indicative for phosphorus in plants treated with arsenic compounds.

8.3.3. Screening Tests for Potassium in Plant Tissue

The potassium supply in the soil is an extremely important parameter for plant growth. If the supply of potassium in the soil decreases, its concentration also declines in the plant sap, causing a series of disorders. Potassium deficiency in grass causes migration of this ion from older tissues to younger ones. The older leaves near the base of the stem turn brown and gradually die. Conversely, in soybeans the younger leaves will become more deficient in potassium than the older ones. The lower leaves will then be richer in potassium than will those toward the top.

The potassium content of the plant tissue is estimated either by squeezing the sap onto a filter paper strip impregnated with potassium reagent or by dipping the reagent strip into a fine suspension prepared from plant tissue ground in a blender.

Potassium reagent paper is prepared by impregnating the paper with dipicrylamine. This bright-yellow, fairly strong acid dissolves in bases such as sodium carbonate and produces orange-red solutions.

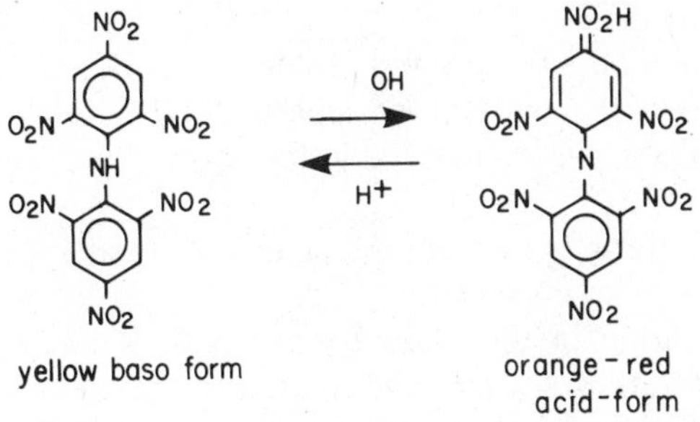

yellow baso form ⇌ (OH / H+) orange-red acid-form

The presence of potassium ions disturbs this equilibrium. A red precipitate of potassium dipicrylaminate appears which is distinctly resistant to dilute acids. The selective test for potassium is based on this reaction (21).

The spot test for potassium has been applied for field use. Test papers were produced in three different concentrations of the reagent system and their sensitivities for potassium calibrated.

PROCEDURE. Place the test paper along the midrib of the leaf for corn, at a node near the middle or upper part of the plant for oats, or on the enlarged base of the petiole in a leaf from the top of the plant for soybeans. Place one jaw of a pair of pliers on one of the test spots, and the other underneath the appropriate portion of the plant. Squeeze the pliers until the plant sap wets the spot. Repeat the procedure on the other two spots. Allow the plant juice to react for ~30 sec. Treat the spots with one to two drops of 0.5 *N* hydrochloric acid solution. The presence of potassium is indicated by the persistence of a reddish-orange color on the spot area, whereas in the absence of potassium a lemon-yellow color appears. If all the spots turn lemon yellow, the plant examined suffers from a major potassium deficiency. When one spot shows a persistence of reddish-orange color and the two others change to yellow, a moderate deficiency is indicated, one still curable by application of potassium fertilizer. If two spots are reddish orange, the potassium supply is adequate for field crops, such as corn, small grain, soybeans, and pasture. An excellent potassium supply is shown by the persistence of an orange-red color at all three spots.

PREPARATION OF THE TEST STRIPS. Prepare solution A by dissolving 0.60 g dipicrylamine and 0.6 g sodium carbonate in 16 ml deionized water, stirring, boiling, cooling, filtering, and making up to a volume of 25 ml. Solution B is prepared by dilution of 8 ml solution A to 25 ml with deionized water. Solution C is prepared by dilution of 10 ml solution B to 15 ml with deionized water.

Make the test papers by dropping solutions A, B, and C onto Whatman no. 1 filter paper strips at locations separated by ~1 cm each. Air-dry the spots. Such reagent papers are stable for about 1 year after preparation.

Indirect testing for potassium deficiency is feasible through detection of iron in the nodes of corn stalks. In cases of potassium deficiency there is iron accumulation in the plant tissues (22). It has been shown that the iron concentration is inversely related to the total potassium in the leaf, nodal, and internodal tissue.

PROCEDURE. Cut the stalk with a clean, sharp knife, and place some drops of potassium thiocyanate on the nodal area, followed by a drop of 6 *N* hydrochloric acid. Potassium deficiency is indicated by the appearance of a red color.

REFERENCES

1. W. Sabbe, *JAOAC* **63,** 763 (1980).
2. D. E. McCoy and S. J. Donohue, *Commun. Soil Sci. Plant Anal.* **10**(4), 631 (1979).
3. A. Hoffer, *Better Crops* **29,** 16 (1945).
4. M. L. Jackson, *Soil Chemical Analysis,* Prentice Hall Inc. Englewood Cliffs, N.J. (1958).
5. E. Truog, *Wisconsin Agricultural Experiment Station Bulletin 312,* (1920).
6. National Plant Food Institute data, Washington, D. C.
7. A. J. Ohlogge, *Purdue Agricultural Experiment Station Bulletin 584,* Purdue University, West Lafayette, Indiana (1954).
8. S. R. Olsen, C. V. Cole, F. S. Watanale, and L. A. Dear, *U. S. Dept. Agr. Circ.* **939** (1954).

9. S. R. Olsen, *Soil Sci. Soc. Am. Proc.* **17,** 20 (1953).
10. H. A. Lunt, "The Morgan Soil Testing System," *Connecticut Agricultural Experiment Station Bulletin 541,* (1950).
11. E. Jungreis, unpublished studies (Jerusalem).
12. J. Hunt, Y. Wai, A. Barnes, and D. J. Greenwood, *J. Sci. Food Agric.* **30**(4), 343 (1979).
13. G. Stanford and J. Hanway, *Soil Sci. Soc. Am. Proc.* **19,** 74 (1955).
14. C. H. Spurway and P. Lawton, *Michigan Agricultural Experiment Station Technical Bulletin 132,* (1945).
15. H. V. Young and R. F. Gill, *Anal. Chem.* **23,** 751 (1951).
16. H. E. Shoemaker, Ph.D. Thesis, The determination of acidity in Ohio soils using lime addition, base titration and buffer equilibrium method, Ohio State University (1959).
17. E. Mehnert and R. Hudec, *Monatschr. Veterin.* **35**(3), 98 (1980).
18. C. D. Welch, unpublished observations, North Carolina Agricultural Experiment Station,
19. R. H. Bray, *Soil Sci.* **60,** 219 (1945).
20. A. Thornton, *Purdue Agricultural Experiment Station Bulletin 204,* Purdue University, West Lafayette, Indiana (1939).
21. N. S. Poluektoff, *Mikrochemie* **14,** 265 (1933–34).
22. A. Hoffer, *Purdue Agricultural Experiment Station Bulletin 298,* Purdue University, West Lafayette, Indiana (1930).

CHAPTER

9

RAPID SCREENING TESTS FOR FOOD ADULTERATION AND FOOD COMPOSITION

9.1. GENERAL

The partial substitution of one product for another in which the substituted material is inferior to the original product is defined as adulteration. Foods may be adulterated in a variety of ways. Mineral oil is sometimes substituted for vegetable oil; saccharin is sometimes used instead of sugar. Starch is sometimes used to extend powdered cocoa, and butter can be replaced by margarine. One of the most common food adulterants is water, which also lowers the nutritional value of the food. Because the law defines exactly the minimum butterfat concentration in milk, water is often added to milk until the fat concentration reaches the minimum allowed value.

Modern agriculture requires the use of chemicals to protect against insect and microbial damage. The resulting pesticide residues can cause health problems. Food merchandisers also face the problem of food deterioration and preservation. Chemical additives solve part of the problem, but new ones emerge because of the possible toxicity of these additives.

Some of the proven toxic materials can be detected by simple screening tests, although most of the analytical procedures that comply with the various safeguards of the Food, Drug, and Cosmetic Act are of complex nature.

In a number of selected examples, however, there is a definite possibility of using simple, unsophisticated tests. The objective of such tests is to provide quality control with a minimum expenditure of resources. Sometimes it is wasteful to devote resources to achieving precision or accuracy beyond that necessary.

The qualitative detection of nonpermissible food preservatives definitely has indicative value. For example, simple spot tests that detect certain volatile acids enable one to determine whether tuna fish is decomposing. The qualitative screening of food products for the presence of mammalian urine and feces contamination is of diagnostic importance for

Public Health Service officials. Economic "cheating" involving adulteration or misbranding, watering of milk, partial substitution of margarine for butter, adulteration of vanillin with the cheaper coumarin, and use of mineral instead of vegetable oils can be detected by inexpensive, rapid screening methods. It should be kept in mind, however, that even in the relatively few examples in which a "yes" or "no" answer has decisive value, these tests are to be considered only preliminary examinations, and in most cases confirmation by an exact analytical procedure is required.

9.2. SCREENING TESTS FOR SUGARS IN FOOD

The natural sweetening agents in food are sugars, simpler carbohydrates having a sweet taste. Most food analysts deal mainly with two monosaccharides, glucose and fructose, and three disaccharides, sucrose, lactose, and maltose. Most "sugar" is the hydrolysis product of sucrose, glucose, and fructose in an equimolar mixture. Milk is the only natural food product that contains lactose. Maltose is formed by the partial hydrolysis of starch.

A number of tests for particular carbohydrates have been proposed based on the reactivity of the carbonyl group, on oxidative splitting of vicinal glycol groups, and on the color reactions between the degradation products (of reactions with strong acid) and organic compounds. With simple screening tests, we may determine, for example, whether a milk product contains only the natural lactose or was sweetened with sucrose.

There are reducing and nonreducing sugars. All of the monosaccharides and most of the disaccharides are able to reduce alkaline solutions of many metallic salts, organic compounds such as picric acid, and methylene blue. Osazone formation is another feature of reducing sugars.

9.2.1. The Fehling Test

One of the classical tests for the presence of reducing sugars is the Fehling test, in which cupric ions are reduced in alkaline solution to form a red cuprous oxide (1).

PROCEDURE. Add 1 ml test solution to a mixture of 1 ml Fehling's alkaline solution and 1 ml Fehling's copper sulfate solution. Boil the mixture for 5–10 min. The presence of reducing sugars is indicated by the appearance of a yellow or red precipitate.

PREPARATION OF REAGENTS. Fehling's alkaline solution consists of 173 g sodium potassium tartrate and 50 g sodium hydroxide made up to

500 ml with deionized water. Fehling's copper sulfate solution consists of 34.64 g crystalline copper sulfate made up to 500 ml with deionized water.

9.2.2. Specific Test for D-Glucose (2)

Free glucose in foods is selectively detected by using a coupled enzyme system containing glucose oxidase and a leucochromogen such as *o*-tolidine (see Section 3.2).

9.2.3. Tests for D-Galactose

D-Galactose-containing sugars and D-galactose can be detected in food by an enzyme system containing galactose oxidase and a leucochromogen (3, 4).

$$\text{D-galactose} + O_2 \xrightarrow[\text{oxidase}]{\text{galactose}} \text{D-galacto-hexo dialdose} + H_2O_2$$

$$H_2O_2 \xrightarrow{\text{peroxidase}} \text{Oxygen} + \text{water}$$

$$\text{Oxygen} + \text{colorless leucochromogen} \rightarrow \text{colored oxidized chromogen}$$

Worthington Biochemical Corporation (Freehold, New Jersey) markets a test kit for qualitative and quantitative estimation of galactose in food, called Galactostat, that contains galactose oxidase, peroxidase, and a chromogen.

PROCEDURE. Add to 1 ml neutral test solution 2 ml Galactostat reagent. Mix and warm for 5 min at 37°C. The presence of galactose is indicated by the appearance of a red color. The test is positive for galactose, galactosamine, galactose glycosides, and galactose-containing polysaccharides.

An alternative test for galactose utilizes the oxidation of this sugar to mucic acid with nitric acid. Galactose-producing di-, tri-, and polysaccharides, such as lactose and raffinose, also produce mucic acid as an oxidation product with nitric acid.

PROCEDURE. Place the test solution in a beaker and add 20 ml conc. nitric acid. Evaporate slowly on a hot plate until the volume is reduced to about one-sixth of the original. After 24 hours, examine the beaker. The appearance of a fine, white precipitate (mucic acid) indicates the presence of galactose or a galactose-producing sugar. A semiquantitative estimation is possible by multiplying the weight of the mucic acid by 1.33.

9.2.4. Test for Pentose

Pentoses, whether as free sugars or as glycosides, are detectable in food with the orcinol reaction (5).

PROCEDURE. Heat 5 ml of the reagent (0.2% orcinol in 30% hydrochloric acid containing ferric ions) and a few drops of the test solution to boiling. The presence of pentoses is indicated by the appearance of a green color.

9.2.5. Test for Keto Hexoses (6)

PROCEDURE. Mix the test solution with 10 ml 1 *N* hydrochloric acid and ~10 mg resorcinol. Heat the mixture on a water bath for 15 min. The presence of ketoses, either free or as glycosides, is indicated by the appearance of a red color.

9.2.6. Test for Differentiation Between Keto and Aldo Hexoses (7)

The reaction in the following test is of unknown mechanism.

PROCEDURE. Place the test solution in a microcrucible with several drops of the reagent solution and heat the mixture over a free flame. In the presence of keto hexoses, a blue color appears. Aldo hexoses produce a red color, but only after longer boiling.

PREPARATION OF REAGENT SOLUTION. Dissolve 4 g urea and 0.02 g stannous chloride in 10 ml hot 40% sulfuric acid.

9.3. SCREENING TESTS FOR ARTIFICIAL SWEETENING AGENTS

The principle artificial sweetening agents used in food are saccharin, cyclamate, dulcin (4-ethoxyphenylurea), and P-4000 (5-nitro-2-propoxyaniline). The use of dulcin and P-4000 has been prohibited for many years in the United States.

9.3.1. Tests for Saccharin

The simplest screening test for saccharin involves its extraction and tasting of the residue.

PROCEDURE. Place the test solution or an aqueous extract of the sample into a separatory funnel, acidify with hydrochloric acid, and extract with three portions of diethyl ether. Combine the ether extracts and wash with 5 ml water. Evaporate the ether extract, and taste the residue. The detection limit by taste is 20 ppm saccharine for a sample size of 50 ml.

In another test, the saccharin is converted to salicylic acid by fusion with sodium hydroxide, and the acid is detected by the ferric chloride reaction.

PROCEDURE. Place the aqueous extract of the sample into a separatory funnel, acidify with hydrochloric acid, and extract with diethyl ether. Evaporate to dryness in a nickel dish, add 10 ml 10% sodium hydroxide solution, bake for 1 hour on a hot plate, and fuse over a Bunsen-burner flame. After leaching with water, acidify the solution in a separatory funnel and extract it with ether again. Evaporate to dryness and add a few drops of ferric chloride solution. The appearance of a deep-violet color indicates the presence of salicylic acid, and thus of saccharin.

Saccharine in food may be detected by its conversion into phenolsulfonphthalein (8). The latter is red in alkaline medium.

PROCEDURE. Evaporate an aqueous extract of the food to dryness. Add a few drops of sulfuric acid and a few grains of phenol and heat for 10 min at 180°C. After cooling, neutralize with solid sodium carbonate. The presence of saccharin is indicated by the appearance of a red color. The limit of detection is 0.4 mg saccharin/liter.

9.3.2. Detection of Dulcin

PROCEDURE. Make a 100-ml food sample alkaline with 10% sodium hydroxide, if necessary. Extract with two to three 50-ml portions of diethyl ether. Divide the ether extracts equally between two porcelain evaporating dishes, and evaporate at room temperature. Dry the residues for 10–15 min in an oven at 105–110°C. To the first evaporating dish add nitric acid and one drop of water. Formation of a brick-red or orange precipitate indicates the presence of dulcin. Carry out a confirmatory test on the residue in the second evaporating dish as follows: Expose it to hydrochloric acid fumes for 5 min; then add one drop of anisaldehyde. The presence of dulcin in this dish will be indicated by the appearance of a deep-red color.

9.3.3. Detection of P-4000 in Food

PROCEDURE. Bring ~200 ml liquid food or an aqueous extract of ~100–200 g of solid food to pH 8 by addition of 10% sodium hydroxide. Extract successively with three 50-ml portions of petroleum ether. Wash the combined extracts with water, add 4 ml dilute hydrochloric acid (1 : 1), and evaporate the solvent. Add a small piece of mossy tin and let stand on a steam bath for 5 min. Decant the liquid into a test tube and add bromine dropwise. In the presence of P-4000 a rose-red to deep-burgundy color appears that disappears with addition of excess bromine water.

9.3.4. Screening Test for Cyclamate (9)

PROCEDURE. To ~100 ml of aqueous extract of food in a beaker add 2 g barium chloride. Let stand for 5 min and filter. Acidify the filtrate with 10 ml conc. hydrochloric acid and add 0.2 g sodium nitrite. In the presence of cyclamate, a white precipitate appears.

9.3.5. Qualitative Screening for Artificial Sweeteners Using Thin-Layer Chromatography

Qualitative identification of the artificial sweeteners saccharin, cyclamate, dulcin, and P-4000, can be carried out by a simple and rapid method. (10). The four sweeteners are separated and identified by thin-layer chromatography on silica gel using a solvent system of butanol, ethanol, ammonia, and water. Saccharin is detected as a fluroescent spot under shortwave ultraviolet radiation. Successive spraying with solutions of 5% bromine in dimethylformamide–alcohol (1 : 1) and 2% *N*-1-naphthyl ethylenediamine dichloride in alcohol reveals cyclamate as a bright pink spot, dulcin as a brownish or blue spot, and P-4000 as a brown-pink spot. Identification limits are 40 ppm for saccharin, 500 ppm for cyclamate, 300 ppm for dulcin, and 20 ppm for P-4000.

9.4. DETECTION OF ADULTERATION OF EGG YOLK BY TURMERIC

Some bakers use turmeric to color bread and rolls yellow. This yellow color is very similar to the hue imparted to baked goods by egg yolk. Examination of bread, rolls, salad dressings, and pickles for this fraud is relatively simple with a modified spot test using a boric acid reagent (11).

PROCEDURE. Evaporate an aliquot of butanol extract to dryness. Dissolve the residue in a few drops of ethanol and spot Whatman no. 1 filter paper with this solution. Dry the paper for ~2 min in an oven preheated to 100°C; then add several drops of boric acid reagent to the yellow spot. The presence of turmeric is indicated by the appearance of red rosocyanin.

PREPARATION OF THE SAMPLE EXTRACT.

Salad dressings and pickles: Add 50 ml butanol with stirring to 2 g salad dressing or crushed pickles. Let stand for 15 min, and filter through Whatman no. 42 filter paper.

Bread and rolls: Add 10 g crumbled bread to 50 ml butanol solution, stopper the flask, shake, and let stand for 15 minutes. Filter through Whatman no. 42 filter paper.

PREPARATION OF BUTANOL SOLUTION. Shake 850 ml of *n*-butanol with 170 ml deionized water in a separatory funnel, and discard the excess water.

PREPARATION OF BORIC ACID REAGENT. Dissolve 1 g boric acid and 5 ml conc. hydrochloric acid in 95 ml ethanol. Dry over anhydrous sodium sulfate and filter.

The principal pigment of turmeric is curcumin. Therefore, a 3-ppm curcumin solutions are prepared for comparison purposes. Curcumin (30 mg) is dissolved in 100 ml ethanol, and 5 ml of this alcoholic solution is diluted to 500 ml in butanol solution.

The time required for this spot test is ~1 hour.

9.5. QUALITATIVE SCREENING FOR NATURAL AND ARTIFICIAL FOOD COLORS

Coloring matter in foodstuff may be divided into three major groups: coal-tar dyes, natural organic coloring materials, and pigments and lakes. A simple screening can answer the question of whether the coloring used is a coal-tar dye or naturally present organic matter.

Of about 2000 available coal-tar dyes, only 11 are permitted by law to be used in the food industry. All these dyes are water soluble, and all are substantive dyes, which means that they do not require addition of a mordant to dye wool or silk. They are all acid dyes, so they dye animal fibers from an acid bath.

The qualitative testing of food colors should be carried out in the following order. First the presence of coal-tar dyes should be checked for, because these dyes are used mostly in the food industry. In the absence of coal-tar dyes, the coloration probably results from the presence of natural organic coloring matter. In the event that the tests for coal-tar dyes and natural dyes are both negative and the food is colored, the possibility of the presence of a pigment or a lake should be considered.

9.5.1. Differentiation Between Coal-Tar Dyes and Natural Organic Coloring Matter (9)

Coal-tar dyes normally dye animal fibers readily, producing bright colors. By placing the dyed fiber in a solution in which a reaction opposite to that in which dyeing took place occurs, the coal-tar color is easily and entirely removed. The stripped coal-tar dye generally can redye animal fibers.

Natural organic coloring materials, on the other hand, do not render a bright color to animal fibers. The color is not removed entirely by stripping. The natural organic colorings (with the exceptions of orchil and cudbear) are not able to redye fibers after stripping.

PROCEDURE. Boil wool or silk fibers in an acidic or basic solution of the dye. After removal and thorough washing, strip the color in a bath opposite in reaction (if dyed from an acidic solution, strip in dilute ammonium hydroxide, and vice versa). Remove the stripped fiber and change the reaction of the solution to that of the original dyeing. Boil a fresh piece of fiber in it for 10–20 min, remove, and wash.

Simple screening tests for individual coal-tar dyes are possible if the food color consists of a single dye. The pH at which a dye changes color can be used for its selective detection. A single coal-tar dye is used to dye animal fibers, and the fibers are then treated with sulfuric acid, hydrochloric acid, sodium hydroxide, and ammonium hydroxide solution. The color changes produced are indicated in Tables 1 and 2.

9.5.2. Color Tests for Individual Organic Coloring Matter

A very quick test for natural coloring matter is based on the fact that these materials do not dye wool or silk as readily as do the coal-tar dyes. The natural coloring materials frequently contain more than one colored substance.

The natural colors of food are extracted by different organic solvents under specified acidity conditions. Carotene, xanthophyll, green chlorophyll, and the coloring substances of tomatoes and paprika are extracted

Table 1. Detection of Permitted Coal-Tar Food Dyes[a]

Dye	Reagent			
	1:1 HCl	Conc. H_2SO_4	40% NaOH	NH_4OH
Amaranth	Slightly darker	Violet to brownish	Brown to orange-red	Little change
Erythrosin	Orange-yellow	Orange-yellow	No change	No change
Ponceau SX	Pepper red	Pepper red	Orange-yellow	Orange-yellow
Tartrazine	Slightly darker	Slightly darker	Little change	Little change
Sunset yellow FCF	Slightly redder	Slightly redder	Browner	No change
Guinea green B	Pale orange-yellow	Yellowish brown	Decolorized	Decolorized
Light green SF yellow	Pale orange-yellow	Yellowish brown	Decolorized	Decolorized
Fast green FCF	Orange	Green to brown	Blue	Blue
Brilliant blue FCF	Yellow	Yellow	No change	No change
Indigotin	Slightly darker	Darker	Greenish yellow	Greenish blue
Acid violet 6B	Light brownish yellow	Dark brownish yellow	Yellow	Bluish

[a] Data from ref. 12. Table indicates the color change that occurs in animal fibers dyed with the indicated dye when treated with the indicated reagent.

by diethyl ether in neutral media. Diethyl ether extracts from slightly acidic solutions the coloring substances of alkanet, annatto, turmeric, sandalwood, fustic, Persian berries, and brazilwood. The extraction of the coloring substances of logwood, orchil, saffron, and cochineal from weakly acidic media is carried out with amyl alcohol. The red coloring materials of most common fruits, called anthocyanins, and caramel are extracted with amyl alcohol.

PROCEDURE (12, 13). Evaporate to dryness the ether extracts of the colors, take up the residue in a little warm alcohol, and dilute in water. Dilute the amyl alcohol extract with gasoline and strip the color with

Table 2. Detection of Nonpermitted Coal-Tar Food Dyes[a]

Dye	Reagent			
	1:1 HCl	Conc. H_2SO_4	40% NaOH	NH_4OH
Rhodamine B	Orange	Yellow	Bluer	Bluer
Orchil	Red	Reddish brown	Violet	Violet
Bordeaux B	Violet	Blue	Brick red	Little change
Ponceau 6R	Violet	Violet	Brown	Orange-red
Safranine	Greenish blue	Green	Red	Red
Bismarck brown	Darker red	Browner	Yellower	Yellower
Crocein orange	Orange-red	Orange	Slightly darker	No change
Orange SS	Cherry red	Cherry red	Slightly yellow	No change
Aniline yellow	Violet-red	Orange-yellow	Little change	No change
Fluorescein	Little change	Little change	Green fluorescent	Green fluorescent
Brilliant yellow S	Violet-red	Violet-red	Little change	Little change
Turmeric	Red	Reddish brown	Orange	Orange
Naphthol green B	Yellowish	Brownish yellow	No change	No change
Malachite green	Almost decolorized	Almost decolorized	Decolorized	Decolorized
Patent blue A	Pale orange-yellow	Green-brown	Little change	Little change
Methyl violet	Yellowish	Yellowish	Decolorized	Almost decolorized
Nigrosine soluble	Dull bluish	Dull greenish	Brownish red	Pale reddish
Auramine	Decolorized	Almost decolorized	Decolorized	Paler

[a] Data from ref. 12. Table indicates the color change that occurs in animal fibers dyed with the indicated dye when treated with the indicated reagent.

deionized water. Carry out the following spot tests on the aqueous solutions of the coloring substances:

1. Add to 1 drop of the test solution in a depression of a white spot plate:

a. Four drops of conc. hydrochloric acid. All the redwoods (brazilwood, sandalwood, camwood, and barwood) and logwood will turn deep red, turmeric will turn orange-red, and the flavone colors will turn an intense yellow. No appreciable change in color will be recorded with other natural coloring matter.
b. One drop of 10% sodium hydroxide solution. The presence of logwood is indicated by the appearance of a violet to violet-blue color; redwoods give a violet-red coloration; anthocyanins of red fruit colors change to green and quickly turn brown by oxidation; alkanet exhibits a deep blue color, and orchil a blue coloration; and cochineal changes to violet, turmeric to orange-brown, and the flavone colors of fustic to bright yellow.
c. A small crystal of sodium hyposulfite. Logwood, orchil, and anthocyanins are quickly decolorized, and the colors of logwood and orchil return by reoxidation. All other coloring substances are little affected.
d. One drop of freshly prepared 0.5% ferric chloride solution. The colors of logwood and redwoods change to dark violet, and the flavone colors of fustic to olive green. No marked change is registered with other natural coloring materials.
e. 5% uranium acetate solution. A quickly fading violet appears in the presence of logwood. Alkanet is indicated by a color change to yellowish green, cochineal turns green, and the flavone colors of fustic turn orange.

2. To five drops of the sample solution add one drop of 10% ammonium aluminum sulfate solution in a depression of a white spot plate. The logwood and redwoods slowly turn rose red; all other substances are barely affected.

3. Evaporate to dryness in a porcelain dish a portion of the aqueous sample. After cooling, add one or two drops of conc. sulfuric acid. Observe the color change immediately after the addition of the acid and disregard the subsequent color changes. The appearance of a violet-blue color indicates the presence of alkanet; a blue color results when saffron,

annatto, carotene or xanthophyll is present; turmeric leads to a red color; and a red color changing to yellow is characteristic of the presence of logwood.

9.6. SCREENING FOR ADULTERATION OF DARK-COLORED FRUIT JUICES (14)

Dark-colored juices such as those of the black cherry, morello cherry, Montmorency cherry, red raspberry, and blackberry are sometimes adulterated with elderberry juice or grape skin extract, or with each other, depending on current prices. The qualitative detection of adulteration is based on the fact that these juices show simple patterns of one or two major bands following paper chromatography. These bands represent fruit anthocyanins. Since most adulterants have more complicated anthocyanin patterns, the adulterants can be readily detected even at low concentrations.

Black cherry and blackberry show only one band, at about Rf 0.3. Black raspberry and morello cherry each give two major bands, the former at Rf 0.3 and 0.35 and the latter near 0.25 and 0.33. The addition of 20% black raspberry to morello cherry is shown by a foreign band between the two main morello cherry bands. The chromatograms have to be sprayed with phosphomolybdic acid to bring out this foreign band. On the other hand, addition of 20% blackberry juice to morello cherry is not detectable under visible light, since the blackberry band largely overlaps with one of the morello cherry bands. The presence of blackberry is detected under long-wave ultraviolet light. Under such conditions, blackberry exhibits a strong blue fluorescence absent in morello cherry.

The main adulterants are grape skin extract and elderberry juice. The anthocyanin bands of elderberry are at Rf 0.15 and 0.18. Thus the presence of this adulterant is detected quite easily. Grape skin extract has anthocyanin bands of about equal color intensity. Since most fruit juices tested show most of their color in the central spot, adulteration by grape skin extract can be detected by the presence of much more intense low and high spots.

PROCEDURE (15). Develop a qualitative and ascending paper chromatogram on Whatman no. 1 paper by applying streaks 1 inch above the bottom of the paper. Pour developing solvent (*n*-butanol–acetic acid–water, 6 : 1 : 1 by volume) to approximately half an inch depth in the chromatographic developing tank. Develop the chromatogram overnight at room temperature. Remove and air dry the paper. Spray with 2% aqueous phosphomolybdic acid reagent and observe the color bands under visible light and under long-wave ultraviolet light.

9.7. TESTS FOR NONPERMITTED FOOD PRESERVATIVES

Objectionable food preservatives such as hydrogen peroxide, salicylic acid, boric acid, borates, formaldehyde, and sulfites may be screened for using simple spot tests (16).

9.7.1. Detection of Peroxide in Milk

Peroxide can be detected by the starch–iodide test.

PROCEDURE. Mix equal volumes of a milk sample and 1% potassium iodide solution dissolved in 2% starch paste. The presence of peroxides is indicated by the appearance of a blue color.

9.7.2. Tests for Salicylic Acid

Two tests are used for qualitative screening for salicylic acid and salicylates: the ferric chloride test and the Jorissen test.

For the ferric chloride test, nonalcoholic liquids may be extracted directly. Alcoholic liquids are first made alkaline, evaporated to one-third volume, and filtered.

PROCEDURE. Place 50 ml of the sample or the same volume of the aqueous extract in a separatory funnel. Acidify with 5 ml dilute (1 : 3) hydrochloric acid solution and extract with 50 ml diethyl ether. Wash the ether extract twice with 5 ml water. Evaporate the ether and add a drop of 0.5% neutral ferric chloride solution. The appearance of violet ferric salicylate indicates the presence of salicylic acid.

Because the ferric chloride test is not completely specific for salicylates, the specific Jorissen test, carried out on the residue left after evaporation of the diethyl ether, can be used to confirm the results.

PROCEDURE. Dissolve in hot water the residue obtained after evaporation of the ether extract obtained as in the ferric chloride test. To 10 ml of the solution add four to five drops of 10% potassium nitrite solution, four to five drops of 50% acetic acid solution, and one drop of 1% copper sulfate solution. Boil for 30 sec and allow to stand for 2 min. The presence of salicylic acid is indicated by the appearance of a red color.

9.7.3. Screening Test for Boric Acid and Borates

Free boric acid produces a deep-red compound when evaporated with tincture of curcuma (turmeric), changing yellow diferuloylmethane into the isomeric red-brown rosocyanine (17). The rosocyanine turns blue to greenish-black under alkaline conditions. This reaction renders the test specific for boric acid.

PROCEDURE. Prepare turmeric test paper by soaking filter paper in an alcoholic extract of ground turmeric. Dry the paper and cut it into strips.

Acidify 100 ml of the sample with 7 ml conc. hydrochloric acid, and dip the turmeric test paper into the solution. Remove the test paper and dry it. The presence of borax or boric acid is indicated by the appearance of a red color that changes to a dark blue-green when held over an open bottle of ammonia.

In the presence of oxidizing materials, make the test solution alkaline with lime water. After evaporation to dryness, char the organic substance over a low red flame. After cooling, digest the residue with 10–15 ml 25% hydrochloric acid. Make sure that the sample is distinctly acidic. Dip the turmeric test paper into the solution, and observe the color change as above.

A common interference in this test is caused by ferric ion, which also reacts with turmeric to form a dark-brown product; this color, however, does not change to blue-green with the ammonia-fume treatment.

9.7.4. Screening Test for Formaldehyde in Foodstuff

Nowadays formaldehyde is very rarely used as a preservative. It is occasionally encountered in milk and in fish products. Formaldehyde may be formed by the partial oxidation of methyl alcohol.

The food sample should be prepared for testing by grinding 200–300 g of the solid or semisolid material with 100 ml of water. Transfer the slurry to a short-necked 500-ml or 800-ml Kjeldahl flask, and after acidifying it with phosphoric acid, distill 40–50 ml through a condenser. Liquid foods, such as milk, should be prepared for testing by distillation of 100–200 ml of a phosphoric-acid-treated sample.

PROCEDURE A. To a test tube add 1 ml of the distillate and 5 ml chromotropic acid reagent (~500 mg chromotropic acid dissolved in 72% sulfuric acid). Place the tube in a boiling water bath for 10 min. In the presence of formaldehyde, a bright violet color appears.

PROCEDURE B (9). Treat 15 ml milk or other liquid food material or 15 ml of the distillate with 1 ml 1% phenylhydrazine hydrochloride solution. Add a few drops of 1% ferric chloride solution and hydrochloric acid. In the presence of formaldehyde a red color develops that with time changes to orange-yellow.

PROCEDURE C (18). Dilute 5 ml conc. sulfuric acid with 1 ml water. Transfer 3 ml of the diluted acid to a test tube, and drop into it several centigrams of potassium bromide cystals. After shaking, overlay with 1 ml of the milk sample without mixing. In the presence of formaldehyde, even in concentrations as low as 1 ppm in milk, a violet zone develops at the interface of the two liquids.

9.7.5. Screening Test for Sulfur Dioxide or Sulfite in Food

Sulfites inhibit enzymatic oxidative discoloration in various food products. The U. S. Food, Drug, and Cosmetics Act allows the use of sulfites as preservatives in dehydrated fruits and vegetables, but prohibits their addition to meats and foods recognized as sources of vitamin B1.

The problem of illegal addition of sulfites to ground meat necessitates the use of an inexpensive screening test for effective detection by law-enforcement agencies. This preservative restores the red color of old and dull-colored meat, giving a false appearance of freshness. Samples that show positive results using this quick screening test in the field must be submitted for quantitative confirmatory determination in the laboratory.

PROCEDURE (19). Place half a teaspoonful of ground meat on impermeable white Benchkote paper, and add 0.5 ml malachite green reagent solution. Mix vigorously for 2 min with a spatula, turning the mass frequently. In the presence of sulfites the dyestuff is decolorized after a few minutes.

PREPARATION OF THE REAGENT SOLUTION. Dissolve 200 mg malachite green in 1 liter water. The solution is stable for several weeks.

Tablets for instant preparation of 15-ml reagent solutions are marketed by La Motta Chemical Products Co., Chesterton, Maryland. The limit of detection of this test is 25 ppm sulfur dioxide.

9.7.6. Detection of Benzoic Acid and Its Salts in Food

Benzoic acid and its sodium salt are used as antimicrobial preservatives in a great variety of foods. Added benzoates must be declared on the label of the food package, because most states in the United States allow not more than a 0.1% content of this preservative.

PROCEDURE. To the test solution add one or two drops of 10% sodium hydroxide and evaporate to dryness. To the residue add 5–10 drops of conc. sulfuric acid and a small crystal of potassium nitrate, and heat for 10 min in a glycerin bath at 120–130°C. Decompose any ammonium nitrite by boiling. After cooling, add one drop of freshly prepared ammonium sulfide without mixing. A red-brown ring is formed in the presence of benzoic acid. This ring is composed of ammonium diaminobenzoate. On mixing and boiling the solution turns greenish yellow, because the diaminobenzoate decomposes. This behavior differentiates benzoic acid from salicylic acid and cinnamic acid, which also form the red-brown ring.

9.8. SIMPLE SPOT TESTS FOR OXIDATIVE RANCIDITY DETECTION

Certain classical screening tests detect rancidity in fats before the spoilage can be detected organoleptically.

One of the oldest field tests (20) is based on the reaction of oxidized fat with phloroglucinol in concentrated hydrochloric acid medium. Epihydrin aldehyde CH_2—CH_2—CHO (with O bridging the two CH_2 groups) present in spoiled fat forms a red color. The impurity is proportional to the extent of oxidative rancidity.

PROCEDURE. Mix 1 ml melted fat with 1 ml conc. hydrochloric acid in a test tube together with 1 ml of a freshly prepared 1% diethyl ether solution of phloroglucinol. Formation of a pink color indicates slightly oxidized fat, whereas a red color indicates definitely rancid fat.

All oxidized fats give positive peroxide reactions, since peroxides are the first compounds formed during the oxidative process. A sensitive spot test for organic peroxides is based on the fact that they oxidize ferrous salts to ferric ones. These then react to form Wurster red with *N,N*-dimethyl-*p*-phenylenediamine.

$$H_2N\text{-}C_6H_4\text{-}N(CH_3)_2 + [O] \longrightarrow HN{=}C_6H_4{=}\overset{+}{N}(CH_3)_2 + H_2O$$

PROCEDURE (21). Place a drop of the fat test solution dissolved in 1 : 1 glacial acetic acid–chloroform on filter paper and treat the spot in succession with a drop of 0.1% ferrous ammonium sulfate and a drop of the reagent. A pink color develops in the presence of rancidity.

PREPARATION OF THE REAGENT SOLUTION. Freshly prepare a 0.1% solution of *N,N*-dimethyl-*p*-phenylenediamine in a 5 : 5 : 1 mixture of glacial acetic acid, chloroform, and water.

9.9. SCREENING TESTS FOR ANTIOXIDANTS

Fat autooxidation processes are greatly slowed down by the presence of substances called antioxidants. These substances give up hydrogen atoms more readily than do fatty acids. The transfer of the hydrogen atom from the antioxidant breaks the reaction chain of the oxidation.

Most of the naturally occurring antioxidants show weak antioxidative properties, and consequently a number of very effective artificial, nontoxic, and heat-stable antioxidants were developed. Their concentration in foods is somewhat limited by the Ford Amendment to the Food and Drug Law.

Four artificial antioxidants are used as additives to retard oxidation and inhibit rancidity in oils and fats. These are butylated hydroxyanisole (BHA), butylated hydroxytoluene (BHT), propyl gallate (PG), and nordihydroguaiaretic acid (NDGA). The use of rapid screening procedures is feasible for the detection and differentiation of these compounds.

9.9.1. Test for Propyl Gallate (PG)

PROCEDURE (22). Dissolve ~30 g melted fat or oil in ~60 ml petroleum ether, and transfer to a 250-ml separatory funnel. Shake with 15 ml water for 1 min. Let the phases separate and drain the aqueous phase into a 125-ml separatory funnel. Repeat the extraction of petroleum ether with two additional 15-ml portions of water. Retain the petroleum ether solution for further tests.

Extract the combined aqueous solution with 15 ml diethyl ether and discard the aqueous phase. Evaporate the ether extract to dryness, dissolve the residue in 4 ml 50% alcohol and 1 ml ammonium hydroxide. The appearance of a rose color that fades after a few minutes indicates the presence of PG.

9.9.2. Test for Nordihydroguaiaretic Acid (NDGA)

PROCEDURE. Extract the petroleum ether solution preserved from the procedure above with 20 ml acetonitrile. Drain the acetonitrile layer into a separatory funnel. Repeat the extraction with two additional 30-ml portions of acetonitrile and discard the petroleum ether. Add to the combined acetonitrile extracts 400 ml water and 2–3 g sodium chloride and shake for 2 min with 20 ml petroleum ether. Allow the layers to separate and drain the diluted acetonitrile into a second separatory funnel. Extract the dilute acetonitrile with two additional 20-ml portions of petroleum ether and remove the diluted acetonitrile solution for further extraction. Keep the combined petroleum ether portions for BHA and BHT tests.

Extract the diluted acetonitrile for 2 min with 50 ml 1 : 1 diethyl ether–petroleum ether mixture. Separate the layers, discard the acetonitrile, and evaporate the ether portion to dryness. Dissolve the residue in 4 ml ethanol and add 1 ml 1% barium hydroxide solution. Formation of a rapidly fading blue color signals the presence of NDGA.

9.9.3. Test for Butylated Hydroxyanisole (BHA)

PROCEDURE. Evaporate one-third of the combined petroleum ether portion preserved in the preceding procedure for BHA and BHT tests. Dissolve the residue in 25 ml alcohol and add 25 ml water. Add 1 ml Ehrlich reagent and 1 ml 1 *N* sodium hydroxide and mix. The presence of BHA is indicated by the appearance of a red-purple color.

PREPARATION OF EHRLICH REAGENT. Mix 1 part 0.5% aqueous solution of sodium nitrite with 100 parts 0.5% solution of sulfanilic acid dissolved in 5% hydrochloric acid.

The solutions of sodium nitrite and sulfanilic acid should be kept separately refrigerated and mixed freshly according to the above proportions when required.

9.9.4. Test for Butylated Hydroxytoluene (BHT)

PROCEDURE (23, 24). Shake 10 g fat vigorously with 5 ml acetonitrile and allow the phases to separate. Shake the decanted upper layer with 5 ml dianisidine solution and 2 ml 0.3% sodium nitrite solution. An orange coloration appears even in the absence of BHT. Then shake the solution with 2 ml chloroform. The appearance of a pink color in the chloroform phase indicates the presence of BHT. The limit of detection is 10 ppm BHT.

PREPARATION OF DIANISIDINE SOLUTION. Dissolve 200 mg reagent in 40 ml methanol and dilute to 100 ml with 1 *N* hydrochloric acid.

9.10. DETECTION OF ANTIBIOTICS IN FOOD

Antibiotics are sometimes ideal food preservatives, because they do not change the flavor, odor, or color of the food and are relatively harmless to humans.

Chlortetracycline (aureomycin®) is added by commercial fisherman to the ice in which ocean fish are packed in order to reduce bacterial decomposition. The permitted tolerance is 5 ppm in raw fish. Sometimes this antibiotic is added to the refrigerated brine in which fish are stored. Chlortetracycline and oxytetracycline are permitted as bacteriostatic agents in raw poultry. Cooking completely destroys the antibiotic, so none is left when the poultry reaches the dining table.

A rapid and reliable spot test identifies and distinguishes between chlortetracycline and oxytetracycline.

PROCEDURE (25). Pipet ~10 ml Sakaguchi reagent into a 9-cm Petri dish. Place a no. 60 sieve over the dish. Sprinkle ~0.5 g ground sample on the sieve with a spatula and gently tap it to obtain a good particle distribution over the surface of the liquid. Place under a stereoscopic microscope and examine with transmitted light at about 15× magnification. As the particles of antibiotic slowly dissolve, diffusing chlortetracycline turns intense purple, and oxytetracycline intense red. The colors fade in 5–10 min.

PREPARATION OF SAKAGUCHI REAGENT. Dissolve 5 g boric acid in 150 ml water and add 350 ml sulfuric acid. Store in a refrigerator.

Traces of antibiotics can be detected by means of their inhibitions of the souring of milk when yogurt cultures are added (26). Acidity changes are detected with the indicator bromocresol purple. The detectable amounts are 0.00001 i.u. penicillin/ml, 5 μg streptomycin/ml, and 30 μg chlortetracycline/ml.

9.11. TEST FOR PENICILLIN IN MILK

The treatment of a cow for a mastitis infection may lead to contamination of the milk with penicillin. Milk is not allowed to be marketed if the milk gives a penicillin reaction, because sensitization of consumers to penicillin or allergic reactions in people sensitive to penicillin might result from its ingestion.

A highly specific, although not very sensitive, chemical reaction is applied for penicillin detection in the 50-ppm range.

PROCEDURE (27). Place five pieces of filter paper on top of five 250-ml beakers. Place two drops of 1% potassium hydroxide solution in anhydrous methyl alcohol in the center of the first paper, and follow immediately with 10 drops of 1% anhydrous ferric chloride dissolved in anhydrous methyl alcohol saturated with hydroxylamine hydrochloride. Sprinkle the machine-ground feed or the milk to be examined evenly over the wet area of the paper. Remove the excess feed by turning the paper over on the beaker. Repeat the procedure in succession with the four remaining filter papers, and observe the clean sides of the filter papers for the appearance of pinkish spots (for solid feeds) or a continuous pinkish color change (for liquid food). Dry the papers completely, remove the remaining solid feed from the paper, and observe the paper against a light source to count the number of spots. Take the mean value of the number of spots. Prepare positive blank tests using penicillin crystals or a feed known to contain penicillin.

9.12. QUALITATIVE TESTS FOR FURAZOLIDONE, TYLOSIN, AND ZOALENE® IN FEEDS (28)

Rapid field methods are needed to screen for furazolidone, tylosin, and Zoalene in mixed feeds. A selective spot test exists for the detection of Zoalene in the presence of other drugs.

The following tests are based on the intense blue color formed by furazolidone and the bright green color formed by Zoalene particles as they dissolve in a solution of dimethylformamide and potassium hydroxide. A pink-orchid color is formed by tylosin when it is left standing for several minutes in distilled water. Color observation in these tests is greatly helped by the use of a stereoscopic microscope.

PROCEDURE A. Into each of three depressions of a white spot plate place nine drops of dimethylformamide and one drop of 4% potassium hydroxide solution in alcohol. Add ~0.1 g finely ground feed material to each solution using the tip of a spatula, and observe with a stereoscopic microscope. The presence of furazolidone causes production of an intense blue color, with a detection limit of 0.0025%. Zoalene produces a bright green color, with the same detection limit. The color of minute particles of Zoalene fades rapidly. The color of larger particles persists for 3–5 min.

PROCEDURE B. Place a piece of dry filter paper at the bottom of a Petri dish and sprinkle ~0.05 g of finely ground feed over the paper. Treat the paper under a fume hood with 2–4 ml ethylenediamine so as to wet the entire paper and sample. Under examination with stereoscopic microscope using 10× magnification, Zoalene particles can be seen to develop a bright purple color, whereas furazolidone turns deep red.

PROCEDURE C. Place ~0.3 g of finely ground feed in a Petri dish and moisten with 4–5 ml water until a paste-like consistency is achieved. Examine with the stereoscopic microscope (10× magnification) against a white background. The presence of tylosin is indicated by the appearance of pink or orchid particles.

9.13. RAPID TESTS FOR THE DETECTION OF PESTICIDES IN FOODSTUFFS (29)

Insecticides, fungicides, miticides, rodenticides, and herbicides are applied by food producers to control impairment of food growth by insects, fungi, viruses, bacteriomites, rodents, and weeds. The use of any of the pesticides may create a residue problem in foodstuffs. Some of these pesticides can be detected by very simple screening methods.

The following chlorinated insecticides are commonly used for fruit and vegetable crop protection: chlordane, aldrin, dichlorodiphenyltrichloroethane (DDT), dieldrin, endrin, heptachlor, lindane, methoxychlor, dichlorodiphenyldichloroethane (DDD), and toxaphene.

Simple qualitative spot tests are still conveniently used to detect and identify the commonly used polychloroinsecticides. Some of the methods are specific enough to be applied to mixtures and simple enough for use in screening (30).

9.13.1. Test for Chlordane

Chlordane is identified by the reaction of some of its components with ethyleneglycol monoethyl ether and ethanolic potassium hydroxide. The *n*-hexane extract of the sample reacts with the reagents to form a wine-red color. No interference by other polychlorinated insecticides is recorded.

9.13.2. Test of Methoxychlor

The following test for methoxychlor was devised by the Grasselli Chemicals Department, E. I. DuPont De Nemours & Co. Inc., Wilmington, Delaware. Methoxychlor is dehydrochlorinated with ethanolic potassium hydroxide. The resulting 1,1-dichloro-2,2-*bis*-(methoxyphenyl) ethylene reacts with 85% sulfuric acid to produce a red coloration.

PROCEDURE. Place a drop of the diethyl ether extract of the sample in a depression of a white spot plate. Treat the sample with a drop of saturated potassium hydroxide solution in ethanol followed by a drop of 85% sulfuric acid. The presence of methoxychlor is indicated by the appearance of a red color. The test can detect less than 1 μg methoxychlor without interference from other polychlorinated insecticides that may be present.

9.13.3. Test for Toxaphene

Toxaphene is a chlorinated camphene that is widely used as an insecticide in agricultural formulations.

A rapid screening test (31) detects toxaphene selectively. The only commonly used polychloroinsecticide is Strobane®, which contains a similarly chlorinated camphene.

PROCEDURE. Carry out the test in a glass-stoppered Erlenmeyer flask. Add 50 ml *n*-hexane to the sample to dissolve the toxaphene. Allow the solution to settle completely, then transfer a 1-ml aliquot to a Pyrex test tube containing 5 ml reagent-grade pyridine. Treat this solution with 0.5 ml 0.1 *N* methanolic potassium hydroxide solution, and place the tube for 15 sec into a boiling water bath. Remove the tube and quickly place it in an ice bath for 1 min. The presence of toxaphene is indicated by the appearance of a pink to rusty-red color within 10 sec after the application of heat.

9.13.4. Tests for DDD (Dichlorodiphenyldichloroethane)

After dehydrochlorination, DDD produces a red color when treated with a mixture of sulfuric and nitric acids.

PROCEDURE A. Extract the sample with *n*-hexane, and take an aliquot of it into a Pyrex test tube. After evaporation to dryness, add three drops 0.5 *N* ethanolic sodium hydroxide solution, and again evaporate to dryness in a hot water bath. After cooling, add 2 ml sulfuric acid–nitric acid mixture. In the presence of DDD a bright red color develops that lasts ~1 min.

PREPARATION OF SULFURIC ACID–NITRIC ACID MIXTURE. Add one drop of conc. nitric acid to 50 ml conc. sulfuric acid.

The only polychlorinated insecticide that interferes with this test is methoxychlor. Its interference may be recognized, however, by the persistence of the red product for more than 10 min. The product caused by the presence of DDD fades after 1 min. Addition of several drops of water destroys the DDD color, whereas the color produced by methoxychlor remains. The sensitivity of the method is ~3–5 μg DDD.

An alternative test for DDD is not affected by methoxychlor.

PROCEDURE B. Extract the sample with benzene and transfer a portion of the extract to a Pyrex test tube. After addition of two to three drops of ethanolic sodium hydroxide solution, evaporate to dryness in a hot water bath. Cool and add four drops of reagent-grade acetic anhydride and 2 ml conc. sulfuric acid. The presence of DDD is indicated by development of an orange color.

DDT produces a light-orange color with this treatment. Addition of nitric acid to the test tube causes DDD to become red and DDT, pale green.

9.13.5. Test for DDT (Dichlorodiphenyltrichloroethane)

After dehydrochlorination, DDT reacts with a sulfuric acid–nitric acid mixture to form a characteristic green color.

PROCEDURE. Extract the sample with *n*-hexane, and transfer a portion of the extract to a Pyrex test tube. After adding three drops of 0.5 *N* ethanolic sodium hydroxide solution, evaporate to dryness in a hot water bath. After cooling, redissolve the residue in four to five drops of carbon tetrachloride. Add 2 ml sulfuric acid–nitric acid mixture (see previous section), shaking carefully to mix. After 5 sec of shaking a green color develops in the presence of DDT that persists for ~1 min.

DDD and methoxychlor weakly interfere with this test. Both interferences may be eliminated easily. Both interferences produce red-colored products. The color caused by the presence of DDD disappears within 5 sec after addition of the acid, whereas the color produced by methoxychlor in amounts of 3 mg or less is masked by the color produced by DDT. The limit of sensitivity of this test is 50 μg DDT. By cooling the acid mixture in an ice bath before carrying out the test, the sensitivity is much enhanced and color-fading delayed.

An alternative spot test specifically identifies DDT with xanth-hydrol (32).

PROCEDURE. Add to 2 ml 0.4% xanth-hydrol solution in anhydrous pyridine 400 mg potassium hydroxide and bring the mixture slowly to a boil. It first becomes yellow, with blue-green streaks rising from the potassium hydroxide, and finally becomes blue. On addition of a trace amount of DDT the solution is decolorized, and then becomes red in a few seconds.

9.13.6. Tests for Heptachlor

PROCEDURE A. Extract the sample with benzene and add to a portion of the extract five drops of redistilled aniline and 2 drops 0.1 *N* methanolic potassium hydroxide solution. Place the test tube in a hot water bath for 15 sec. After removal, add 1 ml reagent-grade pyridine. Reheat the test tube in a hot water bath for 10 sec. A dark green color develops within 1–3 min of the addition of pyridine in the presence of heptachlor.

The reaction is entirely specific to hepatchlor, but large quantities of other polychloroinsecticides interfere by inhibiting the color formation.

When large quantities of other polychloroinsecticides are present, use of the following procedure is advisable. Although rather insensitive (detection limit of 1 mg), it can be used in the presence of any or all of the other chlorinated insecticides.

PROCEDURE B. Extract the sample with benzene and transfer a portion to a Pyrex test tube. Add 1 ml 0.1 *N* ethanolic potassium hydroxide and warm the tube for 30 sec in a hot water bath. Add 1 ml benzene. In the presence of heptachlor a pink to purple color results.

9.13.7. Tests for Aldrin, Dieldrin, and Endrin

Spot tests for aldrin, dieldrin, and endrin detect these polychlorinated compounds by treatment with fuming sulfuric acid in xylene. All three insecticides form red-colored products, but differentiation among them is feasible by utilizing the difference in the rates of reaction. This test is selective for these compounds, and none of the other polychloroinsecticides interfere. The test is specifically recommended for the identification of dieldrin, whereas for aldrin and endrin alternative tests are available.

PROCEDURE. Extract the sample with reagent-grade xylene. Transfer two drops of the extract to a glass-stoppered test tube, and add 1 ml fuming sulfuric acid mixture. Shake carefully for ~30 sec. In the presence of dieldrin an intense red color develops immediately on addition of the acid. The limit of sensitivity is 0.05 mg dieldrin. In the presence of aldrin, 10 sec after the test tube is shaken a pink color develops that gradually becomes intensely red with continued shaking. The limit of sensitivity is

0.02 mg aldrin. In the presence of endrin, a red color develops slowly after ~15 sec of shaking. The limit of sensitivity is 0.05 mg Endrin.

PREPARATION OF FUMING SULFURIC ACID MIXTURE. Mix 1 part 30% fuming sulfuric acid with 4 parts conc. sulfuric acid.

Selective Test for aldrin in the Presence of dieldrin and/or endrin

PROCEDURE. Extract the sample with benzene and transfer two to three drops to a glass-stoppered test tube. Add 2 ml fuming sulfuric acid and shake. Add two to three drops of reagent-grade xylene and continue shaking for ~30 sec. In the presence of more than 2 mg aldrin a pink to red color develops.

Selective Test for endrin in the Presence of dieldrin and/or aldrin

PROCEDURE. Extract the sample with *n*-hexane and transfer an aliquot of the extract to a Pyrex test tube. Add 2 ml conc. sulfuric acid and shake vigorously. Heat the tube in a boiling water bath for 30–40 sec. After cooling the tube, observe the color. In the presence of endrin or chlordane a pink to red color appears. Add one drop of a 1 : 1 mixture of sulfuric and nitric acids and shake. The presence of endrin is indicated by the appearance of a blue-green color. The limit of sensitivity is 0.2 mg endrin.

Test for dieldrin and endrin

A reaction with elemental sulfur and ethanolic sodium hydroxide can be used to detect dieldrin and endrin. The test does not differentiate between these two insecticides.

PROCEDURE. Extract the sample with benzene and transfer an aliquot of the extract to an Erlenmeyer flask fitted with a water-cooled reflux condenser. Evaporate the benzene by heating the flask (without the condenser attached) on a steam bath, and add 20 ml 0.5 *N* ethanolic sodium hydroxide to dissolve the residue. Add ~30 mg elemental sulfur and attach the condenser. Heat the solution on a hot plate just to boiling and allow to digest for 15 min. Remove from the hot plate and leave at room temperature for 20–30 min with occasional swirling. The presence of dieldrin and/or endrin is indicated by the appearance of a pink to orange-red color.

Large quantities of chlordane and heptachlor produce brown colors. All other polychloro insecticides give colorless solutions.

9.14. RAPID TESTS OF PHOSPHOROUS INSECTICIDE RESIDUES IN FRUITS AND VEGETABLES

Using thin-layer chromatography, nanogram concentrations of organophosphorous compounds are detectable by cholinesterase inhibition methods (33, 34).

Organophosphorous insecticide residues can also be detected by another rapid field method (36, 37).

PROCEDURE. Extract the organophosphorous insecticide residues from leaves or soil dust. Place the extract into a test tube containing a 10% acetone solution of 4-(4-nitrobenzyl) pyridine and 0.4% oxalic acid. Heat the tube for 3 min at 150°C and add 2.5 ml 12% triethylamine solution in aqueous 15% sodium chloride. Visually compare the resulting red color with standards.

A wide range of organic insecticide residues react. As little as 0.04 μg insecticide residue/cm can be detected on the surface of a citrus leaf.

The detection of illegal treatment of apples and pears with arsen-containing compounds (e.g., methylarsine *bis*-dimethyl-dithiocarbamate) is possible by extraction of these compounds from the surface of the fruits with a 1 : 1 ethyl acetate–acetone mixture, mineralization of the extract with sulfuric acid and hydrogen peroxide, and detection of arsen by the familiar Gutzeit test (38).

9.15. TESTS FOR FUNGICIDES IN FOOD PRODUCT RESIDUES

Certain fruit and vegetable crops are protected with organic foliage fungicides such as captan, diclone, ferbam, maneb, nabam, thiram, zineb, ziram, and manzoceb.

Most of the fungicides are metallic ethylene *bis*-dithiocarbamates. These chemicals are relatively nontoxic. However, the ethylene *bis*-dithiocarbamates can be contaminated under certain conditions with, and/or be precursors of, ethylene thiourea, a carcinogen. Therefore, tolerance limits for these fungicides are set to protect the health of consumers. A harvested crop containing above-tolerance dithiocarbamate residues might be condemned as unfit for human consumption. Therefore a simple on-site determination is necessary in order that the grower may determine the approximate dithiocarbamate content of the residue and, according to the results of these screening tests, postpone harvesting until the crop is safe from a residue-level standpoint.

Simple color spot tests are used for qualitative differentiation among these compounds. Maneb [(ethylene *bis*-(dithiocarbamate) manganese)] is distinguished from zineb [(ethylene *bis*-(dithiocarbamate) zinc)] and mancozeb [a coordination product of the zinc and manganese ethylene *bis*-dithiocarbamates] by treatment with dithizone (39). This test was modified and extended for the simple screening and differentiation of a series of dithiocarbamate fungicides (40).

These fungicides are actually identified by the formation of colored chelates between the metal participating in the compound and dithizone. Sodium hydroxide is added to release the ionic form of the metal. Zinc ion is then detected by the formation of its purple-red inner complex salt with dithizone, whereas manganese ion is detected by the rapid autooxidation of manganese hydroxide.

PROCEDURE. Disperse ~0.5 g of sample in 5 ml distilled water containing two drops of sodium dioctyl sulfosuccinate solution. Place a drop of the dispersion on Whatman no. 1 filter paper. The undissolved particles should remain in the center of the spot. Dry the spot with a hair-drier and treat the center of the spot with a drop of acid dithizone reagent. Observe the colors at the spot and the surrounding ring, and compare with known standards.

Treat the powder spot with two to three drops of 1 *N* sodium hydroxide and immediately observe any transient colors produced; observe again after 15 min. Air-dry the spot, apply acid dithizone reagent, and again note the color changes.

White zinc hyroxide precipitation on addition of sodium hyroxide confirms the presence of zineb; this enables differentiation between maneb and maneb–zineb mixtures. Maneb alone forms a dark-brown spot, which is the air oxidation product of manganese hydroxide. A maneb–zineb mixture, on the other hand, exhibits a transient pink color and a dark brown end product. The annulus of the white central spot formed by zineb is pink, whereas the rings surrounding the brown central spots produced from maneb and maneb–zineb mixtures are peach colored. This intense pink annulus surrounding a brown central spot results from the complexing of the dithizone with the zinc ions liberated from zineb by the action of sodium hydroxide. Maneb does not give a similar reaction, so differentiation between the two is possible.

PREPARATION OF ACID DITHIZONE REAGENT. Dilute 2 ml neutral dithizone and 0.25 ml glacial acetic acid to 10 ml with chloroform.

For the identification of mancozeb, use the following test.

PROCEDURE. Disperse 0.1 g sample in 5 ml chloroform, spot on Whatman no. 1 filter paper, air-dry the paper, and treat with neutral dithizone reagent. Immediately observe the central spot. In the presence of mancozeb, the yellow powder spot rapidly changes color to an intense purple-pink.

PREPARATION OF NEUTRAL DITHIZONE REAGENT. Prepare a 0.1% solution of diphenylthiocarbamate in chloroform.

Another screening test (41) is based on generation of carbon disulfide generation from dithiocarbamate fungicides into the head-space gas above the contents of a septum-sealed reaction flask at elevated temperatures. The carbon disulfide produced is withdrawn with a syringe and reacts with a secondary aliphatic amine and a cupric salt to produce yellow copper dimethyldithiocarbamate.

PROCEDURE. Take 30 g of chopped representative fruit sample, add to it with a sampling scoop ~0.5 g stannous chloride, and bring to 140 ml with 4 *N* hydrochloric acid. Seal the sample with a silicon rubber septum under an aluminum seal cap by using a hand crimper, and place the sealed sample in a 600-ml beaker of boiling water. Continue heating on the boiling water bath for 1 hour. Remove a 9-ml portion of the head-space gas with a needle and syringe containing 1 ml color reagent. Agitate strongly for 1 min and compare the yellow color produced with a comparator color block.

PREPARATION OF COLOR REAGENT. Dissolve 24 mg cupric acetate monohydrate in 50 ml diethanolamine and 450 ml 95% ethanol.

9.16. DETECTION OF EDTA-TYPE COMPOUNDS IN BEER

Traces of EDTA and related compounds are added to beer to improve its resistance to oxidation and to increase its shelf-life. The MAC of EDTA in beer is 25 ppm as the anhydrous calcium disodium salt.

A rapid and simple screening test is based on the color change of red zinc dithizonate to the bluish-green dithizone in the presence of EDTA-type compounds. The red zinc dithizone complex is dissolved in a water-miscible organic solvent. Use of this solvent turns the two-phase dithizone method into a simple one-phase procedure (42).

PROCEDURE. Place 10 ml beer in a large (25 × 200 mm) test tube. Add 10 ml color reagent and mix. In the presence of EDTA compounds various shades of green develop. In the absence of such compounds the color will be pink.

PREPARATION OF COLOR REAGENT. Dissolve 50 mg dithizone in 100 ml ethylene glycol monoethyl ether (solution A). Prepare a 1000-ppm zinc solution by dissolving 4.398 g zinc sulfate heptahydrate in 1000 ml water. Dilute this solution 100-fold with ethylene glycol monoethyl ether to produce a 10-ppm zinc solution (solution B). Transfer 40 ml ethylene glycol monoethyl ether to a 100-ml volumetric flask, add 2 ml solution A and 5 ml solution B. Dilute with ethylene glycol monoethyl ether to 100 ml. Prepare fresh solution before each use.

The sensitivity of this test is ~3–5 ppm EDTA. The method responds to all the amino acid compounds related to the EDTA molecule (EDTA salts, nitrilotriacetic acid, diethylenetriaminepentaacetic acid, diaminocyclohexanetetraacetic acid).

9.17. TEST FOR ENZYMATIC CHILLPROOFING AGENTS IN BEER

Trace quantities of proteolytic enzymes are added to beer as chillproofing agents to keep the beer clear even when exposed to very low temperatures.

PROCEDURE (43). Place ~100 ml degassed beer into a 150-ml beaker and adjust the pH to 6.4 ± 0.1 with 1 *N* sodium hydroxide solution. Transfer a 50-ml aliquot to a large test tube containing 250 ± 30 mg substrate mixture. Mix the contents and place in a 60°C water bath. The presence of chillproofing agents is indicated by the gradual appearance of a suspension, clouding, and finally coagulative settling of casein. Beers without chillproofing enzymes remain clear. For semiquantitative estimation purposes, the time needed to reach initial coagulation can be measured.

PREPARATION OF SUBSTRATE MIXTURE. Grind together 50 g skimmed milk powder, 5 g L(+)-cysteine HCl.H_2O, 4.4 g sodium biphosphate, 2.5 g sodium chloride, and 1.8 g citric acid monohydrate. Store at 0–4°C and warm to room temperature before use.

9.18. SIMPLE TEST FOR ADULTERATION OF CRABMEAT WITH COD

The substitution of codfish for relatively expensive crabmeat in deviled crab is considered adulteration by the Food and Drug Administration.

This adulteration is detectable by a simple macroscopic test. Codmeat, when held in the path of a focused beam of light, displays a brilliant colored pattern, whereas crabmeat does not exhibit this phenomenon.

All the types of fish tested (flounder, catfish, red snapper, speckled and white sea trout, and mullet) have shown these colored patterns, which are due to refraction, whereas two kinds of shellfish (crab and shrimp) do not exhibit this refractive property.

PROCEDURE (44). Between two microscope slides compress a smear of the product being tested and hold in the path of a light beam from a B & L Nicholas Illuminator. The presence of cod and other fish meat is indicated by the appearance of all the colors of the spectrum. Shellfish or shellfish products not mixed with cod do not show these brilliant patterns. The test has been standardized for authentic pieces of cod and crab, and is feasible even in the presence of several ingredients, as in such homogeneous mixtures as deviled crabs or shellfish cakes.

9.19. TEST FOR SAFROLE IN BEVERAGES, COUGH SYRUP, AND SASSAFRAS TEA

Safrole (4-allyl-1,2-methylene dioxybenzene) was added for many years to carbonated beverages as a flavoring agent. Since 1960 the U. S. Food and Drug Administration has advised against adding this agent to candies, cosmetics, cough syrups, and various beverages because it is considered a weak hepatic carcinogen. Thus, a need arose for a simple qualitative screening test for the detection of safrole in food products.

The test is based upon the development of a greenish-yellow and blue ring at the interface between concentrated sulfuric acid and a solution containing gallic acid and safrole. The detection limit of this test is 5 ppm safrole.

PROCEDURE FOR BEVERAGES (45). Distill 250 ml decarbonated beverage to which 10 ml ethanol has been added. Collect 50–70 ml distillate in a receiving flask containing 10 ml ethanol. Add distilled water to make up to 100 ml. Extract 50 ml of this solution twice with 25-ml portions of diethyl ether. After separation, add to the ether extract 10 ml ethanol and carefully evaporate off the ether. Transfer the ethanolic solution to a test tube, add 1 ml 6% (w/v) freshly prepared gallic acid solution in ethanol, and mix. Holding the test tube at an acute angle, carefully add 3 ml conc. sulfuric acid down the wall of the test tube. Observe the color of the interface against a white background. The presence of methylenedioxy groups is indicated by the immediate development of a yellowish-green ring followed by a lower blue ring.

Although the test is not specific for safrole, and reacts with any chemical compound containing the methylenedioxy group, the probability of such compounds other than safrole being present in root-beer-flavored foods is extremely low.

PROCEDURE FOR COUGH SYRUP, MEDICINAL TONICS, ROOT BEER OIL, AND SASSAFRAS TEA. Dissolve 10–50 ml of the sample in 250 ml distilled water and proceed as above.

9.20. TESTS FOR MAMMALIAN FECES AND URINE RESIDUES IN FOOD

Foods are frequently contaminated with foreign matter that appears to be mammalian feces. Food-processing plants and warehouses are subject to occasional invasions of rats or mice. These rodents can attack a wide variety of foods.

The detection of rodent-contaminated lots, urine-contaminated grains, and mammalian feces in foods is a crucial task for law-enforcement officials. A number of spot tests are applied by sanitary inspectors to detect and prove the presence of these kinds of contamination.

9.20.1. Color Test for the Presence of Mammalian Feces (46)

The intestinal tract of most mammals contains alkaline phosphatase isoenzymes. The spot test for mammalian feces is based on the fact that this enzyme at a certain pH and at a certain temperature splits phosphate radical from phenolphthalein diphosphate to produce a reddish free phenolphthalein. A pH 9.5 borate–carbonate reaction buffer optimizes the response of the enzyme and the substrate. This buffer prevents the hydrolysis of the phosphate by acid phosphatase.

PROCEDURE. Transfer the suspect material to a moistened spot of filter paper in a Petri dish. Cover with a small piece of aluminum foil and crush. Air-dry the paper and the crushed particles. Cut out the stained area and transfer the paper and adhering particles to a 4-ml cup containing 1 ml gelled work test medium (WTM). Add an additional 1 ml WTM and place in a 40–41°C water bath. The appearance of a red color near the particles indicates the presence of mammalian feces.

PREPARATION OF WORK TEST MEDIUM. Prepare only small amounts of this reagent before use. Take equal volumes of the stock test reagent and distilled water in separate beakers. Prepare under heating and stirring a 2% agar dispersion by adding agar to the distilled water. Heat further under stirring until the agar starts to foam. At this point add the stock test reagent and stir for about a minute. Pour 1-ml portions of WTM in cups and cool to 40°C. WTM may be kept for ~48 hours at 40–41°C if covered with foil and plastic.

PREPARATION OF STOCK TEST REAGENT. Dissolve 9.5 g borax and 3.14 g anhydrous sodium carbonate in 500 ml water. Add 0.47 g phenolphthalein diphosphate and add with stirring 1 ml 1 M magnesium chloride solution.

PREPARATION OF MAGNESIUM CHLORIDE SOLUTION. Dissolve 0.406 g magnesium chloride hexahydrate in 1 liter distilled water.

9.20.2. Tests for the Detection of Mammalian Urine Residue in Food

The simplest test for the detection of urine contamination uses a portable ultraviolet light.

PROCEDURE. Place the ultraviolet light within a few inches of a bag of food suspected of being contaminated with urine under the usual semi-dark conditions of a warehouse. Mammalian urine stains exhibit strong fluorescence. After such a simple screening the suspect stains are generally cut out and submitted to further examination. (47).

This test is not selective, since many other natural and artificial materials fluoresce under ultraviolet light.

An alternative test for food contaminated by mammalian urine is carried out with Urograph® strips (General Diagnostics Division, Warner-Chilcott, Morris Plains, New Jersey).

The test strip is constructed from filter paper. The lowest part is impregnated with phosphate-buffered urease. The next band up contains potassium carbonate. This area is bordered by a plastic barrier. Above the barrier, an indicator area containing bromocresol green and tartaric acid undergoes color changes indicative of the presence of urine (48).

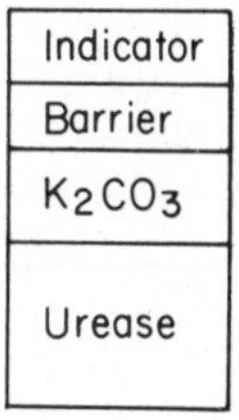

Indicator
Barrier
K_2CO_3
Urease

Urograph R

The mechanism of the test is as follows: The sample solution migrates upward through the paper. Urea, a component of mammalian urine, is converted in the urease band to an ammonium salt, which liberates ammonia gas in the potassium carbonate area. The plastic barrier prevents further migration of the liquid phase. The indicator band, consisting of bromocresol green in tartaric acid, binds the ammonia. The ammonium tartrate formed causes a color change in the indicator band. The extent of the color change is proportional to the amount of urea present in the sample solution.

PROCEDURE. Place the lower end of the Urograph strip in contact with a small amount of the test solution (suspected contaminated area dissolved in one to two drops of water) in a small stoppered test tube. In the presence of urine residues a blue-green band appears in the indicator band within 5 min.

The detection limit is 0.005 ml urine.

An overwhelming majority of positive results for urine contamination obtained with these strips by sanitary inspectors have been confirmed by alternative exact chemical analytical methods.

9.20.3. Tests for Urine-Contaminated Grain

The following test is based on the detection of ionic sodium, a component of mammalian urine. Grain is spread evenly in a single layer and sprayed with zinc uranyl acetate or magnesium uranyl acetate. Examined under ultraviolet light, the contaminated grain kernels exhibit a greenish fluorescence. The suspect kernels may be removed and extracted for subsequent confirmatory tests.

PROCEDURE (49). Spread ~50 g grain in a single layer on a shallow enamel tray or on a piece of wax paper supported by a tray. Spray evenly and lightly with the reagent. Examine under short-wave ultraviolet light. Transfer any fluorescent particles to a depression of a spot plate. Extract the kernels with one or two drops of water for several minutes. Evaporate a drop to dryness on a microscope slide. Add a drop of dilute acetic acid (1 : 2) and a tiny crystal of xanthydrol. In the presence of urea characteristic crystals of dixanthylurea form. These are best observed at 60× magnification with a wide-field stereoscopic microscope.

PREPARATION OF ZINC URANYL ACETATE REAGENT. Dissolve 10 g uranyl acetate by warming in 6 g 30% acetic acid, and diluted with water to 50 ml (solution A). Stir 30 g zinc acetate together with 3 ml 30% acetic acid, and dilute with water to 50 ml (solution B). Mix equal volumes of A and B. Add 5 ml glycerol to 95 ml of the reagent mixture, mix, and filter through washed filter paper (50).

Another spot test for the detection of urine contamination of food and feeds is based on the formation of diphenylcarbazide when urea is heated along with pheylhydrazine. Since diphenylcarbazide forms inner complex salts with several metal ions, addition of ammoniacal nickel salts can be used to identify urea.

PROCEDURE (51). Extract an ~30-g food sample with 80% ethanol. Heat the sample–alcohol mixture for 2–3 min at 50°C. Filter the extract through Whatman no. 1 filter paper. Spot one drop of the extract onto a fresh piece of Whatman no. 1 filter paper, treat the spot with one drop of phenylhydrazine in ethanol, and heat the paper for 5 min at 200°C in a forced draught oven. (The filter paper may turn brown at this stage, but subsequent treatment will clarify the reaction picture.) Cool the paper and add one to two drops of ammonium hydroxide (1 : 1) and one to two drops of 10% nickel sulfate solution. The presence of urea (urine) is indicated by the appearance of a red-violet ring.

Contamination with 11% urea can be detected. Samples of wheat, pulses, and wheat bran spiked with urea have been subjected to this test, and the detection of urea was very good.

Detection of Wheat Grains Contaminated with Urine by the Urease–Bromothymol Blue Reaction in Agar Medium

Urease, an enzyme present in soya beans, rapidly brings about complete hydrolysis of urea at room temperature. The amount of ammonia liberated by this hydrolysis can be estimated conveniently with the bromothymol blue indicator (BTB).

PROCEDURE (52). First test the sample for free alkali using test agar containing silver nitrate solution (45–48°C) by pounding a portion of agar in a shallow dish. Totally immerse the seeds when the agar begins to congeal (36–38°C). Free alkali will change the color of BTB, obviating any further testing for urea. If the free alkali test is negative, repeat the test using silver-nitrate-free agar. Grains or seeds contaminated with urine develop blue colors around the kernels. The color change of the indicator is yellow → green → blue, depending on the concentration of ammonia produced. The reaction usually requires 1–3 min to achieve a detectable color. The reaction time is inversely proportional to the concentration of urea.

PREPARATION OF TEST AGAR. To 300 ml cold water add 0.75 g bacteriological-grade agar and 0.3 g sodium benzoate under vigorous stirring. Heat to boiling on a hot magnetic stirring plate for 1 min. Cool to 45–48°C and add the BTB solution. Adjust the pH to 5.5 (yellow-green color) using dilute sodium hydroxide or dilute sulfuric acid as necessary. Add the urease suspension to the pH-adjusted, 45–48°C agar and readjust the pH if necessary. Divide the agar into two equal portions. To one portion add 0.5 ml 0.1 *N* silver nitrate solution. To the other add ~0.2 ml 0.1 *N* sulfuric acid solution to adjust to the pH of the first portion.

PREPARATION OF BROMOTHYMOL BLUE (BTB) INDICATOR SOLUTION. Dissolve 5 mg BTB in 10 ml water. Add one drop phosphoric acid (1 : 9) to dissolve BTB completely. Change the color of the indicator solution to dark green by dropwise addition of 0.1 *N* sodium hydroxide solution (pH 5.8–6.0).

PREPARATION OF UREASE SUSPENSION. Grind 0.3 g urease in a small mortar, add a few milliliters water, and continue grinding. Slowly dilute to 15 ml with stirring.

9.20.4. Test for Bird Excrement in Food Products

Bird excrement contains nitrogenous metabolic waste material, largely as uric acid. Hence, the selective determination of uric acid will qualitatively reveal contamination by bird excrement.

The test is indicative for both uric acid and xanthine derivatives. The test becomes specific for uric acid treating the oxidation product with a strong alkali. Uric acid furnishes a purple color after oxidation by nitric acid followed by strong alkali treatment, whereas caffeine, theobromine, and similar methylated compounds do not react in the same fashion.

PROCEDURE (53). Place the suspected material into a depression of a spot plate that has been preheated to ~100°C. Add a drop of conc. nitric acid (1 : 1) at the side of the spot plate depression so that it runs in to wet the particles. Evaporate to dryness in 30 sec to 1 min, and continue heating for a further 1–3 min. Cool the spot plate. The presence of uric acid is indicated by the appearance of an orange to deep-red spot, the intensity of which is proportional to the concentration of uric acid present. For confirmation, streak the colored area with a glass rod wetted with 50% sodium hydroxide solution. In the presence of uric acid an intense purple-violet color appears almost immediately.

A modification of the above test for uric acid can be applied to particles of less than 1 mg.

PROCEDURE (54). Place a strong magnifying glass on a metal surface heated to ~110–120°C. Place the suspected particles on an 18-mm no. 2 cover glass, add 5–10 μl nitric acid, evaporate to dryness, heat in an oven for 5–7 min at 135–140°C, cool, and observe under the magnifier. The presence of uric acid is indicated by the appearance of a yellow-orange to orange-red ring. For confirmation of the result, place a small drop of 50% sodium hydroxide on the edge of the cover glass with a 1-mm-diameter glass rod. Wipe the rod, and transfer a small portion of the drop to the edge of the baked residue. The presence of uric acid is indicated by the development of a purple-violet color.

9.21. IDENTIFICATION TESTS FOR THE COMMON ADULTERANT ACIDS IN VANILLA

The two most commonly used adulterant acids in vanilla are citric acid and tartaric acid. The identification test for these substances is based on the fact that the calcium salts of hydroxy acids containing a combination of at least two carboxyl groups, one hydroxy group, and a total of at least four such groups are selectively precipitated in strong ammonium hydroxide.

All organic acids present in vanilla are precipitated with neutral lead acetate. The quantity of insoluble lead salt formed gives the so-called lead number.

PROCEDURE (55). Add one drop of sulfuric acid to 50 ml USP-strength vanilla and dilute to 125 ml with ethanol. After mixing, filter through a folded piece of filter paper and add 5 ml of 1 *N* lead acetate to the clear filtrate. Collect the precipitate on a thin layer of celite in a Buchner funnel and air-dry until peeling takes place. Transfer the contents of the funnel to a small beaker and smear to a paste with a stirring rod. Disperse the paste in ~25 ml water and bubble hydrogen sulfide through the suspension for ~5 min. Remove the lead sulfide by vacuum filtration, and test the filtrate with hydrogen sulfide for residual lead. Evaporate the filtrate to ~2 ml on a hot plate, using a fine jet of air to suppress boiling. Transfer the residue to a 15-ml centrifuge tube and dilute to 3 ml with the rinsings. Add 5 ml 50% (w/v) calcium chloride solution. A positive test for adulteration is indicated by the formation of a white precipitate.

By further spot tests, the organic acids precipitated can be differentiated.

PROCEDURE. Separate out the white precipitate by centrifugation, discard the supernatant, and again disperse the precipitate in ethanol. Centrifuge, discard the liquid again, and detect the particular organic acids as follows:

Oxalic acid: To a portion of the residue add several drops of 0.5 N hydrochloric acid. Mix for ~1 min. Incomplete dissolution shows the presence of calcium oxalate.

Citric acid: Heat another portion of the residue in a microcrucible with four drops of thionyl chloride until self-sustaining fuming takes place. Treat with 28% ammonium hydroxide and evaporate the excess ammonia. After cooling, add six drops of sulfuric acid and heat until acid fumes are detected with litmus paper. Make the residue alkaline in a depression of a spot plate with 28% ammonium hydroxide. Examine the spot under long-wave ultraviolet light in a dark room. The appearance of a blue-green fluorescence quenched by acidification is proof of the presence of citric acid.

Tartaric acid: Add a small portion of the white precipitate to a solution of a few crystals of resorcinol in several drops of sulfuric acid, and heat the mixture. The formation of a characteristic rose color indicates the presence of tartrate.

Saccharic acid: Disperse a part of the white residue in 1 ml water in a centrifuge tube. Dissolve the precipitate in two drops of sulfuric acid while heating in a boiling water bath. Add 1 ml 10% oxalic acid and reheat the tube in the bath for 10 min. After cooling, neutralize with 5% barium hydroxide using one drop of 1% phenolphthalein indicator. Discharge the color with a minimal amount of acid, centrifuge, and after separation, evaporate the supernatant to ~1 ml. Filter and evaporate to dryness in a microcrucible. Add a drop of 85% phosphoric acid, and cover the crucible with a piece of filter paper impregnated with aniline acetate [10% (w/v) in 10% acetic acid (v/v)]. Heat the covered crucible on a hot plate. The presence of saccharic acid is indicated by the formation of a pink to rose-red color on the filter paper.

Detection of Coumarin in Vanillin Extract

Coumarin is the chief flavoring component of the tonka bean. It is a 1,2-benzopyrone, and resembles vanilla extract in general appearance and odor. Prior to 1953 coumarin flavorinig was used in foods, either alone or blended with vanillin. The Food and Drug Administration discovered in that year some toxic properties of coumarin, and thereafter the use of coumarin in food has been regarded as adulteration.

The spot test for coumarin is based on its conversion to a fluorescing alkali salt of *o*-hydroxy cinnamic acid (56).

PROCEDURE (57). Pipet 5 ml of sample into a 100-ml volumetric flask and dilute to 100 ml. Pipet 2 ml of this solution into a second 100-ml volumetric flask, add 2 ml 0.1 *N* sodium hydroxide solution, and dilute to 100 ml. Examine ~20 ml of this solution in a small beaker under ultraviolet light in a darkened room. In the presence of even low concentrations of coumarin (0.01%) a brilliant green fluorescence develops within 5 min.

9.22. TESTS FOR CYANIDES AND CYANOGENETIC GLYCOSIDES

9.22.1. Test for Cyanide in Almond Extract

Highly toxic hydrocyanic acid can form by the hydrolysis of the cyanogenetic glucoside amygdalin in almond extract. The familiar Prussian blue reaction identifies traces of hydrocyanic acid.

PROCEDURE (58). Add 5 ml sample solution to three drops of freshly prepared 3% ferrous sulfate heptahydrate solution and 1 drop of a 1% solution of ferric chloride. Add 10% sodium hydroxide solution drop by drop until no further precipitation occurs. Add dilute (1 : 4) hydrochloric acid until the precipitate dissolves. The presence of hydrocyanic acid is shown by formation of the characteristic Prussian blue color.

9.22.2. Tests for Cyanogenetic Glycosides in Animal Feeds

The presence of cyanogenetic glycosides in animal feeds can be critical if the plant material is stored in a humid environment for an extended period.

The test for cyanogenetic glycosides in feeds is based on the familiar Gutzeit reaction (59).

PROCEDURE (60). Chop up a sample of plant material and place it in a test tube. Moisten with water. In the case of relatively dry material, hand a moistened test paper in the tube, taking care not to let it touch the sample material. Add a few drops of chloroform and stopper the tube tightly. Cyanogenetic glycosides are indicated by the gradual appearance of an orange to brick-red coloration on the test paper.

PREPARATION OF TEST PAPER. Dip filter paper strips into 1% citric acid and dry. Dip the dried strips into 10% sodium carbonate solution and dry again. Store the papers in a stoppered bottle.

9.23. SIMPLE TESTS FOR THE QUALITY AND COMPOSITION OF FLOUR

Exact analytical data are required to characterize the nutritional quality of flour. The protein, carbohydrate, oil, mineral, and vitamin content of the flour have to be assayed by complex analytical procedures. However, some gross adulteration, the presence of various bleaching agents, and the addition of different kinds of improvers can be detected by unsophisticated qualitative tests.

9.23.1. Test for Soy Flour Content in Wheat Flour

Simple field tests are able to ascertain the presence of soy flour in food. A rapid and simple procedure is based on the fact that the enzyme urease normally occurs in soybean flour. In this method the urease activity is estimated by the pH change resulting from incubation of soybean meal with urea in a buffer solution under specific conditions.

PROCEDURE A (61). Place an ~0.5-g sample in a small test tube and mix with 5 ml 2% urea solution. Immerse a strip of red litmus or bromophenol paper into the liquid, mix, stopper, and let sit for 3 hours at 40°C. Examine the paper for color changes resulting from the liberation of ammonia. In the presence of soy flours in larger than trace quantities the litmus paper or bromophenol paper turns blue.

The shortcoming of this simple test is that it is not reliable for products that were previously heated beyond the temperature of urease inactivation.

An alternative rapid test (62) that is not restricted by heat-dependent enzyme activity estimates the quantity of soy flour in wheat flour by the canary-yellow fluorescence of soybean particles as viewed under ultraviolet light with no magnification.

PROCEDURE B. Either spread the sample on a microscope slide or place it in a depression of a spot plate. View the sample under a stereomicroscope at low magnification (18×, 24×, or 48×) in the dark under an ultraviolet light with a predominant wavelength of 360 nm. In the presence of soybean particles an intense fluorescence occurs in the yellow range of the spectrum. This fluorescence is also visible with the unaided eye, but it is far more visible under low-power magnification.

The smallest quantity of soyflour that can be detected with this technique is 0.01%, and the test is independent of previous heat treatment. Various quantities of soyflour can be semiquantitatively estimated by counting the fluorescent particles under low-power magnification. This estimation presumes that the particle size distribution of soybean flour in the test material the distribution of soy flour particles in the wheat flour are reasonably uniform.

9.23.2. Detection of Rye Flour in Wheat Flour

Expensive flour is commonly adulterated with cheaper flours. The presence of rye flour is detected by a simple method.

PROCEDURE (63). Weigh 5 g flour into a 50-ml centrifuge tube, add 20 ml 70% ethyl alcohol, and shake continuously for 15 min. Cool in an ice–salt bath at −3°C for 10 min, continuously stirring the viscous mass with a glass rod during cooling. Centrifuge for 5 min, then decant the supernatant fluid, which if cloudy must be filtered clear. Add 0.5 ml 1 *N* alcoholic sodium hydroxide solution. The presence of rye flour is indicated by the formation of a pronounced cloudiness or even a precipitate. Pure wheat flour gives a clear solution.

As little as 10% rye flour can be detected by this test.

9.23.3. Test for Bleached Flours

Whiteness is a primary characteristic of bread and cake flour, although the flours used for pastas should have a yellowish appearance to indicate the presence of egg.

Bleaching is practiced extensively in the milling of white flour. Nitrogen peroxide, nitrogen trichloride, chlorine dioxide, and benzoyl peroxide are all used to this end. Chlorine, nitrosyl chloride, and chlorine dioxide have both bleaching and maturing actions (i.e., they have an artificial aging effect due to inhibition of the active proteolytic enzymes present in flour).

Maturing agents are added only if baking tests show that the flour needs such treatment. In some crop years no maturing treatment is necessary.

The presence of oxidizers and improvers has an indicative value in the flour industry; therefore tests for their presence have diagnostic meaning even if not providing any quantitative results. With the exception of potassium bromate and azodicarbonamide, there are no quantitative restrictions on use of such additives.

Test for Chlorine Bleaches (64)

PROCEDURE. Extract ~30 g flour with 50 ml petroleum ether. Decant through a filter, catching the solvent in an evaporating dish, and evaporate to dryness. Dip the hot end of a copper wire (previously heated in a nonluminous Bunsen-burner flame until fully oxidized to black cupric oxide) into the small amount of oily residue in the evaporating dish. Place the wire again in the flame. The presence of chlorine bleaching agents is indicated by the appearance of a green color in the flame.

Test for Bromates and Iodates in White or Whole Wheat Flour

The use of bromates as maturing agents in flour is permitted in the United States and Canada.

PROCEDURE (65). Sieve ~4 g flour over a white pan and cover with a freshly prepared mixture of equal volumes of 1% potassium iodide and dilute (1 : 7) hydrochloric acid. The presence of bromates, iodates, and persulfates is indicated by the formation of black or purple specks.

The following test can be used to confirm the presence of iodates in flour.

PROCEDURE. Sprinkle ~1 g flour evenly over the bottom of a Petri dish and cover with a mixture of equal volumes of 1% potassium thiocyanate solution and dilute (1 : 7) hydrochloric acid. The presence of iodates is signaled by the appearance of black and purple specks in the sample.

Test for Persulfates

Neutral or weakly acidic solutions of alkali persulfates turn blue on addition of benzidine due to the formation of a quinoidal oxidation product of benzidine. Alkali peroxides, perborates, chlorates, perchlorates, bromates, iodates and nitrates and hydrogen peroxide do not react under the test conditions.

PROCEDURE (64). Place ~10 g flour on a glass plate. Slowly pour over the surface a solution of 0.5 g benzidine in 100 ml ethanol. Bluish-black specks appear in the presence of persulfates.

Test for Benzoyl Peroxide in Flour

PROCEDURE (66). Shake ~1 g flour with 3.5 ml petroleum ether in a test tube. After settling, add 1.5 ml 1% dipara-diamino-phenylamine sulfate in ethanol. In the presence of benzoyl peroxide a green color appears in the petroleum ether phase, whereas blue crystals in the sediment show the presence of persulfates.

Test for Nitrogen Oxides in Flour

PROCEDURE (66). To an Erlenmeyer flask add ~20 g flour and 20 ml nitrite-free water. Heat the Erlenmeyer flask to 40°C. in an incubator, stopper, and shake vigorously for 5 min. Then digest at 40°C for 1 hour, shaking every 10 min. Filter through a nitrite-free filter and add to the filtrate 2 ml sulfanilic acid solution (0.5 g sulfanilic acid in 120 ml 20% acetic acid) and 2 ml α-naphthylamine hydrochloride solution (0.2 g reagent dissolved in 150 ml 20% acetic acid). Shake, and allow to stand for 1 hour. The presence of nitrite nitrogen is indicated by the formation of a pink to reddish-purple color. Visual comparison of the color intensity against a series of standards containing known amounts of nitrites enables semiquantitative estimation of nitrogen oxides in flour.

9.24. TEST FOR DIFFERENTIATION BETWEEN BUTTER AND MARGARINE

The difference between the critical temperatures of dissolution (CTDs) in an alcohol–isoamyl alcohol mixture of butterfat and margarine fat is ~24°C. The rapid field test (67) that exploits this difference for the differentiation of butter from margarine has been adopted as an official test by the AOAC.

PROCEDURE. Melt the butter at ~60°C until the moisture and curd separate completely. Filter the clear supernatant oil through dry filter paper placed in a hot filter funnel. Refilter if the filtrate is not perfectly clear. Fill a Pyrex tube calibrated at 2 ml and 4 ml to the 2-ml mark with the filtered oil. Add the mixed alcohol reagent to the 4-ml mark. Using a 0–100°C graduated thermometer as a stirring rod and heat the tube on a microburner until the mixture becomes clear and homogeneous. Do not boil. Remove from the heat and continue to stir until the mixture becomes turbid. Record the temperature of the first appearance of the turbidity. The CTD range for various brands of butter is 42–53°C, whereas the CTD range for margarine samples falls between 66 and 75°C.

PREPARATION OF MIXED ALCOHOL REAGENT. Redistill isoamyl alcohol and collect the distillate between 128 and 132°C. Using pipets for measurement, mix 2 volumes of the 95% ethyl alcohol with 1 volume of the redistilled isoamyl alcohol, and keep in a tightly stoppered container.

9.25. TESTS FOR MILK ADULTERATION

9.25.1. Watering

The adulteration of milk by watering is usually detected by determining the ordinary fat and, nonfat ratio or by performing a freezing-point test with the Fiske cryoscope. The cryoscope method is capable of detecting as little as 3% added water in milk.

This method is based on the fact that the freezing point of milk is relatively constant at −0.550°C. The accurate measurement of the freezing point offers an excellent and simple method for the detection of added water in milk.

The Fiske cryoscope (Advanced Instrument Inc., Newton Highlands, Massachusetts) consists of a cooling bath, an air agitator, a thermistor probe, a seeder rod, a Wheatstone bridge, and a galvanometer measuring circuit. Freezing-point values are read from a temperature dial that balances the bridge.

PROCEDURE. (68). Prechill the sample tube containing 2 ml sample by immersion in an ice bath to two-thirds of its length. Locate the thermistor in the center of and 10 mm from the bottom of the tube. After seeding, a long, steady temperature plateau must be attained. Take the freezing point as the position where the galvanometer spot ceases to move to the right. The presence of added water is indicated if the freezing point is above −0.530°C.

9.25.2. Test for Gelatin in Milk

The lowering of the nonfat solid content of watered milk is sometimes counterbalanced by adulterant addition of gelatin to the milk. Qualitative detection of gelatin in milk is indicative of milk adulteration.

PROCEDURE (68). Add to 10 ml sample 10 ml acid mercury nitrate. Shake the mixture, add 20 ml water, shake again, let stand for 5 min, and filter. In the presence of large quantities of gelatin, the filtrate will be opalescent. Add to an aliquot of the filtrate an equal volume of a saturated aqueous solution of picric acid. The presence of gelatin will be indicated by yellow precipitation, whereas trace amounts will lead to cloudiness. The gelatin–picric acid precipitate tends to remain in suspension, settling very slowly and sticking to the walls and bottom of the container. When gelatin is present in relatively large concentrations (1% or more), precipitation is rather quick.

PREPARATION OF ACID MERCURY NITRATE SOLUTION. Dissolve metallic mercury in twice its weight of nitric acid and dilute to 25 times the resulting volume with water.

9.25.3. Test for Adulteration of Milk by Addition of Extraneous Ammonium Sulfate

Ammonium sulfate is favored as an adulterant of milk because it increases both the nonfat solid content and the nitrogen content as determined by the Kjeldahl procedure. A rapid field test for the detection of extraneous ammonium sulfate detects such adulteration.

PROCEDURE (70). Add to 1 ml milk in a test tube 0.5 ml each of 2% sodium hydroxide, 2% sodium hypochlorite, and an aqueous solution of 5% phenol. Heat the mixture in a boiling water bath for ~20 sec. The presence of extraneous ammonium sulfate is indicated by the appearance of a rapidly deepening blue color.

As little as 0.1% ammonium sulfate can be detected with this test, even in the presence of other adulterants such as urea, glucose, and sucrose.

9.26. SIMPLE TESTS OF MEAT PRODUCTS

9.26.1. Tests for Starch

In many Continental-type sausages the use of carbohydrates is prohibited altogether. In the United Kingdom it is usual to manufacture fresh sausage containing relatively large amounts of carbohydrates, whereas in the United States carbohydrate content is restricted to 3.5%

The qualitative test for starchy flour in chopped meat, sausage, and deviled meat is based on the familiar iodine–starch reaction.

PROCEDURE (71). Boil ~5–6 g of the sample for 2–3 min. After cooling, test the supernatant liquid with some drops of iodine–potassium iodide solution (0.5 g iodine and 1.5 g potassium iodide dissolved in a very small amount of water and diluted to 25 ml). The formation of an intense blue color indicates the presence of starch. Even in the absence of added starchy flour, a weak positive reaction may appear due to the presence of spices.

A simple semiquantitative test for starchy flour in meat products measures volumetrically the precipitated starchy material. This method cannot be used in the presence of cellulosic material.

PROCEDURE (72). Into a 100-ml graduated ASTM conical tube graduated from 0 to 3 ml in 0.1-ml units, place 10 g of sample, and add 50 ml 8% alcoholic potassium hydroxide solution. Digest the mixture on a steam bath for 1 hour. Dilute to 100 ml with ethyl alcohol and mix. Let stand for 1 hour, occasionally rotating the tube. Read the volume of the sediment after 1 hour. In the presence of spice oils only, more than 0.5 ml precipitate shows the content of added starchy flour. The precipitate volume read indicates the same percentage of added starchy flour present.

If the contaminated meat product contains also dried skim milk or dried corn syrup, the lactose or maltose has to be removed before the test by shaking the sample in a 100-ml centrifuge tube with two 50-ml portions of warm water, centrifuging and decanting after each shaking.

9.26.2. Test for Dried Milk ro Dried Skim Milk in Meat Products

In addition to cereals, vegetable starch, and starchy vegetable flour, dried milk or dried skim milk may be present as an added ingredient meat product. The presence of dry skim milk in sausage can be qualitatively detected by a test for lactose. The question of whether a sausage meets kosher requirements, naturally cannot be answered by a chemical test. If, however, lactose is detected in the meat product, the product is definitely not kosher; a negative answer, however, does not prove the opposite.

PROCEDURE (73). To 10 g of sample in a small beaker, add 20 ml hot (70–90°C) water. Shake vigorously and filter the solution. Add to 4 ml filtrate in a test tube three to four drops of 5% methyl amine hydrochloride solution and boil for 30 sec. After removal from the flame, add three to five drops of 20% sodium hydroxide solution and shake for 10 sec. The presence of lactose is indicated by the immediate appearance of a yellow color that slowly changes to carmine.

9.26.3. Rapid Test for Urea and Ammonium Salts in Meat Products

The crude protein value of meat products is usually determined by the Kjeldahl method. This value is based on the assumption that the nitrogen is derived from meat protein. The Kjeldahl procedure, however, does not distinguish between meat protein and other nonmeat sources of nitrogen, such as urea and ammonium salts. Addition of these materials to meat products is considered adulteration, and hence their simple detection has great importance.

PROCEDURE (74, 75). Extract a sausage sample with 1 *N* sodium hydroxide, and dip a urea test strip (see Section 3.13) into the extract. The serum urea value obtained from the scale is converted to the concentration in the meat by a given factor.

9.26.4. Detection of Excessive Amounts of Nitrite in Meat Products

The following test is carried out with Nitratesmo test paper (Macherey-Nagel).

PROCEDURE (76). Press a sample of sausage between two sheets of Plexiglas and insert the test paper between the two sheets so as to absorb some of the expressed juice. Dip the test strip into hydrochloric acid (1 : 1). The presence of nitrite is indicated by the development of a yellow color.

As little as 200 ppm nitrite can be detected with this test.

9.27. TESTS FOR EDIBLE OILS

Rapid qualitative testing for different edible oils is a very old practice.

9.27.1. Teaseed Oil in Olive Oil

Olive oil is sometimes adulterated with the cheaper teaseed oil. A specific color reaction (77) detects as little as 5% teaseed oil.

Treating the suspected oil with acetic anhydride in the presence of concentrated sulfuric acid leads to the formation of deep green pigment as seen with reflected light (brown with transmitted light) in the presence of teaseed oil. Pure olive oil treated similarly becomes green in both reflected and transmitted light.

PROCEDURE. Put exactly 0.8 ml acetic anhydride and 1.5 ml chloroform together with 0.2 ml sulfuric acid in a test tube. Mix and allow to cool to room temperature. Add seven drops of the oil sample to be tested, mix, and cool again. If the mixture is cloudy, add further drops of acetic anhydride until the solution clears completely. After 5 min add 10 ml absolute ether and mix immediately by inverting once. In the presence of teaseed oil, a brown solution is produced that changes to intense red within about a minute. The red color reaches a maximum and fades within minutes. Pure olive oil initially forms a green solution on addition of ether. This green color fades slowly to brown-gray. Mixtures of teaseed oil and olive oil show characteristic teaseed-oil colors proportional in intensity to the quantity of teaseed oil present.

9.27.2. Test for Cottonseed Oil

A very old test dating back to 1897 is still used in the selective identification of cottonseed oil (78, 79). The only interfering vegetable oil known is kapok oil. The sensitivity of the test enables detection of cottonseed oil in concentrations as low as 1%. Fat from pigs fed on cottonseed meal will also show a positive reaction.

PROCEDURE. Mix equal volumes of Halphen reagent (carbon disulfide containing 1% elementary sulfur mixed with an equal volume of amyl alcohol) and of the oil sample in a test tube. Heat to about 75°C in a water bath until carbon disulfide is expelled. Then heat the tube in a saturated brine bath or a glycerin bath at 110–115°C for ~1–2 hours. The rapid appearance of a red color indicates large quantities of cottonseed oil, while lower concentrations (~1%) will show the red reaction only after 2 hours of heating. A semiquantitative estimate can also be established by running simultaneously a series of known dilutions of cottonseed oil along with the sample. Visual comparison of the resulting color intensities enables rough quantitation.

9.27.3 Tests for Sesame Oil

PROCEDURE A (80). Prepare furfural reagent by diluting 2 ml furfural in 100 ml ethanol. Mix 0.1 ml of this reagent with 10 ml of the oil sample and add 10 ml hydrochloric acid. After shaking for about 15 sec, let stand for ~10 min until the emulsion separates. A negative test for sesame oil is indicated by the absence of a pink or crimson color in the lower layer. If a color does appear, add 10 ml water and shake again. If after separation of the layers the red color persists, sesame oil is present. If the color disappears, sesame oil is absent. As little as 0.5% sesame oil can be identified. Obtaining a semiquantitative estimation is possible by simultaneously running known standards of sesame oil; the results can then be visually compared with the outcome of the sample test.

PROCEDURE B (66). Heat the fat in a 35% ethanolic ammonia solution in a water bath until the ethanol and ammonia evaporate. Shake the residual solution with a 1% solution of sucrose in 35% hydrochloric acid. In the presence of sesame oil, a pink coloration develops in the acid layer.

9.27.4. Test for Mineral Oil in Fats and Vegetable Oils

PROCEDURE (80). Add 1 ml potassium hydroxide solution (3 N) and 25 ml ethyl alcohol to a 1-ml sample of oil or melted fat in an Erlenmeyer flask. Reflux with an air condenser for ~5 min until saponification is completed. Add 25 ml water and mix. The presence of at least 0.5% mineral oil is shown by a distinct appearance of turbidity.

9.27.5. Detection of Rapeseed Oil as an Adulterant in Almond or Olive Oil

Rapeseed oil is produced from various species of *Brassica*. Although it is used as a salad oil and shortening in Europe, it is not used for food in the United States. Adulteration based on addition of rapeseed oil can be qualitatively detected on the basis of the high erucic acid content of rapeseed oil.

PROCEDURE (81, 82). Place the oil sample in a screw-capped tube with 3.5 ml ethanol and heat to 70°C to obtain a clear solution. Then transfer the tube to a water bath at 30°C and note the time it takes for the solution to turn completely opaque. The higher the erucic acid (and the rapeseed oil) content of the oil, the shorter the time required for the solution to turn opaque. Calibration graphs are rectilinear.

9.27.6. Detection of *Argemone Mexicana* Seeds Among Mustard Seeds

Edible vegetable oils are sometimes adulterated with *Argemone mexicana* seed oil. The latter seeds can be identified by their rough surface as seen under magnifying glass. For further evidence, a spot test can be carried out that identifies the presence of the toxic alkaloids sanguinarine and dihydrosanguinarine in the oil.

PROCEDURE (83). Place a single seed between the two halves of a folded piece of filter paper and press with a paperweight to cause the seed to burst, leaving an oily spot on each half of the paper. Soak one spot with hydrochloric acid and test with Dragendorff's reagent (see preparation in Section 4.4.1) and the other with nitric acid. If the first spot becomes orange-red and the second becomes orange-yellow to crimson, the seed is that of *Argemone mexicana*. Mustard seeds do not give these reactions.

9.27.7. Spot Test for Heated Oils

A rapid and simple spot test monitors the accumulation of oxidative products in heated oils and prevents the sale and distribution of food with a high concentration of thermally oxidized products. These oxidative products, including polymeric triglycerides, cyclic acids, and free fatty acids, may induce various symptoms including organ enlargement, growth inhibition, necrosis of the liver, and extensive cellular damage.

The spot test described (84) assesses the degree of thermal oxidation of a fat on the basis of a color reaction of free fatty acids with bromocresol green incorporated into a gel. The colors developed by several oils range from blue to an intense yellow.

PROCEDURE. With a glass rod, place a drop of the oil sample on the dried blue silica-coated reagent slide. Observe the color change within 15 sec. Fresh eating oils are indicated by a blue color, whereas heated oils give various shades of pale green to pale yellow according to their fatty acid concentrations.

PREPARATION OF THE REAGENT SLIDE. Add to a big beaker containing 100 ml of a methanol–water (1 : 1) mixture 0.1 g bromocresol green and 90 g silica gel G type 60. Stir the suspension and adjust its pH to 7.3 with 0.1 *N* sodium hydroxide. Immerse a precleaned glass slide in the suspension, swirl, remove, and air-dry for 20 min on a paper towel, then dry in an oven at 100°C for 1 hour. Keep the slides in a closed container.

9.28. TESTS FOR ENZYME ACTIVITY IN FOOD

Enzymatic activity is used as a criterion for food quality. Qualitative and quantitative analysis characterizes the baking quality of wheat, the brewing quality of barley, and the keeping quality of eggs. The acceptablility of frozen and dehydrated fruit and vegetables depends on their enzyme activities. Dried or frozen vegetables develop an undesirable odor and taste if not heat pretreated. The food industry uses a number of measures to prevent this haylike odor, which is caused by the presence of enzymes. Steam treatment and the application of sulfites inhibit enzyme activity, as does packaging in a nitrogen atmosphere.

Peroxidase is one of the most heat-resistant plant tissue enzymes. It is thermally inactivated at about 72°C. The adequacy of the heat pretreatment should be tested by rapid screening tests for residual peroxidase activity after blanching. Residual peroxidase activity has been used as a criterion for the adequacy of the thermal process in acidic fruits such as apples, pears, and cherries. Blanching temperatures are usually in the 77–88°C range for peroxidase inactivation. Since the taste and texture of some vegetables are adversely influenced by treatment at such temperatures, in these cases catalase activity is used as the criterion for sufficiency of the blanching. The enzyme lipoxidase also has to be thermally inactivated, because this enzyme causes the undesirable flavor changes in fat-containing vegetables such as dried beans and peas. Rapid tests for inactivation of lipoxidase are thus very useful.

9.28.1. Test for Peroxidase

PROCEDURE (85). Grind in a blender a representative vegetable sample of 100–200 g, adding to each gram of sample 3 ml water, and filter through a cotton milk filter. Dilute 2 ml filtrate in 20 ml distilled water in a test tube. Add 1 ml 0.5% guaiacol solution in 50% ethanol and 1 ml 0.08% hydrogen peroxide solution. Prepare a color comparison tube by diluting 3 ml filtrate to 22 ml with distilled water without adding the reagents. Invert the tube and observe the development of any color that differs from the color of the comparison tube. Inadequate blanching is indicated by a color difference between the two tubes. If no such color difference shows up in 3.5 min, the peroxidase activity is considered negative, which means that the product was adequately blanched. Color development after 3.5 min is considered a negative outcome.

A rapid semiquantitative screening test for peroxidase (86) has been developed based on the reaction of the previous procedure. The time for the development of a colored spot on a paper disc is measured, and a linear calibration curve is obtained by plotting the log of the development time against the log of the concentration of peroxidase.

PREPARATION OF CALIBRATION CURVES. Place a piece of Schleicher & Schuell no. 57 GH paper disc on a clean glass plate and drop 0.05 ml of a standard diluted solution of peroxidase onto the disc. Add a drop of a

0.5% solution of guaiacol in 50% ethyl alcohol and a drop of 0.5% hydrogen peroxide solution. Measure the time with a stopwatch from the moment the drop of reagent touched the paper disc until the first color appears. The log value of the time in seconds plotted against the log value of the enzyme concentration (as percentage by weight) gives a linear relationship.

For the determination of peroxidase in food processing industries, test paper strips named Peroxtesmo KO are marketed by Macherey-Nagel (Düren, F. R. G.).

9.28.2. Tests for Catalase

Catalase is an enzyme that decomposes hydrogen peroxide into water and molecular oxygen. The estimation of the amount of oxygen liberated is the basis of the rapid qualitative and semiquantitative tests of catalase activity. These quick tests have been successfully used in vegetable freezing plants.

PROCEDURE (87). Grind 100 g of cooled blanched vegetable or thawed frozen vegetable in a blender. Quickly transfer 1 g of the mashed material to a mortar containing 1 g sand and 0.6 g calcium carbonate. Add 10 ml water and grind again. Transfer 4–5 ml of this mixture to a fermentation tube, add 8 ml 3% hydrogen peroxide (not containing a preservative), and shake gently for 3 min. Compare the volume of oxygen gas developed with that from a similar test carried out on fresh or unblanched vegetable. If the volume of gas is greater than that from the fresh or unblanched sample, the catalase test is considered positive. The gas evolution shown by the blank should be less than 0.1 ml.

Other qualitative and semiquantitative tests are based on the same oxygen liberation by the enzyme. These tests do not measure the released oxygen volume as such, but utilize the buoyant effect of the liberated oxygen on paper discs saturated with the enzyme solution. The following test is based on recording the time required for the disc to rise to the surface of a solution of hydrogen peroxide.

QUALITATIVE PROCEDURE (86). Dip a Schleicher & Schuell no. 57 GH paper disc into a sample solution or sample slurry. Alternatively, moisten the disc by pressing it against the bruised tissue of the food product. Drop the disc into a test tube containing 5 ml 3% hydrogen peroxide solution. The presence of catalase enzyme activity is indicated by the flotation of the disc within a short time period. A blank should be run to establish the time required for the disc to rise in the absence of catalase.

A semiquantitative estimate can be obtained from the same test by preparing a calibration curve.

PROCEDURE. Saturate Schleicher & Schuell no. 57 GH paper discs with various dilutions of catalase enzyme solutions by dipping them into the liquid. Drop them immediately into test tubes containing 5 ml 3% hydrogen peroxide solution. Measure the time between the moment the disc touches the surface of the hydrogen peroxide when dropped into the tube and the moment when it again reaches the surface. Plot the logarithm of the concentration of the enzyme versus the logarithm of this time.

Liquidize a solid sample in a blender, adding a known amount of distilled water, and filter the solution. Proceed as in the qualitative test, above. The catalase content is semiquantitatively estimated by locating the time required for the sample disc to float on the calibration curve.

Typical time periods required for the disc to rise to the surface are 2.7 sec for 0.0100% catalase solution, 9.3 sec for 0.00250% catalase solution, 21.6 sec for 0.001% catalase solution, and 98 sec for 0.00010% catalase solution.

9.28.3. Test for Lipoxidase Activity

Lipoxidases are active at low temperatures, and are partly responsible for the deterioration of stored frozen foods. They catalyze the oxidation of fatty acids with two or more methylene-interrupted double bonds to peroxides. This enzyme occurs also in flour, and may oxidize carotenoid pigments during peroxidation of polyunsaturated fatty acids. Bleaching of flour during natural aging can be caused by lipoxidase activity. Rancidity in cereals and cereal products may also be the consequence of lipoxidase enzyme activity.

PROCEDURE (88). Grind 5 g sample to no. 20 mesh and stir for 10 min with 50 ml distilled water. Centrifuge, and add to the supernatant fluid 5 ml 6.4% calcium chloride solution. Mix, and centrifuge again. The addition of this salt precipitates inactive proteins.

Add 5 ml pH 5.5 acetate buffer to 100 ml distilled water and 1 ml cottonseed oil substrate. Add 1 ml enzyme extract at zero time, and remove 2-ml aliquots from the reaction mixture, which must be kept at room temperature (21 ± 2°C), at 8, 16, and 32 min to test the peroxide formation. Place these aliquots into test tubes containing 25 ml acidified ethyl alcohol. Add 1 ml ferrous ammonium sulfate solution, and 1 ml 20%

ammonium thiocyanate solution. Mix, and observe the color formation within a 3-min period. Prepare a blank solution by adding 2 ml of the buffered water and substrate mixture to 25 ml acidified ethyl alcohol. Comparison of the color of the blank with the color developed by the sample will indicate the lipoxidase activity present in the sample. The colors of the 8-, 16- and 32-min aliquots not being darker than that of the blank indicates 95% or better lipoxidase inactivation. At 90% inactivation, a darker pinkish-red color can be observed in the 32-min aliquot.

Preparation of Substrate. Commercial cottonseed oil (0.02–0.04% free fatty acid, zero peroxide value, and bland odor and flavor) should be used as the lipoxidase substrate. Dissolve 1 g of the oil in 100 ml of a 1 : 1 mixture of acetone and 95% ethyl alcohol. Prepare the substrate fresh, and store below 0°C.

Preparation of Acetate Buffer (pH 5.5). Mix 8.8 ml 0.2 *M* acetic acid with 41.2 ml 0.2 *M* sodium acetate and dilute to 100 ml.

Preparation of Ferrous Ammonium Sulfate Solution. Dissolve 0.125 g reagent-grade ferrous ammonium sulfate in 100 ml 3% hydrochloric acid.

Preparation of Acidified Ethyl Alcohol. Add 4 ml conc. hydrochloric acid to 500 ml 95% ethyl alcohol.

REFERENCES

1. H. Fehling, *Ann. Chem.* **72,** 106 (1849).
2. A. Cantafora, V. D. Ayala, and D. Rodini, *Riv. Soc. Ital. Sci. Alim.* **7,** 131 (1978).
3. G. Avigad, D. Amiral, C. Asensio, and B. L. Horecker, *J. Biol. Chem.* **327,** 2736 (1962).
4. D. Amaral, L. Bernstein, D. Moore, and B. L. Horecker, *J. Biol. Chem.* **238,** 2281 (1963).
5. M. Bial, *Dtsch. Med. Wochenschr.* **28,** 253 (1903)
6. N. Seliwanoff, *Ber.* **20,** 181 (1881).
7. F. Feigl, *Spot Tests in Organic Analysis,* 7th ed., Elsevier, Amsterdam (1966), p. 341.
8. W. C. Kao and H. F. Chang, *Anal. Abstr.* **43,** 189 (1982).
9. H. O. Triebold and L. W. Aurand, *Food Composition and Analysis,* D. Van Nostrand Co., New York (1963).
10. T. Korbelak, *JAOAC* **52,** 487 (1969).
11. A. B. Karasz, F. de Cocco, and L. Bokus, *JAOAC* **56,** 626 (1973).
12. Official Methods of Analysis of the Association of Official Analytical Chemists, 9th ed. Association of Official Analytical Chemists, Arlington, Virginia (1960), p. 580.

13. A. Perditz and I. Ekar, *Z. Lebensm. Unters. Forsch.* **163,** 100 (1977).
14. J. Fitelson, *JAOAC* **51,** 937 (1968).
15. J. Fitelson, *JAOAC* **50,** 293 (1967).
16. *Official Methods of Analysis of the Association of Official Analytical Chemists,* 13th ed., Association of Official Analytical Chemists, Arlington, Virginia (1980), pp. 325, 343.
17. F. Feigl and V. Anger, *Spot Tests in Inorganic Analysis,* 6th ed., Elsevier, Amsterdam (1972), p. 136.
18. C. C. Fulton, *Ind. Eng. Chem. Anal. Ed.* **3,** 199 (1931).
19. E. Kaplan, *JAOAC* **44,** 485 (1961).
20. N. Kerr, *Ind. Eng. Chem.* **10,** 471 (1918).
21. A. Vioque and E. Vioque, *Grasas Aceitas* **13,** 211 (1962).
22. S. A. Scheidt and H. W. Conroy, *JAOAC* **49,** 489 (1966).
23. F. Mordret and N. L. Barbanchon, *Rev. fr. Cps. gras* **22,** 387 (1975).
24. G. Giovetti, *Ind. Alim. Pinerolo* **15,** 128 (1976).
25. C. T. Smith, *JAOAC* **51,** 750 (1968).
26. F. Gorner, H. Simkovica, and G. Majlatova, *Prum. Protavin* **29,** 700 (1978).
27. J. P. Barrette, *JAOAC* **48,** 1071 (1965).
28. L. A. Barefield, *JAOAC* **56,** 762 (1973).
29. F. L. Hart and H. J. Fisher, *Modern Food Analysis,* Springer-Verlag, Berlin (1971).
30. D. P. Johnson, *JAOAC* **39,** 490 (1956).
31. D. P. Johnson, *JAOAC* **38,** 156 (1956).
32. H. Auterhoff and G. Ahlers, *Arch. Pharm. Ber.* **306,** 631 (1973).
33. S. Khamova and K. Bokarev, *Biol. Abstr.* **63,** 66159 (1977).
34. E. M. Lores, D. W. Bristol, and R. F. Mosenman, *J. Chromatogr. Sci.* **16,** 358 (1978).
35. B. Berck, Y. Iwata, and F. A. Gunther, *J. Agric. Food Chem.* **29,** 209 (1981).
36. F. A. Gunther, Y. Iwata, E. Papadopolou, B. Berck, and C. A. Smith, *Bull. Environ. Contam. Toxicol.* **24,** 209 (1981).
37. H. Berg and H. Sperlich, *Mitt. Bl. GD Chem. Fachgr. Lebensmittelgericht. Chem.* **28,** 298 (1974).
38. J. M. Walker, *Labor. Pract.* **17** (1970).
39. A. Stevenson, *JAOAC* **55,** 939 (1972).
40. D. E. Ott and F. A. Gunther, *JAOAC* **65,** 909 (1982).
41. I. Stone, *JAOAC* **47,** 714 (1964).
42. I. Stone, *JAOAC* **52,** 88 (1969).
43. C. C. Freeman, *JAOAC* **51,** 509 (1968).
44. C. Genest, D. Morison Smith, and D. G. Chapman, *JAOAC* **44,** 631 (1961).
45. J. J. Trasher and J. S. Gecan, *JAOAC* **64,** 196 (1980).
46. W. S. Cox, *JAOAC* **45,** 655 (1962).
47. S. J. Koziol, Distributed by General Diagnostic Division, Warner-Chilcott, Morris Plains, New Jersey (1960).
48. See ref. 16, p. 813.
49. G. E. Keppel, *JAOAC* **45,** 657 (1962).

50. B. R. Roy and P. K. Bose, *JAOAC* **55,** 664 (1972).
51. J. J. Trasher, *JAOAC* **55,** 76 (1962).
52. J. J. Trasher, *JAOAC* **45,** 659 (1962).
53. J. J. Trasher, *JAOAC* **47,** 516 (1964).
54. R. M. Roberts, *JAOAC* **38,** 941 (1955).
55. F. Feigl, H. E. Feigl, and D. Goldstein, *J. Am. Chem. Soc.* **77,** 4162 (1955).
56. See ref. 49, p. 307.
57. F. L. Hart and H. J. Fisher, *Modern Food Analysis,* Springer-Verlag, Berlin (1971), p. 171.
58. G. Gutzeit, *Helv. Chem. Acta* **12,** 713 (1929).
59. *Official Methods of Analysis of the Association of Official Analytical Chemists,* 12th ed., Association of Official Analytical Chemists, Arlington, Virginia (1975), p. 481.
60. See ref. 49, p. 220.
61. Y. Pomeranz and G. D. Miller, *JAOAC* **43,** 442 (1960).
62. See ref. 58, p. 75.
63. See ref. 58, p. 68.
64. See ref. 49, p. 217.
65. *British Standard Institution* **BS 684,** Section 2.3 (1978).
66. See ref. 58, p. 69.
67. See ref. 49, p. 250.
68. See ref. 49, p. 251.
69. S. B. Mittal and N. K. Roy, *Indian J. Dairy Sci.* **29,** 283 (1970).
70. See ref. 60, p. 329.
71. See ref. 49, p. 381.
72. See ref. 49, p. 382.
73. P. Weigert, H. Thoma, E. Unkauf, and L. Kotter, *Arch. Lebensmittelhyg.* **26,** 145 (1975).
74. K. O. Honikel and H. Sauer, *Fleischwirtschaft* **55,** 1724 (1975).
75. K. Hoffman and E. Bluechel, *Fleischwirtschaft* **54,** 1083 (1974).
76. J. Fitelson, *JAOAC* **19,** 493 (1936).
77. S. Halphen, *Analyst* **22,** 326 (1897).
78. *British Standard Institution* **BS 684,** Section 2.29 (1978).
79. See ref. 49, p. 458.
80. D. Sonanini and S. Weber, *Pharm. Acta Helv.* **50,** 379 (1975).
81. D. I. McGregor, *Can. J. Plant Sci.* **57,** 133 (1977).
82. P. K. Bose, *J. Food Technol.* **11,** 28 (1974).
83. R. Haskel and L. Gray, *Can. Inst. Food Sci. Technol. J.* **14,** 150 (1981).
84. See ref. 58, p. 435.
85. W. M. Hunting, M. Gagnon, and W. B. Esselen, *Anal. Chem.* **31,** 144 (1959).
86. B. E. Proctor, *Food Ind.* **14,** 51 (1942).
87. *Military Specifications MIL-B-43193,* General Services Administration, Washington, D.C. (1964).

INDEX